Surfactants

The proceedings of a meeting held in Bristol, England
in July 1983

Surfactants

Edited by

Th. F. TADROS *ICI plc Plant Protection Division Jealott's Hill Research Station Bracknell, England*

ACADEMIC PRESS, INC.
Harcourt Brace Jovanovich, Publishers
London Orlando San Diego
Austin New York Toronto Montreal
Sydney Tokyo

ACADEMIC PRESS INC. (LONDON) LTD
24/28 Oval Road
London NW1

United States Edition published by
ACADEMIC PRESS INC.
(Harcourt Brace Jovanovich, Inc.)
Orlando, Florida 32887

British Library Cataloguing in Publication Data

Surfactants.
1. Surface active agents
I. Tadros, T.F.
668'.1 TP994

ISBN 0-12-682180-1

LCCCN 83-73319

PRINTED IN THE UNITED STATES OF AMERICA

85 86 87 88 9 8 7 6 5 4 3

PREFACE

Surfactants find application in almost every chemical industry of which the following may be worth mentioning: detergents, paints, dyestuffs, cosmetics, pharmaceuticals, pesticides, fibres, plastics, etc. Moreover, surface active agents play a vital role in the oil industry, for example in enhanced and tertiary oil recovery. They are also occasionally used for environmental protection, e.g., in oil slick dispersants. Therefore, a fundamental understanding of the physical chemistry of surface active agents, their unusual properties and their phase behaviour is essential for most industrial chemists. Moreover, an understanding of the basic phenomena involved in the application of surface active agents, such as in the preparation of emulsions and suspensions, in foams, in microemulsions, in wetting and adhesion, etc., is of vital importance in arriving at the right composition and control of the systems involved.

Due to the widespread use of surfactants in many industrial applications, a residential school was held at Bristol University during 1983 to fulfil some of the above requirements. The scientific content of the course was organized by the Editor and the residential school was sponsored by the Royal Society of Chemistry of Great Britain. The short course was attended by about 80 people who represented various countries. The lecture contents, which were circulated before the course, covered a wide range of topics (see below) and were planned to form a book on the subject, which would be useful to graduate, research and industrial chemists.

The book starts with an introductory chapter by Professor R.H. Ottewill who gave a brief account of the various classes of surface active agents and their solution properties. This is followed with a chapter on the fundamental aspects of surfactant solutions by Dr Couper, with particular attention to the thermodynamics of micellization and adsorption at interfaces. Chapter 3, by Dr Laughlin, deals with the phase equilibria and mesophases in surfactant systems. The structural aspects of micellar systems are dealt with in Chapter 4 by Professor Lindman. Particular attention is paid to the dynamic features of surfactant micelles which could be investigated using self-diffusion studies. These four chapters constitute, perhaps, the fundamental basis of the physical chemistry of surface active agents. They provide the reader with ample references for more detailed information on the various topics.

Chapters 5 to 9 deal with some of the basic principles involved in the application of surface active agents in dispersions. Chapter 5, by Professor Overbeek, lays the foundation for the fundamentals of microemulsion formation and stability. The most recent thermodynamic theories are briefly reviewed and some suggestions for future extensions are made. Chapter 6, by Dr Goodwin, deals with the rheology of concentrated systems and gives an account of the various complex rheological characteristics of such systems. Chapter 7, by Dr Aveyard, deals with the subject of adsorption of surfactants at the air/liquid, liquid/liquid and solid/liquid interface which is the key to understanding how surfactants function as stabilizers for emulsions, foams and suspensions. These last three areas are dealt with in the following two chapters. Chapter 8, by Dr Vincent, deals with emulsions and foams and the role of surfactants in their formation and stability. Chapter 9, by the Editor, gives an account of the role of surfactants in suspension preparation and in the control of their physical stability. A brief account is then given of the influence of surfactants on the flow behaviour (rheology) of suspensions. Chapter 10, by Dr Blake, deals with the role of surfactants in wetting phenomena, spreading of liquids on solids, etc. The author treats the various phenomena at a fundamental level and gives numerous examples to illustrate the application of such basic phenomena.

The last three chapters in the book have been written by industrial chemists to illustrate some of the applications of surfactants in complex industrial processes. Chapter 11, by Dr Neustadter, covers the topic of surfactants in enhanced oil recovery. The author highlights the various complex phenomena involved and the role played by the surfactants in them. This chapter should be considered only as an introduction to this complex field. For more detail other books are available. Chapter 12, by Dr Hancock, gives a comprehensive review of the various types of the commonly used, nonionic surfactants and some of their applications. Particular attention is given to macromolecular surfactants which have now found a great many applications in various industrial fields. Finally, Chapter 13, by the Editor, gives some examples of the application of surfactants to biological systems. Again, this is a vast field and arbitrary examples have been chosen to illustrate the role of surfactants in biological applications.

Thus, the book, which has been produced as a result of the

residential school on surfactants, is by no means a comprehensive text on surfactants. The topics have been carefully chosen to cover a wide range of applications. The book should, therefore, provide a useful text for readers involved with surfactants and their applications.

I would like to thank all the contributors for their care and cooperation in preparing and producing the manuscripts in good time, which made my editing job much easier. I would also like to thank the Royal Society of Chemistry, in particular Dr A.D. Ashmore and Miss Lorraine Hart for organizing the administrative side of the course and their help during the residential school. I would also like to thank the local organizers from Bristol University, namely, Drs Goodwin and Vincent for taking care of all the arrangements at Bristol University. I would also like to thank my secretary, Mrs Irene Gallacher for looking after the timetable of producing the manuscripts and communicating at the right time with the authors. Finally, I would like to thank my wife and children for coping with me during several weekends whilst I was finalizing the text.

December 1983 *Tharwat F. Tadros*

CONTRIBUTORS

AVEYARD R. Department of Chemistry, University of Hull, Hull, England

BLAKE T.D. Kodak Limited Research Laboratories, Headstone Drive, Harrow, Middlesex HA1 4TY, England

de BRUYN P.L. Van't Hoff Laboratory, University of Utrecht, Padualaan 8, 3584 CH Utrecht, The Netherlands

COUPER A. Department of Physical Chemistry, University of Bristol, Cantock's Close, Bristol BS8 1TS, England

GOODWIN J.W. Department of Physical Chemistry, University of Bristol, Cantock's Close, Bristol BS8 1TS, England

HANCOCK R.I. ICI Petrochemicals and Plastics Division, PO Box 90, Wilton, Middlesbrough, Cleveland TS6 8JE, England

LAUGHLIN R.G. Miami Valley Laboratories, The Procter and Gamble Company, Cincinnati, Ohio, USA

LINDMAN B. Physical Chemistry 1, Chemical Centre, PO Box 740, S-220 07 Lund, Sweden

NEUSTADTER E.L. BP Research Centre, Sunbury-on-Thames, Middlesex TW16 7LN, England

OTTEWILL R.H. School of Chemistry, University of Bristol, Cantock's Close, Bristol BS8 1TS, England

OVERBEEK J.T.G. Van't Hoff Laboratory, University of Utrecht, Padualaan 8, 3584 CH Utrecht, The Netherlands

TADROS TH.F. ICI Plant Protection Division, Jealott's Hill Research Station, Bracknell, Berkshire RG12 6EY, England

VERHOECKX F. Van't Hoff Laboratory, University of Utrecht, Padualaan, 8, 3584 CH Utrecht, The Netherlands

VINCENT B. School of Chemistry, University of Bristol, Cantock's Close, Bristol BS8 1TS, England

CONTENTS

Preface v

Contributors ix

CHAPTER 1
Introduction 1
R.H. Ottewill

CHAPTER 2
Thermodynamics of Surfactant Solutions 19
A. Couper

CHAPTER 3
Phase Equilibria and Mesophases in Surfactant Systems 53
R.G. Laughlin

CHAPTER 4
Structural Aspects of Surfactant Micellar Systems 83
B. Lindman

CHAPTER 5
Microemulsions 111
J.T.G. Overbeek, P.L. de Bruyn and F. Verhoeckx

CHAPTER 6
The Microstructure and the Rheology of Surfactant and Related Systems 133
J.W. Goodwin

CHAPTER 7
Adsorption at the Air/Liquid, Liquid/Liquid and Solid/Liquid Interfaces 153
R. Aveyard

CHAPTER 8
Emulsions and Foams 175
B. Vincent

CHAPTER 9
Suspensions 197
Th.F. Tadros

CHAPTER 10
Wetting 221
T.D. Blake

CHAPTER 11
Surfactants in Enhanced Oil Recovery 277
E.L. Neustadter

CHAPTER 12
Macromolecular Surfactants 287
R.I. Hancock

CHAPTER 13
Some Applications of Surfactants in Biological Systems 323
Th.F. Tadros

Index 337

INTRODUCTION

R.H. Ottewill

School of Chemistry, University of Bristol, Cantock's Close, Bristol BS8 1TS, England.

The groups of chemical compounds known as surface active agents are in the most common form constituted of a hydrocarbon portion and a polar or ionic portion as illustrated schematically in Fig. 1. The hydrocarbon portion, which can be linear or branched, interacts only very weakly with the water molecules in an aqueous envircnment. Moreover, the strong interactions between the water molecules arising from dispersion forces and hydrogen bonding act cooperatively to squeeze the hydrocarbon out of the water, hence the chain is usually called hydrophobic. The polar or ionic portion of the molecule, usually termed the head-group, however interacts strongly with the water via dipole-dipole or ion-dipole interactions and is solvated. Consequently the head-group is said to be hydrophilic.

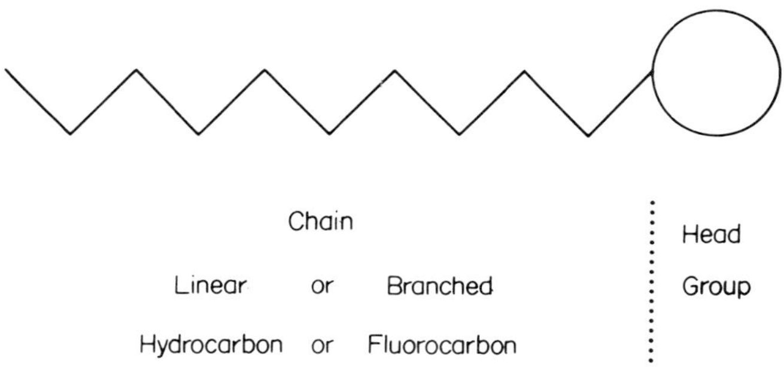

Fig. 1. Schematic diagram of surface active molecule.

SURFACTANTS
ISBN 0-12-682180-1

It is essentially the balance between the hydrophobic and hydrophilic parts of the molecule (ion) which gives the special properties which we associate with surface active agents. In addition to the name surface active agents these materials are often called by other names which include: surfactants (particularly in the USA); association colloids; colloidal electrolytes; paraffin chain salts; amphipathic compounds; tensides (particularly in Europe). It should be noted, however, that whereas most detergents (from the Latin, *detergens* cleansing) are surface active agents not all surface active agents are detergents.

In addition to hydrocarbon chains many surface active agents have been synthesized with fluorocarbon moieties as the hydrophobic, and in this case oleophobic, portion of the molecule.

GENERAL CLASSIFICATION OF SURFACE ACTIVE AGENTS

The classification usually used is that which puts surface active agents into various groups depending on the nature of the head-group, that is, anionic, cationic, nonionic or zwitterionic. Some examples of these materials in common use are:

Anionic

$$C_{12}H_{25}SO_4^-Na^+$$

Sodium dodecyl sulphate, a synthetic material. The name is often abbreviated to SDS.

$$C_{11}H_{23}COO^-Na^+$$

Sodium dodecanoate, a material which can be synthesized. There is however, also a large group of materials of this type formed from natural sources, e.g., the hydrolysis of triglycerides. Salts of carboxylic acids form the group of compounds known as soaps.

Cationic

$$C_{12}H_{25}N^+Me_3Br^-$$

Dodecyltrimethylammonium bromide (DTAB)

$$C_{16}H_{25}N^+Me_3Br^-$$

Hexadecyltrimethylammonium bromide (HTAB). As a group these materials usually have good germicidal properties.

Nonionic

Many of these are based on polyoxyethylene and a typical example is,

$$C_{12}H_{25}[OCH_2CH_2]_6OH$$

Dodecylhexaoxyethylene glycol monoether, often abbreviated to $C_{12}E_6$ to denote hydrocarbon chain length and ethylene oxide chain length.

It should also be noted that with this type of compound the head-group is larger than the hydrocarbon chain, the E_6 portion being ca. 24Å in length in the fully extended form and the C_{12} chain ca. 18Å.

Nonionic surface active agents with small head-groups also exist, some examples being,

$$C_{12}H_{25}SOCH_2CH_2OH$$

Dodecyl sulphinyl ethanol

and

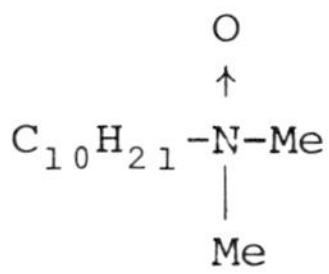

Decyl dimethyl amine oxide.

The latter material is nonionic in aqueous solutions above pH 7 but at lower pH values can protonate to give the cation form, namely,

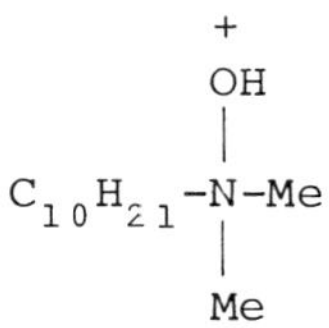

Zwitterionic

A simple example of this type of material is,

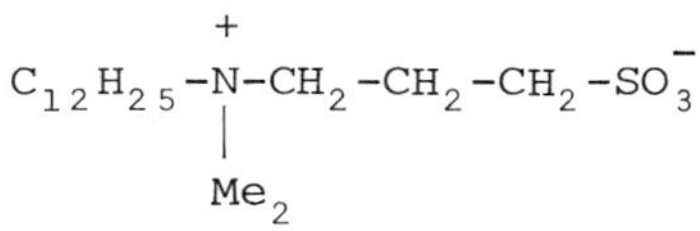

3-dimethyldodecylamine propane sulphonate

Within this group there are a number of important naturally

occurring materials, known as triglycerides, a good example being lecithin, which occurs in the membranes of many animal cells.

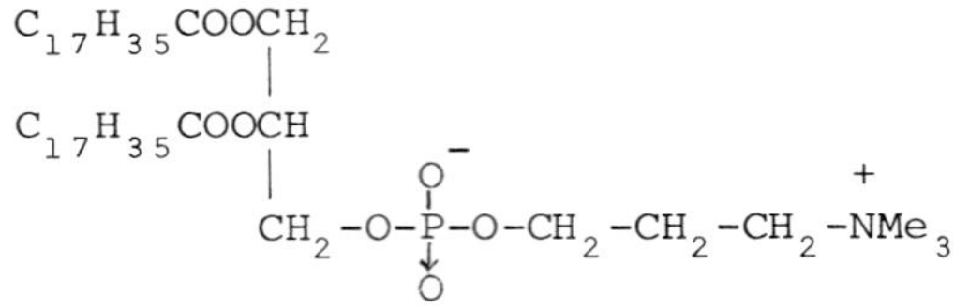

Although surface active, lecithin has an extremely low solubility in water and would in general terms be considered insoluble.

Most, if not all, materials which are considered, in practical terms, to be surface active agents are soluble in water. As a rather rough approximation the compounds with chain length $>C_{16}$ are insoluble in water whilst those with a chain length $<C_{16}$ are water soluble.

The unique property of these materials is their ability to adsorb strongly at various interfaces, for example, air - water, oil - water, water - solid, oil - solid, etc. and to lower the interfacial surface energy, hence the name surface active. Solubility of the material ensures that a reservoir of surface active agent is available to accommodate the large surface area of the various interfaces encountered in practical applications.

In this chapter we will concentrate on the properties of aqueous solutions of surface active agents.

PROPERTIES OF SOLUTIONS OF SURFACE ACTIVE AGENTS

Surface Properties

With pure materials of chain length between about C_8 and C_{16} a curve of surface tension against the logarithm of the molar concentration is obtained having the form shown in Fig. 2*a*. A characteristic feature of the curve is that at a certain concentration a break appears in the curve and from this point up to higher concentrations the surface tension remains essentially constant.

The Gibbs' adsorption equation which has the form,

$$\Gamma_2 = -\frac{1}{RT}\left(\frac{\partial\gamma}{\partial \ell n a_2}\right)_{P,T} \tag{1}$$

where Γ_2 is the surface excess, γ the surface tension and a_2 is the solution activity, will be discussed in detail in the chapter by Couper (1983). It is useful at this juncture, however, to make a number of points. At concentrations just before the break point we have the condition,

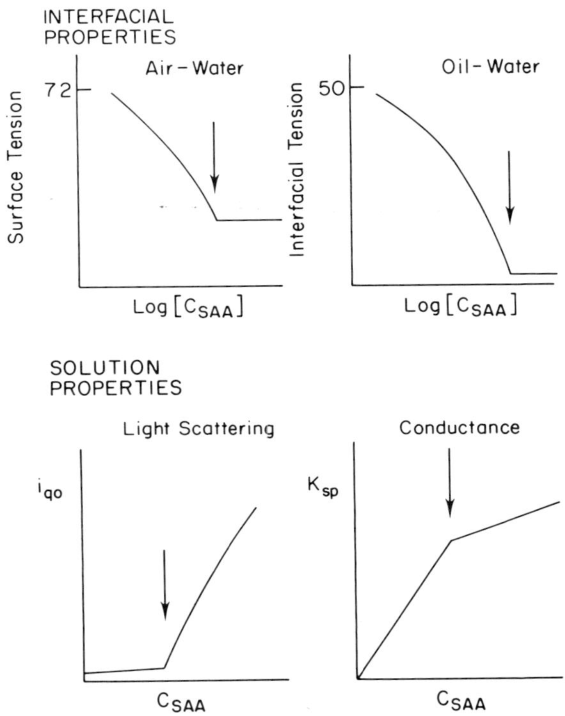

Fig. 2. Interfacial and solution properties of surface active agents:- (a) air-water tension versus log concentration, (b) oil-water interfacial tension versus log concentration, (c) scattered light intensity versus concentration, (d) specific conductance versus concentration.

$$\left(\frac{\partial \gamma}{\partial \ell n a_2}\right)_{P,T} = \text{const.} \tag{2}$$

indicating that saturation adsorption of the surface has been reached. Just above the break point we have,

$$\left(\frac{\partial \gamma}{\partial \ell n a_2}\right)_{P,T} = 0 \tag{3}$$

giving on integration, γ = constant × $\ell n a_2$.

Consequently, since γ is constant in this region this indicates that the activity of the surface active agent in the aqueous phase must remain constant - at least as a simplistic argument. Therefore, even though more material is being added to the solution, its activity in that solution is remaining constant, an effect which can only be explained if the surface active molecules are associating to form units with a low activity in solution.

Typically, hydrocarbon based surface active agents lower the surface tension of water to ca.30-35 mN m^{-1}. Fluorocarbon materials, on the other hand, can lower the surface tension to ca. 17 mN m^{-1}.

The oil - water interfacial tension is also lowered by the addition of a surface active material as illustrated in Fig. 2*b*. With single components values of the order of a few dynes are typical but with mixed systems values of the order 10^{-3} mN m^{-1} can be achieved.

Bulk Properties

Light scattering Measurement of the intensity of the light scattered at 90° to the direction of the incident beam, i_{90}, gives initially at low concentration a very small scattered intensity. However, at higher concentrations at or close to the break point observed in interfacial tension measurements it is found that an increase in scattering occurs and this continues as the concentration is increased. The implication of such measurements, see Fig. 2*c*, is immediately that association must be occurring in the solution phase to give larger units.

Conductance For ionic surface active agents the curve of specific conductance against molar concentration, C, has the form shown in Fig. 2*d*. The curve (schematically idealized) shows a change of gradient at a certain concentration which is again close to, or at, the break point found by interfacial tension measurements.

From these data a plot of molar conductance against $\sqrt{C}$ can be obtained which has the form shown in Fig. 3. A sharp decrease in molar conductance, consistent with a decrease of mobility occurs in the vicinity of the interfacial tension break point.

General Conclusion

Both the interfacial properties and bulk properties of solutions of surface active agents are consistent with the fact that at a certain concentration the surface active ions (molecules) in solution associate to form larger units. These associated units are called micelles. The concentration at which this association phenomenon occurs is known as the critical micelle concentration, usually abbreviated to c.m.c.

Each surface active agent at a given temperature and electrolyte concentration has a characteristic c.m.c. value. Values for a number of materials are listed in Table 1. A detailed compilation of values can be found in the book by Mukerjee and Mysels (1971).

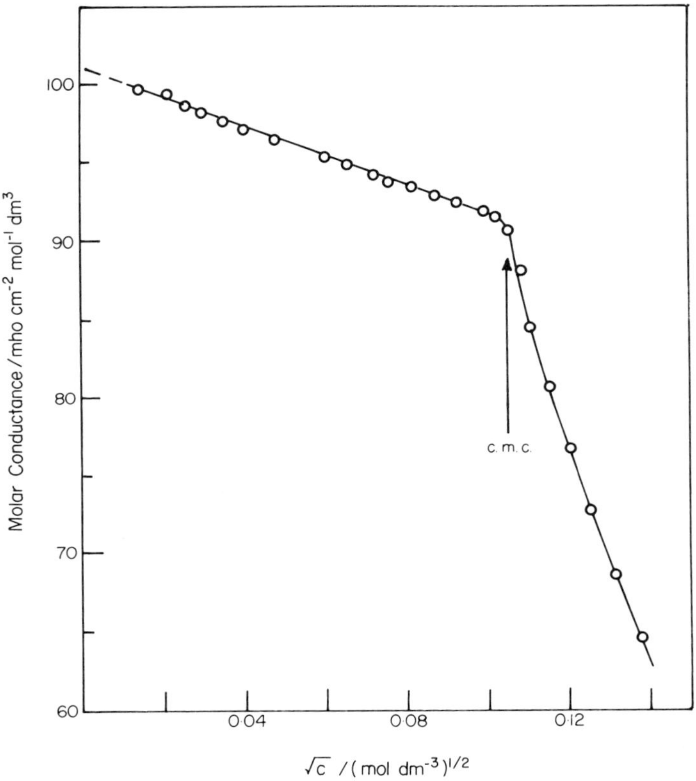

Fig. 3. Molar conductance versus $\sqrt{c}$. *Dodecylpyridinium bromide in water at 25°C (Billett, 1965).*

As a very rough working approximation it is useful to remember that many C_{12} ionic surface active agents have a c.m.c. value of approximately 10^{-2} mol dm^{-3}, and that the addition of two $-CH_2-$ groups lowers the value to approximately 10^{-3} mol dm^{-3}. The c.m.c. values of nonionic surface active agents are usually lower than those of the corresponding chain length ionic materials, for example, dodecylhexaoxyethylene glycol monoether at ca. 10^{-4} mol dm^{-3}.

Solubility of Surface Agents

The solubility behaviour of surface active agents is anomalous in that as the temperature is increased a value is reached at which there is a sudden increase in solubility and the material then becomes very highly soluble (Krafft, 1899). This is illustrated in Fig. 4.

TABLE 1

Critical micelle concentrations for some surface active agents in water at 25°C

Material	c.m.c./mol dm^{-3}
Sodium Octyl-1-sulphate	1.30×10^{-1}
Sodium Decyl-1-sulphate	3.32×10^{-2}
Sodium Dodecyl-1-sulphate	8.39×10^{-3}
Sodium Tetradecyl-1-sulphate	2.05×10^{-3}
Octyl Trimethylammonium Bromide	1.30×10^{-1}
Decyl Trimethylammonium Bromide	6.46×10^{-2}
Dodecyl Trimethylammonium Bromide	1.56×10^{-2}
Octylhexaoxyethylene Glycol Monoether	9.80×10^{-3}
Decylhexaoxyethylene Glycol Monoether	9.00×10^{-4}
Dodecylhexaoxyethylene Glycol Monoether	8.70×10^{-5}

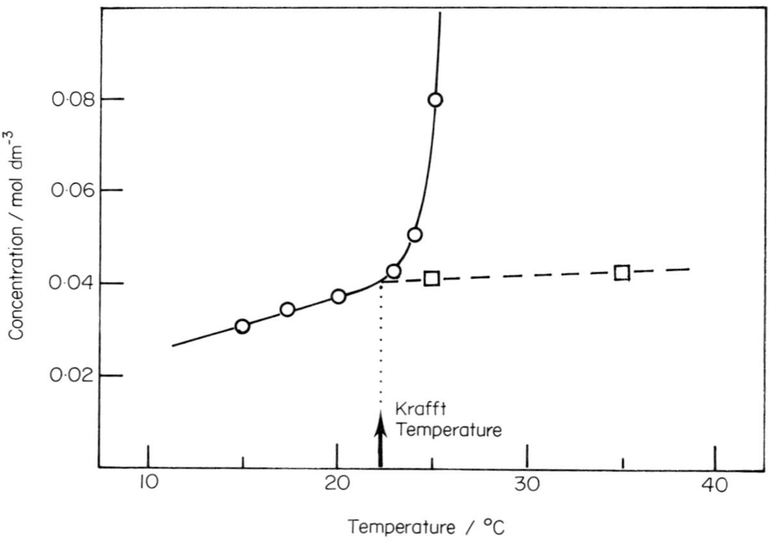

Fig. 4. Solubility versus temperature for sodium decylsulphonate in water showing location of the Krafft point (after Shinoda, 1974); -O-, solubility data; -□-, c.m.c. values.

When the variation of c.m.c. with temperature is plotted on the same curve it becomes clear that at a particular temperature the concentration of the dissolved material becomes equal to the c.m.c. The temperature at which this intersection occurs is known as the Krafft temperature or Krafft point, that is, it is the temperature at which the solubility becomes equal to the c.m.c. Moreover, it is at this temperature that the solubility starts to increase rapidly. As mentioned earlier this implies that the activity of the monomer remains essentially constant at the c.m.c. and it is the micelles which have the increased solubility, although an increase in solubility of the monomer would be anticipated with an increase in temperature - or a decrease according to the compound.

Some Historical Notes

In 1920 McBain and Salmon investigated the osmotic activity of a 1 mol dm^{-3} solution of potassium stearate at 90°C. They found that the concentration of osmotically active material was 0.42 mol dm^{-3} and hence concluded that considerable association had occurred in the solution. McBain suggested that the associated units should be called micelles. In order to explain the drop in molar conductance, which was also observed, he suggested that two types of micelles were formed: spherical ionic micelles and lamellar neutral micelles and represented these schematically in the manner shown in Fig. 5.

Later in a now classical monograph entitled, "Aqueous Solutions of Paraffin Chain Salts", published in 1936, G.S. Hartley suggested that the properties of solutions of surface active agents could be explained by the formation of one type of micelle and he depicted this in the manner shown in Fig. 5.

The micelle which Hartley proposed had the following properties:-

1) the associated unit was spherical with a radius approximately equal to the length of a hydrocarbon chain;
2) it contained in the region of 50 to 100 monomeric units of surface active agent;
3) as a consequence of the high association number the process of micellization occurred over a very narrow concentration range;
4) many of the counter-ions were bound to the charged head-groups composing the micelle surface, thus reducing the mobility of the unit from that expected for a micelle with no counter-ion binding (McBain's model); this factor explained the drop in conductance on association;
5) the interior of the micelle formed by the association of hydrocarbon chains essentially had the properties of a liquid hydrocarbon. This conferred on the micelle the ability to solubilize many organic molecules, e.g., dyes, normally insoluble in water.

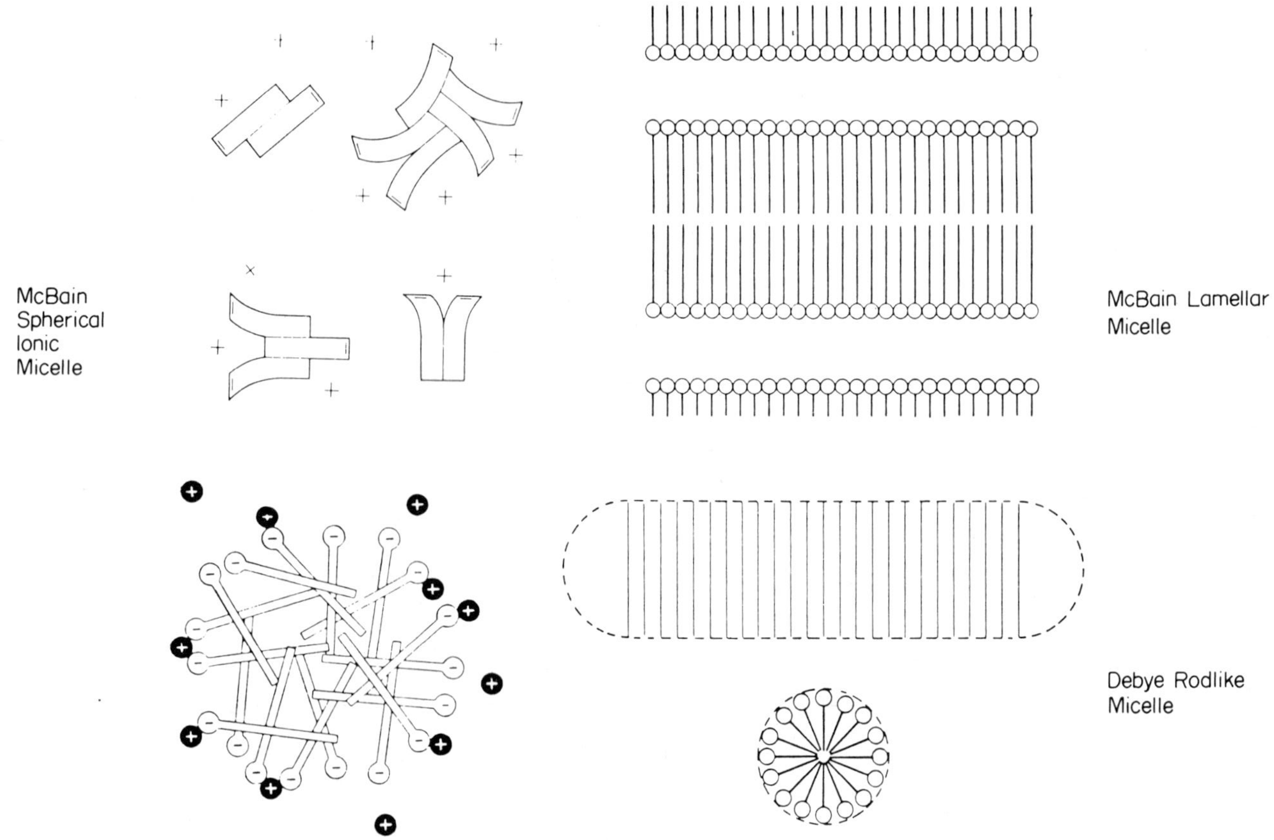

Fig. 5. Various models of micelle structures, McBain spherical ionic micelle, McBain lamellar micelle, Hartley spherical micelle, Debye cylindrical micelle.

The micelle structure proposed by Hartley was correct from many points of view and has provided an excellent foundation for various present-day studies. However, recent small angle neutron scattering studies (Hayter, 1981) indicate that as a consequence of the motion of the molecules the surface may be somewhat rougher than depicted by Hartley. Moreover, many modern techniques such as light scattering and neutron scattering indicate that in my systems the micelles are not spherical. For example, Debye and Anacker (1951) proposed a cylindrical micelle as shown in Fig. 5 to explain their light scattering results on hexadecyltrimethylammonium bromide in water. Evidence for disc-shaped micelles has also been obtained.

A Simple Treatment - Thermodynamics of Micellization

For simplicity, and to avoid the complications of electrical effects we will confine our attention to the case of nonionic surface active agents. The association of N single molecules, S, to form a micelle M composed of N monomers can be represented by the equation,

$$N\ S \rightleftharpoons M \tag{4}$$

hence by the law of mass action we can write for the equilibrium constant K the equation,

$$K = [M]/[S]^N \tag{5}$$

Representing the concentration of monomer as C_s and that of the micelles as C_m but expressing the concentration of surface active agent in monomeric units we obtain,

$$K = C_m/C_s^N \tag{6}$$

where the total concentration of surface active agent is given by $C = C_s + C_m$. The free energy of micellization is given by

$$\begin{aligned} -\Delta G &= RT\ \ell n\ K \\ &= RT\ \ell n\ C_m - NRT\ \ell n\ C_s \end{aligned} \tag{7}$$

or expressing the free energy as that required to insert one monomer unit into the micelle, we obtain,

$$-\Delta G^\circ = -\frac{\Delta G}{N} = \frac{RT}{N}\ \ell n\ C_m - RT\ \ell n\ C_s \tag{8}$$

For many micellar units N will be of the order of 100 and hence, provided this is so, we can write with little loss of accuracy,

$$\Delta G^\circ = RT\ \ell n\ C_s \tag{9}$$

Inherently, by this approach we have treated micellization as a kinetic phenomenon. However, as soon as we make the assumption that N is large we have also assumed that it is equivalent to a phase change (see later, Couper, 1983).

We can now make a further simplifying assumption which will hold for many instances of micellization. It is, that provided above the c.m.c. all the additional surface active agent added forms new micelles then the concentration of monomeric material remains constant at the c.m.c. value, and hence C_S = [c.m.c.]. This provides the very simple and useful equation,

$$\Delta G^\circ = RT \, \ell n \, [\text{c.m.c.}] \tag{10}$$

It should be remembered of course that the numerical value of ΔG° will depend on the choice of concentration units, i.e. the choice of standard states. Values found in the literature are frequently given in concentrations expressed as either mol dm^{-3} or mol fractions. As an example, the c.m.c. for dodecylhexaoxyethylene glycol monoether in water is 8.70×10^{-5} mol dm^{-3} at 25° (Table 1), or as a mol fraction this is,

$$\frac{8.70 \times 10^{-5}}{55.6 + 8.70 \times 10^{-5}} \cong \frac{8.70 \times 10^{-5}}{55.6} = 1.57 \times 10^{-6}$$

Consequently, the values of ΔG° are,

in molar concentration units $\Delta G^\circ = -23.2$ kJ mol^{-1}
in mole fraction units $\Delta G^\circ = -33.1$ kJ mol^{-1}

For the case of ionic surface active agents in water it can be shown by similar, but more complicated arguments (Phillips, 1955) that

$$\Delta G^\circ = RT \, [2 - p/N] \, \ell n \, [\text{c.m.c.}] \tag{11}$$

where p is the effective charge on the micelle, that is with (N-p) counter-ions located on charged head-groups. The presence of any salt must also be taken into consideration.

The Constancy of Monomer Concentration above the c.m.c.

One method of obtaining the concentration of monomer units present above the c.m.c. is to measure, using n.m.r., the chemical shift of an ^{19}F labelled material. The area under the peak can be used as a measure of the concentration of the monomer. Corkill *et al.* (1969) used the material $C_3F_7(CH_2)_5\overset{+}{N}(CH_3)_2(CH_2)_3\,SO_3^-$ and obtained the results shown in Fig. 6 wherein the monomer concentration is shown plotted as a function of total concentration. As can be seen, initially the monomer concentration increases linearly with increase in total concentration and then above the c.m.c. becomes almost constant. More rigorous analyses of micellization phenomena

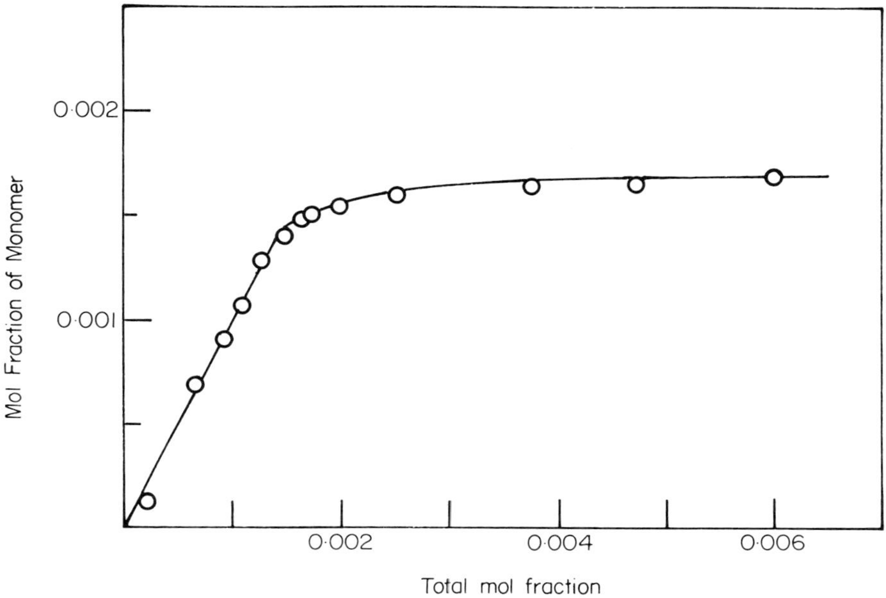

Fig. 6. Mol fraction of monomer versus total mol fraction of surface active agent. Data obtained from n.m.r. experiments using $C_3F_7(CH_2)_5\overset{+}{N}(CH_3)_2(CH_2)_3\ SO_3^-$ *(Corkill* et al., *1969).*

take this into account, for example, the multiple equilibrium model of Corkill *et al.* (1969).

The Effect of Association Number

As originally pointed out by Hartley (1936) the influence of N on the micellization process can be most easily shown by plotting the rate of increase of micellar concentration with total concentration against the total concentration, that is dC_m/dC against C. This is shown in Fig. 7. As can be seen when N = 2 there is a gradual change. For N = 20 it becomes more abrupt and for N = 100 the sudden onset of the micellar phase can be clearly seen.

Enthalpy of Micellization

For a nonionic surface active agent the enthalpy of micellization is given by the equation,

$$-\Delta H^\circ = RT^2 \frac{d\ \ln\ [c.m.c.]}{dT} \qquad (12)$$

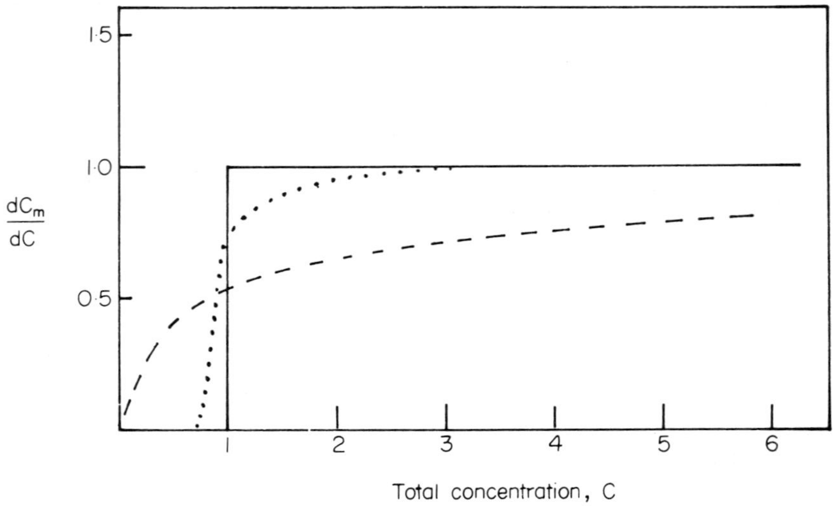

Fig. 7. dC_m/dC versus C for, --- N = 2; .... N = 20; ——— N = 100.

and hence the value of ΔH° can be obtained from the variation of the c.m.c. with temperature. Since as shown earlier ΔG° is also calculable from the same information, the entropy of micellization, ΔS°, or TΔS°, can also be obtained. Some values are listed in Table 2.

An alternative, and preferable, method of determining ΔH° is by calorimetry and some values obtained by this technique are also included in Table 2. It should be noted that they are more precise than those obtained from the variation of the c.m.c. with temperature.

TABLE 2

Micellization of octylhexaoxyethylene glycol

	Thermodynamic parameters			
Temp/ °C	ΔG°/ $kJ\ mol^{-1}$	ΔH°/(1) $kJ\ mol^{-1}$	ΔH°/(2) $kJ\ mol^{-1}$	TΔS°/ $kJ\ mol^{-1}$
25	-21.3 ± 2.1(1)	8.0 ± 4.2	20.1 ± 0.8	41.4 ± 1.0
40	-23.4 ± 2.1		14.6 ± 0.8	38.0 ± 1.0

(1) From variation of c.m.c. with temperature.
(2) Calorimetric determination.

It can also be seen from Table 2 that for the compounds cited, which are not atypical, the $T\Delta S°$ term is positive and larger in magnitude than the $\Delta H°$ term over the range of temperatures quoted.

Micellization as a Kinetic Process

Earlier, in writing the equation for micelle formation as

$$NS \rightleftharpoons M \quad \text{or} \quad N\,S \rightleftharpoons S_N \tag{13}$$

we have implicitly assumed that there will be definite rates for the formation and breakdown of micelles. Using modern rapid reaction techniques such as pressure jump and stopped flow methods it has now been established that the half-life for micellar formation or breakdown is usually in the region of 10^{-3} to 1 seconds, that is, they are fast reactions (Jaycock and Ottewill, 1964; Aniansson *et al.*, 1976; Okubo *et al.*, 1979).

Another important feature of micellar systems is the rate of interchange of the surface active molecules between the micelle and the solution phase. Again modern techniques such as ultrasonic absorption (Rassing *et al.*, 1974; Aniansson *et al.*, 1975) have established that the lifetime of a surface active molecule in a micelle is of the order of 10^{-7} seconds. This emphasizes the dynamic nature of micelles and it is important to realize that these molecules are in continuous motion and that there is constant interchange between micelles and solution. It is even possible that concentration fluctuations in solution would lead to the complete dissolution of micelles in one microregion of the solution while micelles formed in another. This has interesting consequences when organic molecules are solubilized within micelles.

The Two-Component System: Surface Active Agent - Water

The simplest surface active agent solution is that composed of the surface active agent and water, namely, a two-component system. As we have mentioned earlier, as soon as micelles are formed, the solubility of the surface active agents becomes very high. In dilute solutions the micelles are randomly distributed throughout the solution phase. As the concentration is increased, however, changes occur in the system and organized structures are formed. Three types of organized systems are frequently found and these are illustrated in Fig. 8 (Corkill and Goodman, 1969; Ottewill, 1969). They are:-

1) the hexagonal phase (middle) - composed of cylindrical units, hexagonally close-packed;
2) the viscous-isotropic phase - composed of spherical units

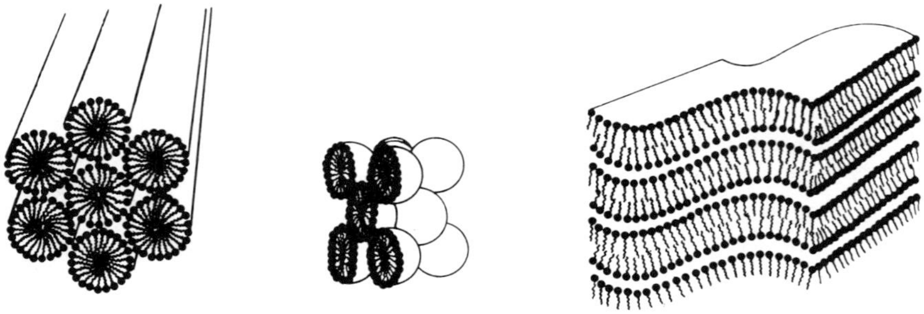

Fig. 8. Schematic representation of structures formed in concentrated solution of surface active agents (after Corkill and Goodman, 1969).

in a body-centred cubic arrangement;

3) the lamellar phase (neat) - composed of sheets of molecules in a bimolecular packing with the head-group exposed to the water layers in between them.

The concentration and temperature domains over which these phases are formed, the so-called lyotropic mesomorphic phases or lyotropic liquid crystals, vary widely for different surface active agents. Changes also occur on the addition of salt to give a three-component system and on the addition of another organic phase. It is always advisable, but often time-consuming, to explore the phase behaviour of water - surface active agent mixtures and the effect of additives on them.

As an example of a two-component phase diagram, Fig. 9 gives the phase diagram for the water - dodecylhexaoxyethylene glycol monoether system (Clunie *et al.*, 1969). It shows the position of the hexagonal, viscous isotropic and lamellar phases. In some literature these will be referred to by the soap-boilers names of middle, viscous-nigre and neat respectively.

An additional feature of this diagram is important. Many aqueous solutions of nonionic surface active agents on heating separate into two liquid phases, one rich in water and the other rich in surface active agent. Consequently, for these systems a lower consolute boundary is present in the phase diagram. In practice this boundary is recognized by the fact that during the heating process, at a certain temperature, the solution suddenly goes cloudy. The temperature at which this occurs is usually referred to as the cloud point. The practical quantity is not usually quoted at a particular composition and hence can be somewhat imprecise. However, it

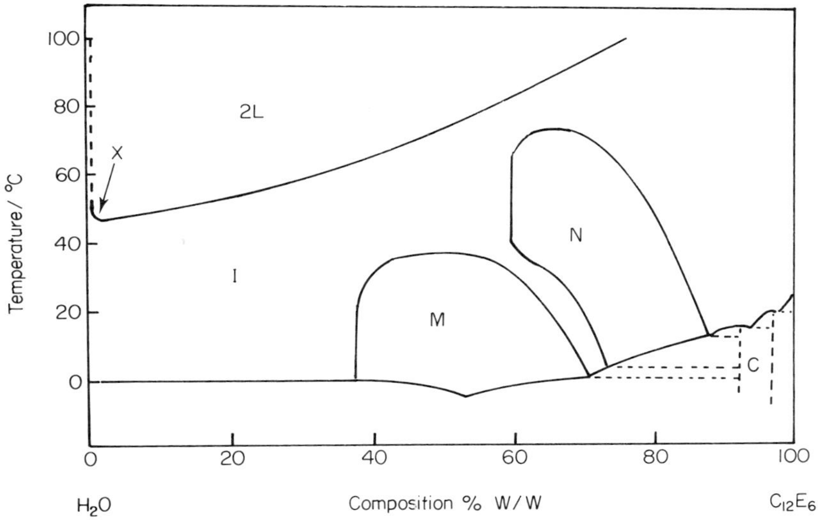

Fig. 9. Phase diagram for dodecylhexaoxyethylene glycol monoether - water mixtures. I = isotropic solution, M = middle phase, N = neat phase, C = crystalline phase, 2L = two liquid phases. X shows location of cloud point.

is usually determined just above the c.m.c. and hence is often close to the minimum in the lower consulute boundary. For many practical purposes this is quite adequate. It must also be remembered that the cloud point can be raised or lowered by the presence of a third component.

CONCLUSION

There are a very large number, ca. 10^3, of surface active agents available on the commercial market, usually under trade names, and not infrequently the same basic material appears from different manufacturers under different names. Useful references in this context are McCutcheon, an annual publication which lists surface active agents under trade name and gives the chemical constitution and the Directory of Surface Active Chemicals compiled by Hollis.

REFERENCES

Aniansson, E.A.G. and Wall, S.N. (1974). *J. Phys. Chem.* 78, 1024.

Aniansson, E.A.G. and Wall, S.N. (1975). *J. Phys. Chem.* 79, 857.

Aniansson, E.A.G., Wall, S.N., Almgren, M., Hoffmann, H. Kielmann, I., Ulbricht, W., Zana, R., Lang, J., and

Tondre, C. (1976). *J. Phys. Chem.* 80, 905.
Billett, D.F. (1965). M.Sc. Thesis, University of Bristol.
Clunie, J.S., Goodman, J.F. and Symons, P.C. (1969). *Trans. Faraday Soc.* 65, 287.
Corkill, J.M. and Goodman, J.F. (1969). *Adv. Colloid Interface Sci.* 2, 297.
Corkill, J.M., Goodman, J.F., Walker, T. and Wyer, J. (1969). *Proc. Roy. Soc.* A312, 243.
Couper, A.C. (1983). This volume, Chapter 2, p.19.
Debye, P. and Anacker, E.W. (1951). *J. Phys. and Colloid Chem.* 55, 644.
Hartley, G.S. (1936). "Aqueous Solutions of Paraffin Chain Salts". Hermann and Cie, Paris.
Hayter, J.B. (1981). *Ber. Bunsenges. Phys. Chem.* 85, 887.
Hollis, G.L. (1979). "Directory of Surface Active Chemicals". 2nd edn.
Jaycock, M.J. and Ottewill, R.H. (1964). *Fourth Int. Congress Surface Activity* 2, 545.
Krafft, F. (1899). *Ber. Deutsch. Chem. Gesell.* 32, 1596.
McBain, J.W. and Salmon, C.S. (1920). *J. Amer. Chem. Soc.* 43, 426.
McCutcheon, "Detergents and Emulsifiers". Allured Publishing Co., New Jersey. Published annually.
Mukerjee, P. and Mysels, K.J. (1971). "Critical Micelle Concentrations of Aqueous Surfactant Systems". National Bureau of Standards, Washington, USA.
Okubo, T., Kitano, H., Ishiwatari, T. and Isem, N. (1979). *Proc. Roy. Soc.* A366, 81.
Ottewill, R.H. (1969). *Ann. Rep. Chem. Soc.* A66, 183.
Phillips, J.N. (1955). *Trans. Faraday Soc.* 51, 561.
Rassing, J., Sams, P.J. and Wyn-Jones, E. (1974). *J. Chem. Soc. Faraday II* 70, 1247.
Shinoda, K. (1974). "Principles of Solution and Solubility". Marcel Dekker Inc., New York.

General References

Elworthy, P.H., Florence, A.T. and Macfarlane, C.B. (1968). "Solubilization by Surface Active Agents". Chapman and Hall, London.
Garrett, H.E. (1972). "Surface Active Chemicals". Pergamon Press, Oxford.
Lindman, B. and Wennerström, H. (1980). "Micelles - Amphiphile Aggregation in Aqueous Solution". Springer-Verlag, Berlin.
Rosen, M.G. (1978). "Surfactants and Interfacial Phenomena". John Wiley and Sons, New York.
Shinoda, K., Nakagawa, T., Tamamushi, B. and Isemura, T. (1963). "Colloidal Surfactants". Academic Press, London.

THERMODYNAMICS OF SURFACTANT SOLUTIONS

A. Couper

Department of Physical Chemistry,
University of Bristol, Bristol, BS8 1TS, England.

INTRODUCTION

The physical interactions which amphiphile molecules undergo at liquid and solid interfaces are almost always rapidly reversible so that a quantitative description of the corresponding equilibria can be given in thermodynamic terms. This chapter is concerned with the properties of surface active agents in solution, but the same principles are generally applicable when a solid adsorbent is introduced except that adsorption may then become much stronger and localized at immobile sites so that it appears to be irreversible on the time scale appropriate to physical adsorption.

Figure 1 illustrates some of the equilibria which occur in solutions of an amphiphile molecule, comprising a hydrophilic "head" and a hydrophobic "tail", in water. Adsorption at an air or vapour/solution interface is very similar to that at an oil/water interface for surfactants which are water-soluble but only sparingly soluble in a hydrocarbon liquid. The same forces which cause adsorption, cause the surfactant to aggregate into micelles when the chemical potential of the monomer in solution has risen to an appropriate level, and the equilibrium at the surface of the hydrocarbon core of the micelle is similar to that of a layer adsorbed at an oil/water interface. Micelles are often small aggregates of nearly spherical shape, but in some cases, often at higher chemical potentials in solution, larger micelles which may be laminar or rod-shaped may be formed, followed by the separation of surfactant-rich phases and eventually of solid crystals. It is convenient to describe the relative tendencies of these processes

SURFACTANTS
ISBN 0-12-682180-1

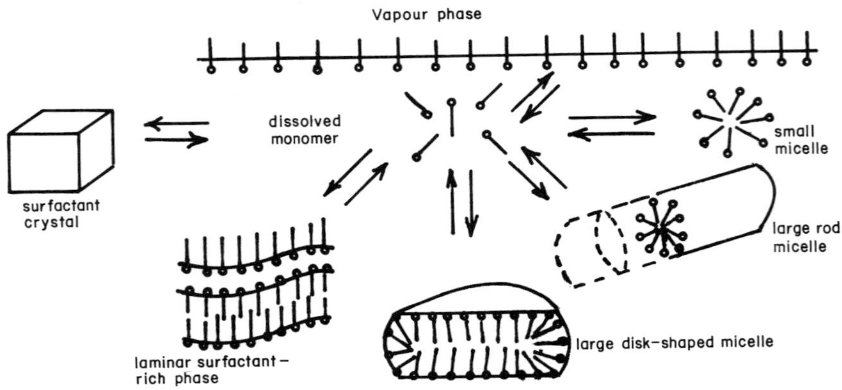

Fig. 1. Possible states of an amphiphile in solution.

to occur in thermodynamic terms.

The solvent is usually water, sometimes with added electrolyte to define the ionic strength or pH, and the technical importance of surfactants designed to operate in water has stimulated extensive study of them, but there have also been studies of adsorption at solid surfaces from non-aqueous media and of the properties of surfactants in such solvents (Jarvis and Zisman, 1960) and even in nonpolar solvents (Singleterry, 1955; Eicke, 1980). It is, however, questionable whether the concept of the micelle should be applied to aggregates of 3 to 7 molecules, in C_6H_6 or CCl_4 (Fendler *et al.*, 1973), and "reversed micelles" of high aggregation number containing an aqueous core, are perhaps more appropriately described as microemulsions.

LIQUID SURFACES

The surface of a liquid is the boundary between two bulk phases, usually between the liquid and air, or its own vapour alone. Similarly, an interface is formed between two "immiscible" liquids, when the requirement that each liquid should be saturated with the other, replaces the requirement for saturation vapour pressure at the air/liquid boundary. Stability of the interface, in the absence of gravitational fields, requires that the free energy should increase if the area of the interface increases, otherwise the interface will become convoluted, increasing the interfacial area, until the liquid evaporates or the "immiscible" phases dissolve in each other. This requirement leads to a convenient definition of the surface tension, γ, at a liquid surface, as the differential of the Helmholtz free energy, A, at constant temperature and composition with the area, $\mathcal{A}$:

$$\gamma = (\partial A/\partial \mathcal{A})_{T,V,n_i} \tag{1}$$

and the work done in extending the surface is given by

$$dA_{T,V,n_i} = \gamma d\mathcal{A} \tag{2}$$

where n_i indicates that the number of moles of each component, i, is kept constant. It is obvious that for a stable interface, γ should have a positive finite value. The Helmholtz free energy for a system which includes a surface, $\mathcal{A}$, can be expressed as:

$$dA = -SdT - pdV + \gamma d\mathcal{A} + \sum_i \mu_i \, dn_i \tag{3}$$

and if the surface only is being considered

$$dA^\sigma = -S^\sigma dT - pdV^\sigma + \gamma d\mathcal{A} + \sum_i \mu_i \, dn_i^\sigma \tag{4}$$

in which the superscript, σ, denotes surface quantities.

MODELS OF THE LIQUID INTERFACE

The region in which the properties of one bulk phase change into those of the other at an interface is thought to be limited to a few molecular diameters only, normal to the interface, and may extend only slightly beyond the monolayer of the largest molecule involved. Two conventions have been applied to describe the energetics of processes occurring at the surface, leading to identical results. These are represented in Fig. 2.

All changes of composition and other properties, such as density, between the two phases α and β are confined between boundaries A-A and B-B, the phases being homogeneous outside this region. The two conventions may then be summarized as follows.

1) The region between A and B can be regarded as a <u>surface phase</u> having finite thickness and volume. Extensive quantities, e.g., A or n_i are total quantities associated with the phase between A-A and B-B, and are denoted by a superscript, s. The description will include terms involving γ and the area, $\mathcal{A}$.
2) The model proposed by Gibbs represents the two phases as meeting at some arbitrary surface S-S, and the amount of adsorption of a component, i, is defined as the <u>surface excess</u> of i, i.e., the amount of i between A-A and B-B in

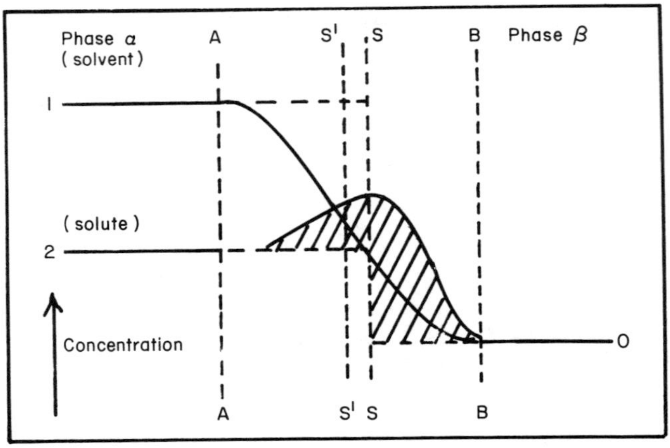

Fig. 2. Model of adsorption in the surface phase.

excess of the amount which would have been present if the two phases had remained homogeneous up to the plane S-S, at which a discontinuous change of properties occurred.

A typical concentration profile is shown in Fig. 2 for two components of which (1) is a solvent and (2) is a solute which is positively adsorbed at an air/solution interface, so that the concentrations in phase β of both components are zero. The surface excess of (2) is shaded in Fig. 2, and since positive adsorption has occurred, the solvent, (1), has been displaced, its concentration profile having moved in the direction of phase α from a position symmetrical about the plane S-S.

In this model the surface excess quantities of (1) and (2) are each arbitrary, depending upon the position assigned to the interface S-S, although the adsorption of (2) relative to (1) is determined by their concentration profiles. The excess amounts of other extensive thermodynamic quantities, e.g., A, H or S, are similarly defined. It is always possible to assign a position to the interface such that the adsorption of one component is zero. In Fig. 2, the surface excess of the solvent, (1), becomes zero if the interface is moved to the plane S'-S', because equal areas are enclosed to the right and left of S'-S' by the concentration profile. A plane such as S'-S', for which the adsorption of one component (usually the solvent) is zero, is called the Gibbs Dividing Surface, and surface excess quantities defined relative to it are denoted by a superscript, σ, as, for example, the Helmholtz free energy in equation (4).

SURFACE TENSION AND SURFACE FREE ENERGY

The surface tension, i.e., the force per unit length tangential to the surface measured in units of mN m^{-1}, is dimensionally equivalent to an energy per unit area measured in units of mJ m^{-2}. For this reason it has often been stated that the excess surface free energy is identical to the surface tension, but this is true only for a single component system, i.e., a pure liquid. Adopting the Gibbs model for which the surface thickness δ, and volume, V^{σ}, are zero, the second term in equation (4) is zero, and integrating at constant temperature, chemical potential and surface tension:

$$A^{\sigma} = \gamma A + \sum_i \mu_i n_i^{\sigma} \qquad (5)$$

Equation (5) may be expressed in quantities per unit area, Γ_i^{σ}, which are surface excess concentrations (mol m^{-2}) giving

$$A^{\sigma}/A = \gamma + \sum_i \mu_i \Gamma_i^{\sigma} \qquad (6)$$

Thus, for a multicomponent system, the surface excess free energy per unit area is equal to the surface tension, γ, only if the total adsorption is zero.

THE GIBBS ADSORPTION ISOTHERM (Gibbs, 1928)

At constant temperature and assuming that V^{σ} is zero (Gibbs convention) or that V^{σ} dp is negligible (surface phase convention), since A^{σ} is a total differential, equation (5) may be differentiated:

$$dA^{\sigma} = \gamma dA + A d\gamma + \sum_i \mu_i dn_i^{\sigma} + \sum_i n_i^{\sigma} d\mu_i \qquad (7)$$

Now, with the assumptions above, the first two terms of equation (4) disappear, and substracting the remaining terms from equation (7) yields:

$$A d\gamma + \sum_i n_i^{\sigma} d\mu_i = 0 \qquad (8)$$

and reducing this to unit area by dividing by A,

$$-d\gamma = \sum_i \Gamma_i d\mu_i \quad . \qquad (9)$$

Equation (9) is the general form of the Gibbs adsorption isotherm relating a decrease of the surface tension, γ, to the amount of adsorption, Γ_i, and a change in the chemical potential of components, i.

The simplest case of this isotherm is a system of two components in which the solute, (2), is surface active, i.e., it is adsorbed at the surface of the solvent, (1), as in Fig. 2.

For such a case, equation (9) may be written

$$-d\gamma = \Gamma_1\, d\mu_1 + \Gamma_2\, d\mu_2 \tag{10}$$

and if the Gibbs dividing surface (S'-S' in Fig. 2) is used, $\Gamma_1 = 0$, and

$$-d\gamma = \Gamma_2^{(1)}\, d\mu_2 \tag{11}$$

where $\Gamma_2^{(1)}$ is the <u>relative adsorption</u> of (2) with respect to (1). The significance of this term can best be understood by referring to the surface phase model, having a finite thickness, in which Γ_1 and Γ_2 are total surface concentrations. The chemical potentials of (1) and (2), μ_1 and μ_2, must be the same in all phases at equilibrium, and for a bulk phase at constant temperature and pressure, the Gibbs-Duhem equation indicates that in a multicomponent system the chemical potentials of components are related to one another:

$$n_1\, d\mu_1 + n_2\, d\mu_2 + n_3\, d\mu_3 + \ldots = 0 \tag{12}$$

so that for a two-component system

$$d\mu_1 = -(n_2/n_1)d\mu_2 \tag{13}$$

in which n_1 and n_2 are the number of moles of (1) and (2).

Equation (10) can now be written

$$-d\gamma = [\Gamma_2^S - (n_2/n_1)\, \Gamma_1^S]\, d\mu_2 \tag{14}$$

The term in square brackets is clearly equivalent to the relative adsorption $\Gamma_2^{(1)}$ of the Gibbs model, and in the case of dilute solutions of surfactants, n_2/n_1 is very small, e.g., for a solution of a surfactant near to its critical micelle concentration (c.m.c.) at 10^{-3} mol dm^{-3} in water with a concentration of 55.5 mol dm^{-3}, $n_2/n_1 \approx 1.8 \times 10^{-5}$. Since the adsorption of one surfactant molecule is unlikely to cause the desorption of more than 10 solvent molecules, $\Gamma_1^S < -10\, \Gamma_2^S$, and the second term in the bracket of equation (14) is $< 1.8 \times 10^{-4}\, \Gamma_2^S$. Any error, therefore, in identifying $\Gamma_2^{(1)}$ with the total surface excess concentration, Γ_2^S is very small. It is clear also that in the Gibbs model, by locating the dividing surface at S'-S', $\Gamma_1 = 0$, and the term in brackets becomes Γ_2.

In the bulk solution at equilibrium with the surface, the chemical potential of solute, μ_2, is given by

$$\mu_2 = \mu_2^{\ominus} + RT\, \ln a_2 \tag{15}$$

in which the standard chemical potential of (2), $\mu_2^{\ominus}$ refers to a solution of unit ideal activity with either $a_2 = c_2 f_2$ or

$a_2 = X_2 f_2$ where c_2 is the concentration of (2), X_2 is its mol fraction and f_2 is the appropriate activity coefficient. Using the differential form of equation (15), $d\mu_2 = RT\,d\,\ell n\,a_2$, equation (11) may now be written

$$\Gamma_2^{(1)} = -\frac{1}{RT}\frac{d\gamma}{d\,\ell n\,a_2} \tag{16}$$

The choice of units to denote composition in the bulk phase is arbitrary. Surfactants are normally used in dilute solutions and their chemical potential is often limited to the activity at or near their c.m.c., which rarely exceeds 10^{-3} mol dm^{-3}, so that these units, or mol kg^{-1} (molality) seem to be a natural choice. However, interest in micellar aggregation and the structure of phases present in mixtures which may contain mol fractions of solvent between 0.75 and zero, makes it convenient in many cases to use mol fraction units which are easily converted into molalities. In solutions of low concentration, however, the difference between the densities of solutions and their solvents is very small so that volume and mass concentrations are almost identical.

THE ORIENTED MONOLAYER

If an hydroxyl group is substituted in the terminal group of a hydrocarbon molecule, the interfacial tension of the oil with water is dramatically lowered from a value near to the difference between the surface tensions of the phases in contact (Antonow's rule, 1907) to a small value which is almost independent of the hydrocarbon chain length (Table 1). The inference is that the polar oil is oriented with the -OH group in contact with the water.

TABLE 1

Interfacial tensions (mN m^{-1}) between water and pure liquids

n-hexane	51.0	n-butanol	1.6
n-octane	50.8	n-hexanol	6.8
h-hexadecane	53.8	n-octanol	8.5

The changes produced in the surface tension of water by various solutes (Pappenheimer *et al.*, 1936) are illustrated in Fig. 3, from which it is evident that increasing the hydrocarbon content from glycine to ε-amino caproic acid reduces $d\gamma/dc$ from +0.92 to -0.25 mN m^{-1} mol^{-1} dm^3. Separating the hydrophilic "head" from the hydrophobic "tail" in the isomer, α-amino caproic acid, makes it "surface-active" with $d\gamma/dc$ = -23 mN m^{-1} mol^{-1} dm^3. Such observations are consistent with

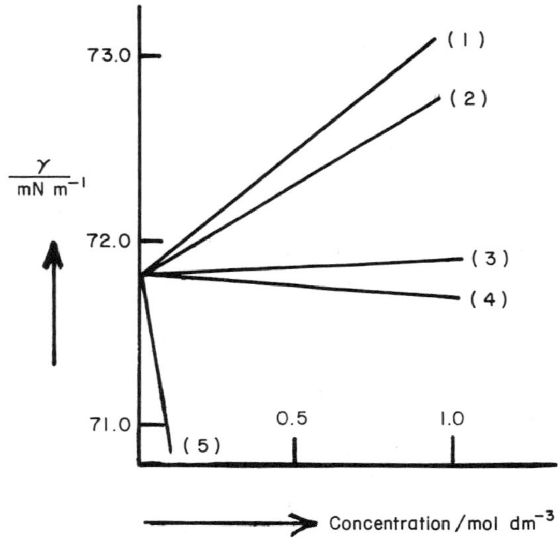

Fig. 3. Surface tensions of aqueous amino-acid solutions (Pappenheimer et al., *1936).*
1 Glycine, $H_2N.CH_2.COOH$ *2 β alanine,* $H_2N.CH_2.CH_2.COOH$
3 β amino butyric acid, $H_2N(CH_2)_3.COOH$
4 ε amino caproic acid, $H_2N(CH_2)_5.COOH$
5 α amino caproic acid, $CH_3(CH_2)_3CH.NH_2.COOH$

an oriented monolayer.

If an adsorbed molecule occupies an area, $\underline{a}$, 1 mol of substance would occupy $N_A.\underline{a}$, where N_A is Avogadro's constant, and equation (16) can be expressed in the alternative forms

$$\Gamma_2^{(1)} = 1/N_A\underline{a} = -(1/RT)\ (d\gamma/d\ \ln a_2) \tag{17a}$$

or

$$1/\underline{a} = -(1/kT)\ (d\gamma/d\ \ln a_2)\ . \tag{17b}$$

For ideal solutions with $f_2 = 1$, a_2 may be replaced by X_2 or c_2 and an alternative form of equations (17a) and (17b) is

$$1/\underline{a} = -(a_2/kT)\ (d\gamma/d\ a_2)\ . \tag{17c}$$

Lengthening the hydrophobic "tail" decreases the solubility of the amphiphile, but it may form an insoluble monolayer either by spreading spontaneously ($C_{16}H_{33}OH$) or using a volatile spreading solvent ($C_{17}H_{35}COOH$). In such cases Γ or $\underline{a}$ is known from the surface area and the amount of substance, and since the amphiphile is insoluble, the spread layer may be separated from the clean surface of the pure liquid substrate by a hydrophobic barrier, and the force per unit length on such a barrier exerted by the surface film is given by $\pi = \gamma_o - \gamma$, i.e., the surface pressure, π, as measured on a Langmuir-Adam trough (Adam, 1941) is equal to the lowering of the surface tension. In differential forms of the Gibbs adsorption isotherm (equations 9, 14, 16 and 17), $-d\gamma$ may be replaced by $d\pi$.

ADSORPTION OF SURFACTANTS AT LIQUID INTERFACES

The term "surfactant" is usually applied to an amphiphile which is water-soluble and strongly adsorbed at the air/water interface, but when adsorption at an oil/water interface takes place it may also be necessary to allow for its solubility in the oil phase, and the term in square brackets in equation (14) would include the negative adsorption of the oil as second solvent.

The hydrophilic "head" group may be unionized (e.g., alcohols or poly(ethylene oxide)-alkane compounds), weakly ionized (e.g., carboxylic acids) or strong electrolytes (e.g., sulphates, sulphonates and quaternary ammonium halides).

Nonionic Surfactants

These surfactants are usually strongly adsorbed at the surface of water from very dilute solutions, having c.m.c. of the order of 10^{-5} mol dm^{-3}, and form closely packed adsorbed layers at even lower concentrations. The activity coefficient of such surfactants is close to unity, and is only slightly affected by variations in ionic strength or pH, so that aqueous solutions behave like the two-component system of equation (10) and the adsorption is accurately described by equations (16) and (17) with concentration, C_2, substituted for the activity, a_2.

1, 1 Electrolyte, e.g., Na^+R^-

In the absence of other electrolytes, the Gibbs isotherm takes the form

$$-d\gamma = \Gamma_{Na^+}\, d\mu_{Na^+} + \Gamma_{R^-}\, d\mu_{R^-} + \Gamma_{H^+}\, d\mu_{H^+} + \Gamma_{OH^-}\, d\mu_{OH^-} + \Gamma_{H_2O}\, d\mu_{H_2O} \qquad (18)$$

and except in the most dilute solutions, the terms in H^+ and OH^- are neglected. The Gibbs-Duhem relationship then reduces to

$$n_{Na^+}\, d\mu_{Na^+} + n_{R^-}\, d\mu_{R^-} + n_{H_2O}\, d\mu_{H_2O} = 0 \qquad (19)$$

which combined with equation (18) gives

$$-d\gamma = (\Gamma_{Na^+} - \Gamma_{H_2O}\, n_{Na^+}/n_{H_2O})\, d\mu_{Na^+} + (\Gamma_{R^-} - \Gamma_{H_2O}\, n_{R^-}/n_{H_2O})$$
$$+ (\Gamma_{R^-} - \Gamma_{H_2O}\, n_{R^-}/n_{H_2O})\, d\mu_{R^-} \qquad (20)$$

Now for the electrical neutrality in the solution and in the surface,

$$n_{Na^+} = n_{R^-} = n_{NaR} \quad ; \quad \Gamma_{Na^+} = \Gamma_{R^-} = \Gamma_{NaR}$$

so that

$$-d\gamma/RT = (\Gamma_{NaR} - \Gamma_{H_2O}\, n_{NaR}/n_{H_2O})\, d\, \ell n\, a_{Na^+}\, a_{R^-} \quad . \qquad (21)$$

But the mean activity $a_\pm = (a_{Na^+}\, a_{R^-})^{\frac{1}{2}}$.

Therefore,

$$-d\gamma/RT = 2(\Gamma_{NaR} - \Gamma_{H_2O}\, n_{NaR}/n_{H_2O})\, d\, \ell n\, a_\pm \qquad (22)$$

and the term in brackets is the relative adsorption, $\Gamma_{NaR}^{(H_2O)}$, of the surfactant.

The factor of 2 in equation (22) arises because both surfactant ion and counterion must be adsorbed to maintain neutrality, and $d\gamma/d\, \ell n\, a_\pm$ is twice as large as for an unionized surfactant.

1, 1 Electrolyte in Presence of Excess Non-adsorbed Electrolyte, e.g., $Na^+R^- + Na^+Cl^-$

If the non-adsorbed electrolyte is present in large excess, an increase in the concentration of Na^+R^- produces a negligible increase of Na^+ so that $d\mu_{Na^+}$ is negligible. Consideration of the Gibbs-Duhem relationship then shows that $d\mu_{Cl^-}$ is also negligible

$$-d\gamma/RT = \Gamma_{R^-}^{(H_2O)}\, d\, \ell n\, a_{R^-} \quad . \qquad (23)$$

The activity coefficient for R^-, f_{R^-}, depends upon the ionic strength of the solution, which is determined by the excess concentration of Na^+Cl^-, and is therefore constant so that

$$-d\gamma/RT = \Gamma_{NaR}^{(H_2O)}\, d\, \ell n\, C_{NaR} \qquad (24)$$

in which $\Gamma_{NaR}^{(H_2O)}$ is the adsorption of the surfactant, and C_{NaR} its concentration in solution. The factor of 2 in equation

(22) has disappeared and the ionic surfactant in excess electrolyte is adsorbed as though it was an unionized molecule.

Weak Electrolyte, e.g., $RH \rightleftharpoons R^- + H^+$

Dissociation of the surfactant can be represented by a dissociation constant, K_a,

$$K_a = C_{R^-} C_{H^+}/C_{RH} \tag{25}$$

and at constant pH, $d \ln C_{R^-} = d \ln C_{RH}$ so that

$$-d\gamma/RT = \Gamma_{RH}\, d \ln C_{RH} + \Gamma_{R^-}\, d \ln C_{R^-} + \Gamma_{H^+}\, d \ln C_{H^+} \tag{26}$$

At constant pH, $d \ln C_{H^+} = 0$, and

$$-d\gamma/RT = (\Gamma_{RH} + \Gamma_{R^-})\, d \ln C_{RH} \quad . \tag{27}$$

Thus the lowering of surface tension depends upon the adsorption of both ionized and unionized forms of the surfactant molecule.

The activity coefficient required for combination with the concentration in equations (22) and (23) is usually determined by experiment, but in many cases application of the Debye-Hückel equation gives values correct within 0.25% for 1:1 electrolytes up to concentrations of 0.1 mol dm^{-3}, and a suitable expression for the activity coefficient, $f_\pm$ in aqueous solution of ionic strength, I, at 25°C is

$$\ln f_\pm = \frac{-1.172\, |Z_+ Z_-|\, I^{\frac{1}{2}}}{1 + 3.3\, \alpha\, I^{\frac{1}{2}}} \tag{22a}$$

where Z_+ and Z_- are the valences of the predominant cations and anions, and α is the mean distance of approach of ions expressed in nm (Boucher *et al.*, 1968).

SURFACE EQUATIONS OF STATE AND ADSORPTION ISOTHERMS

The adsorption, Γ, or area per molecule, $\underline{a}$, can be derived from measurements of the surface tension lowering, π, as described above. An equation relating π to $\underline{a}$ is a <u>surface equation of state</u> and an ideal surface film, behaving as a two-dimensional perfect gas can be represented by the ideal equation,

$$\pi\, \underline{a} = k\, T \tag{28}$$

An equation relating the adsorption, Γ, to the concentration in the bulk phase, c_2, at a given temperature is an <u>adsorption isotherm</u>, and it may be derived from the equation of state by using the Gibbs adsorption isotherm. Determining Γ or $\underline{a}$ from measurements of γ usually involves graphical evaluation of $d\gamma / d \ln c_2$ with substantial errors, so it has been found convenient to test experimental results using the integrated form of the Gibbs adsorption isotherm in the form of a <u>surface pressure equation</u> relating π to c_2 or a_2 directly. From this equation the standard free energy of adsorption may be obtained.

Ideal Surface Equation of State

Combining equations (17c) and (28)

$$\pi = kT/\underline{a} = a_2 \; d\pi/da_2 \tag{29}$$

and on integration

$$\ln a_2 = \ln \pi + I \tag{30}$$

where I is the constant of integration.

Clearly, π/a_2 is constant, and the surface tension decreases linearly with increasing concentration if c_2 is so low that the film obeys equation (28). Γ/a_2 is also constant and represents a constant partition coefficient between surface and bulk phases so that this state of the film is often described as the "Henry's Law Region". Such a region is rarely experimentally attainable.

The constant of integration, I, represents the strength of the tendency of the amphiphile to be adsorbed. If $\Gamma_2^{(1)}$ in equation (11) is replaced by $1/N_A\underline{a}$ and if $\underline{a}$ has the ideal value, kT/π,

$$-d\gamma = d\pi = d\,\mu_2\;\pi/RT \; . \tag{31}$$

If an ideal equation of state can be applied to the surface film, and if unit pressure is adopted as the standard surface state,

$$\mu_2 = \mu_2^{\ominus,s} + RT \ln \pi \tag{32}$$

in which $\mu_2^{\ominus,s}$ is the standard chemical potential of (2) in the surface. Combining equation (32) with equation (15) the standard free energy of adsorption, ΔG_a^o, is given by

$$\Delta G_a^o = \mu_2^{\ominus,s} - \mu_2^{\ominus} = RT \ln (a_2/\pi) \tag{33}$$

ΔG_a^o is obtained, therefore, from the initial slope, $da_2/d\pi$ at very low values of c_2, where the activity coefficient is equal to unity, so that

$$(d\pi/dc_2)_{c_2 \to 0} = \exp(-\Delta G_a^o/RT) \tag{34}$$

Calorimetric measurement of the enthalpy of adsorption at liquid surfaces is impracticable, so this quantity and the standard entropy of adsorption are obtained from adsorption studies over a range of temperatures. Differentiating equation (33)

$$\frac{d\ \ell n\ (d\pi/dc_2)_{c_2 \to 0}}{dT} = \frac{\Delta G_a^o}{RT^2} - \frac{1}{RT}\frac{d(\Delta G_a^o)}{dT} \qquad (35)$$

and since the standard entropy of adsorption, ΔS_a^o, may be defined as $-d(\Delta G_a^o)/dT$,

$$\Delta S_a^o = -\frac{\Delta G_a^o}{T} + RT\frac{d\ \ell n(d\pi/dc_2)_{c_2 \to 0}}{dT} \ . \qquad (36)$$

ΔH_a^o is obtained from ΔG_a^o and ΔS_a^o. The variation of the initial slope of graphs of π vs c_2 with T must be determined.

Choice of Standard States in the Surface

In the ideal equation (28) the standard pressure, $\pi = 1$ mN m^{-1} corresponds to $\underline{a} = 1.38 \times 10^{-2}$T nm^2, which at 298K gives the standard surface area, $\underline{a} = 4.12$ nm^2. This is less convenient for adsorption at solid surfaces where $\underline{a}$ or Γ are measured, because these are dependent on T if $\pi = 1$ mN m^{-1} is the standard state. It would be convenient if a single system of standard states could be adopted so that such thermodynamic quantities as ΔG_a^o could be directly compared for different systems, but for reasons such as that given above, no uniform system has been adopted, and an additional term in (33) for ΔG_a^o is required if results using different standard states are to be compared.

It is desirable also that the standard state in the surface should be as similar as possible to that in the bulk phase, so that a perfect gas pressure of 1 atmosphere in a surface phase of 0.6 nm thickness, and 1 atmosphere in the vapour phase, were adopted for adsorption of vapours at the Hg surface (Kemball and Rideal, 1946). A similar convention (Ward and Tordai, 1946) was adopted for the adsorption of carboxylic acids at the water surface, using a thickness, δ, related to the length of the hydrocarbon chain, and standard concentrations of 1 molecule cm^{-3} or 1 mol dm^{-3} (Davies and Rideal, 1963) were used.

Separation of molecules in the surface by the same distance as in a perfect gas at STP has also been suggested (De Boer, 1968), which applied to the ideal equation (28) gives $\underline{a}$ = 11.16 nm^2 and $\pi = 0.338$ mN m^{-1} (Ross and Chen, 1965). Another proposal, convenient if a Langmuir adsorption isotherm is followed, is half-coverage of the surface, which for surfactants might correspond to $\underline{a} \approx 60\ nm^2$ (Ross and Olivier, 1964).

Non-ideal Surface Films

Deviations from the ideal surface equation (28) occur because the molecules are not dimensionless, but occupy an area, $\underline{a}_o$, which may approximate to that observed in closely packed monolayers at high π. Lateral interactions may reduce the observed pressure, π, by attractive forces between chains at the air/water interface, or may increase π by electrostatic repulsions in the case of ionic surfactants. These deviations may be sufficiently large, even at the lowest measurable concentrations and surface pressures, to render extrapolation to standard conditions a very dubious procedure. In such cases the experimental results can be fitted to suitable equations of state, and for soluble surfactants the integrated form of the pressure equation is often used. Table 2 lists eight equations of state (Parsons, 1955) which have been used in this way to obtain thermodynamic parameters for adsorption.

In addition to the equations listed in Table 2, which are based on physical models of the adsorbed state, experimental results may be fitted to a virial equation,

$$\pi \underline{a} = kT\ (1 + B^*\pi + C^*\pi^2 + D^*\pi^3 + \ldots) \qquad (37)$$

and application of the Gibbs adsorption equation then gives

$$\ell n\ (a_2/\pi) = \Delta G_a^o/RT + B^*\pi + \tfrac{1}{2}C^*\pi^2 + \tfrac{1}{3}D^*\pi^3 + \ldots) \qquad (38)$$

If B^* is independent of temperature, and if C^*, $D^* \ldots \rightarrow 0$, equation (38) reduces to the Volmer equation of state (4 in Table 2).

Interpretation of the various fitting parameters in Table 2 requires the adoption of physical models for the surface film from statistical mechanics or other arguments.

Standard Free Energy of Adsorption

Langmuir (1917) recognized that Traube's rules (1891) that an additional CH_2 group causes a three-fold decrease in the concentration required to produce a given surface pressure, π, could be ascribed to a constant increment in ΔG_a^o per CH_2 group.

If it is assumed that for films formed by members of an homologous series of surfactants, a surface pressure, π, results from a particular value of $\underline{a}$, or Γ, then for two such surfactants, 1 and 2, the activities, a_1 and a_1 at equilibrium with a film of given surface pressure, π, are related by equation (33) in the form

TABLE 2

Equations of state and surface pressure equations

(i) Name	(ii) Equation of state	(iii) Pressure equation	(iv) Adsorption isotherm
1. Ideal	$\pi \underline{a} = kT$	$\ell n\ \pi$	$\ell n(kT/\underline{a})$
2. Freundlich	$\pi \underline{a} = xkT$	$x\ \ell n\ \pi$	$x\ \ell n\ (kT/\underline{a})$
3. Langmuir	$\pi \underline{a} = kT\ \ell n[\underline{a}/(\underline{a}-\underline{a}_o)]$	$\ell n[\exp(\pi \underline{a}_o/kT)-1]$	$-\ell n\ (\underline{a}-\underline{a}_o)$
4. Volmer	$\pi \underline{a} = kT - \underline{a}_o\ \pi$	$\ell n\ \pi + \pi \underline{a}_o/kT$	$\ell n\ [1/\underline{a}-\underline{a}_o) + \underline{a}_o/(\underline{a}-\underline{a}_o)]$
5. Amagat	$\Pi \underline{a} = x\ kT - \underline{a}_o\ \pi$	$x\ \ell n\ \pi + \pi \underline{a}_o/kT$	$x\ \ell n\ [x\ kT/(\underline{a}-\underline{a}_o) + x\ \underline{a}_o/(\underline{a}-\underline{a}_o)]$
6. Van der Waals	$\pi \underline{a} = kT\ \underline{a}/(\underline{a}-\underline{a}_o) + B/\underline{a}_o$		$\ell n\ [kT/(\underline{a}-\underline{a}_o)] + \underline{a}_o/(\underline{a}-\underline{a}_o) + 2B/\underline{a}_o\ kT$
7. Square root	$\pi \underline{a} = kT + C\pi^{\frac{1}{2}}$	$\ell n\ \pi + 2C\pi^{\frac{1}{2}}/kT$	
8. Frumkin	$\pi \underline{a} = (kT\ \underline{a}/\underline{a}_o)\ \ell n\ [\underline{a}/(\underline{a}-\underline{a}_o)] + D/\underline{a}$		$\ell n\ [\underline{a}_o/(\underline{a}-\underline{a}_o)] + 2D/\underline{a}\ kT$
9. Temkin	$\pi \underline{a} = E/\underline{a}$	$2E^{\frac{1}{2}}\ \pi^{\frac{1}{2}}/kT$	$2E/\underline{a}\ kT$

Notes: The expression in column (iii) is the value of $\ell n\ a_2 - \Delta G^o_a/RT$ in terms of the surface pressure, π. The expression in column (iv) is the value of $\ell n\ a_2 - \Delta G^o_a/RT$ in terms of the area per molecule, $\underline{a}$, i.e., is the adsorption isotherm ($\underline{a} = 1/\Gamma$). ΔG^o_a is the standard free energy of adsorption at zero coverage, i.e., $\pi \to 0$, $\underline{a} \to \infty$, except for equations (2) and (5), which do not reduce to the ideal equation under these conditions. Equations (6), (7), (8) and (9) take a similar form at large $\underline{a}$.

$$a_1/a_2 = \exp(\Delta G_a^{o,1}/RT)/\exp(\Delta G_a^{o,2}/RT)$$

$$= \exp\{(\Delta G_a^{o,1} - \Delta G_a^{o,2})/RT\} \quad (39)$$

where $\Delta G_a^{o,1}$ and $\Delta G_a^{o,2}$ are the standard free energies of adsorption of 1 and 2, respectively. Equation (33) refers to an ideal surface equation of state, but any of the equations in Table 2 lead to equation (39) provided that the same equation can be applied to both surfactants.

If the contribution made by the "head" group, $\Delta G_a^{o,h}$ and that of n_c CH_2 groups to the free energy of adsorption, ΔG_a^o is represented by

$$\Delta G_a^o = \Delta G_a^{o,h} + n_c W \quad (40)$$

then, if surfactant 2 has one CH_2 group more than 1,

$$a_1/a_2 = \exp(-W/RT) \quad (41)$$

and putting $a_1/a_2 = 3$ in accordance with Traube's rule gives $W = -2.72$ kJ mol^{-1} per CH_2 group at 25°C. The incremental nature of ΔG_a^o is indicated by the values listed in Table 3.

TABLE 3

Standard free energies, entropies and enthalpies of adsorption at the air/water interface of n-carboxylic acids at 20°C

Acid	n_c	$-\Delta G_a^o$ kJ mol^{-1}	$\delta\Delta G_a^o$	ΔS_a^o JK^{-1} mol^{-1}	$-\Delta H_a^o$ kJ mol^{-1}
propionic	2	6.8		-60.2	24.7
			3.3		
n-butyric	3	10.1		-26.1	18.1
			3.3		
n-valeric	4	13.4		4.6	11.8
			2.1		
n-caproic	5	15.6		27.6	7.1
			3.4		
n-heptanoic	6	19.0		30.3	10.1
			3.7		
n-octanoic	7	22.7			
			2.9		
n-nonanoic	8	25.6			
			4.1		
n-decanoic	9	29.7			

The average increment is -3.26 kJ mol^{-1} per CH_2 group. ΔS_a^o is negative for shorter chains, indicating the loss of degrees of freedom as adsorption takes place, but with increasing chain length, ΔS_a^o becomes less negative and is large and positive for n-heptanoic acid. It is suggested that orientation of water molecules around the hydrocarbon chain lowers the entropy of the molecule in solution, and that adsorption of longer chains results in an increasing gain in entropy.

Between $n_c = 2$ and $n_c = 6$, the average increase in entropy is 22.6 $JK^{-1}mol^{-1}$ per CH_2 group, so that if two water molecules are oriented at each CH_2 group, the entropy of orientation is -11.3 JK^{-1} mol^{-1} per H_2O molecule. This is about half the entropy of ice-formation. Similar results have been obtained for aliphatic alcohols (Posner *et al.*, 1952).

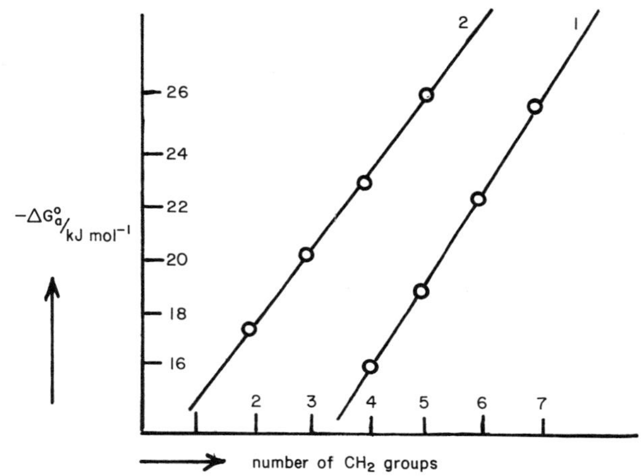

Fig. 4. Standard free energies of adsorption of n-alkanols. 1 at n-dodecane/water interface at 20°C (Mitchell, 1969). 2 at air/water interface at 25°C (Clint et al., *1968).*

Figure 4 shows how additivity rules, analogous to equation (40), can be applied to the adsorption of n-alkanols at the oil/water interface (Mitchell, 1969), compared with the air/water interface (Clint *et al.*, 1968). Similarly, the standard free energies of transfer of some hydrocarbons from water to pure liquid hydrocarbon, ΔG^o_p, illustrated in Fig. 5, show the same linear dependence on the number of carbon atoms (McAuliffe, 1966). A list of the "head" group contributions to the free energy of transfer processes involving an aqueous phase, and incremental values per CH_2 group as in equation (40) is given in Table 4.

The head-group term, $\Delta G^{o,h}_a$, is dependent upon the choice of standard states, but this is unlikely to affect values of W.

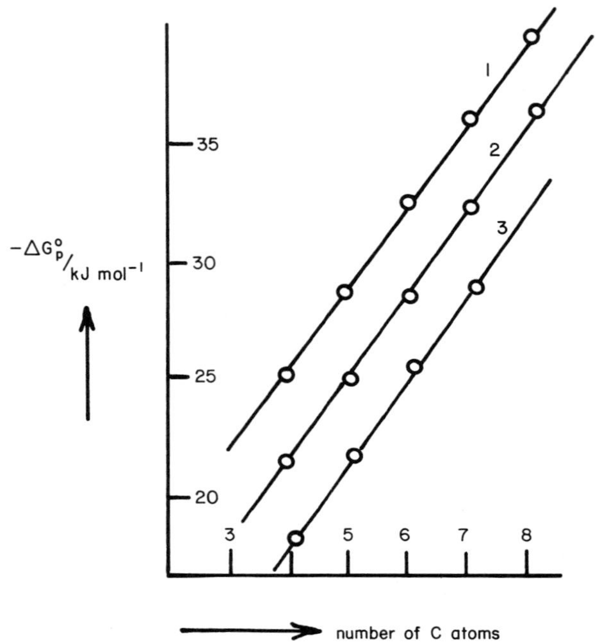

Fig. 5. Standard free energies of transfer from aqueous solution to pure liquid hydrocarbon, ΔG_p^o (Tanford, 1980).
1 Alkanes 2 Alkenes 3 1,ω alkadienes

THE HYDROPHOBIC EFFECT

An early proposal was that the transfer of a hydrocarbon chain from water to a non-aqueous environment was accompanied by the disappearance of an area of oil/water interface, $\delta \underline{a}$, proportional to the number of C atoms, and that the resultant decrease of free energy, δG_a, was given by

$$\delta G_a = \gamma_{12} \, \delta \underline{a} \tag{42}$$

in which γ_{12} is the oil/water interfacial tension. The value of δG_a so predicted often agrees well with values of W.

The separate contributions of enthalpy and entropy to ΔG_a^o is usually obtained from the variation of ΔG_a^o with temperature

$$d(\Delta G_a^o/T)/d(1/T) = \Delta H_a^o \tag{43}$$

but may also be found calorimetrically in the case of transfer between bulk phases. The results of some such studies are

TABLE 4

Standard free energies of adsorption of homologous series of surface active molecules at various interfaces. Free energies of transfer from solution in hydrocarbon liquid to solution in water are given for comparison. Free energies are expressed in the component terms of equation (40).

Substance	Number of C atoms, n_c	$\Delta G_a^{o,h}$/kJ mol^{-1}	W/kJ mol^{-1}	Ref.
	Aqueous solution → liquid hydrocarbon			
Alkanes	4 - 8	-10.23	-3.71	a
Alkenes	3 - 8	-6.31	-3.71	a
Alkadienes	4 - 7	-3.79	-3.61	a
Aliphatic cpds			-3.61	b
n-Alkanols		21.0	-3.3	c
n-Alkanoic acids	8 -22	17.1	-3.47	d
	Water → air/water interface			
n-Alkanols	3 - 7	-2.47	-2.90	e
n-Alkanols	1 - 5	-12.25	-2.75	f
Alkanoic acids	2 - 9	-0.28	-3.26	g
Alkanoic acids			-2.52	b
Acids, alcohols, esters			-2.63	h
	Water → oil/water interface			
n-Alkanol	4 - 7	-2.77	-3.23	i
-COOH		-6.85		j
-OH		-3.36		j
Aliphatic cpds			-3.40	b
Ionic* and nonionic surfactants			-3.40	k
Aliphatic cpds			-3.35 to -3.44	h
	Oil → oil/water interface			
Higher alkanols		-23.3	independent of n_c	l
$-CF_2-$			-5.88	m
	Vapour → air/water interface			
n-Alkane		-2.33	-1.76	n
n-Alkanol		-24.00	-2.25	e

*Ionic surfactants adsorbed from solution at high ionic strength

a. McAuliffe, 1966; b. Davies and Rideal, 1963; c. David *et al.*, 1972; d. Smith and Tanford, 1973; e. Posner *et al.*, 1952; f. Clint *et al.*, 1968; g. Ward and Tordai, 1946; h. Tanford, 1980; i. Mitchell, 1969; j. Langmuir, 1917; k. Haydon and Taylor, 1960; l. Aveyard and Haydon, 1973; m. Klevens and Davies, 1957; n. Cutting *et al.*, 1955.

TABLE 5

Thermodynamic parameters for transfer of hydrocarbons from water to liquid hydrocarbon at 25°C

	ΔG^o	ΔH^o	ΔS^o	ΔC_p^o	$C_p^{o,gas}$
	kJ mol^{-1}	kJ mol^{-1}	$JK^{-1}mol^{-1}$	$JK^{-1}mol^{-1}$	$JK^{-1}mol^{-1}$
C_2H_6	-16.4	10.5	88.2	-	-
C_3H_8	-20.6	7.1	92.4	-	-
C_4H_{10}	-24.8	3.4	96.6	-273	-143
C_5H_{12}	-28.8	2.1	105.0	-403	-172
C_6H_{14}	-32.5	0	109.2	-441	-197
C_6H_6	-19.3	-2.1	58.8	-227	-134
$C_6H_5CH_3$	-22.7	-1.7	71.4	-265	-155
$C_6H_5C_2H_5$	-26.0	-2.0	79.8	-319	-185
$C_6H_5C_3H_8$	-29.0	-2.3	88.2	-395	-

Adapted from Tanford (1980) based mainly on calorimetric data of Gill *et al.* (1976).

given in Table 5. From these it is evident that the principal contribution to the value of ΔG^o is the large positive value of ΔS^o, which increases with hydrocarbon chain length whereas ΔH^o is positive. It is the ordering of water molecules in the vicinity of the hydrocarbon molecule which is relaxed by the removal of hydrocarbon rather than the formation of new H-bonds in the water; the number of such bonds in the presence of hydrocarbon may even slightly exceed that in pure water (Frank and Evans, 1945). This is confirmed by the heat capacity change on transfer from water, ΔC_p^o, which is much larger than the heat capacity in the gas phase, C_p^o, so that it cannot be ascribed to the internal motions of the solute molecules but must result from the change in the structure of the water left behind. Similar results have been obtained from studies of the solubility of n-alkanols in water (Arnett *et al.*, 1969), except that the hydroxyl group forms H-bonds with water which alternate with unbonded regions around the alkyl chains.

An alternative proposal (Aranow and Witten, 1960) is that water molecules surrounding the hydrocarbon chain may restrict the hindered rotation about the C-C bonds. Such a mechanism might contribute about 2.72 kJ mol^{-1} per CH_2 group at 25°C.

MICELLAR AGGREGATION IN DILUTE SOLUTION

An alternative to adsorption as a means of segregating the hydrophobic part of an amphiphilic molecule from an aqueous solvent is the formation of aggregates such that the hydrophobic chains form an inner core with the hydrophilic "head" groups forming an outer surface. These aggregates, called micelles, have a high solubility in water. It was observed (Ekwall, 1927; Jones and Bury, 1927; Preston, 1948) that physical properties of surfactant solutions, which normally depend upon the number of solute molecules per unit volume of solution, often exhibit a discontinuity at a particular concentration, called the c.m.c. above which the vapour pressure and freezing-point, etc., of the solution change very slowly. The simplest model for such behaviour is that above the c.m.c. a phase equilibrium exists with a constant activity of the surfactant in the micellar phase.

Phase Separation Model

This model gives the clearest definition of the c.m.c. and may be formulated by assuming that the chemical potential of the surfactant in the micellar state is a constant at a given temperature and may be adopted as the standard chemical potential, μ_m^o, analogous to the standard chemical potential of a pure liquid or solid. Equilibrium with the surfactant in solution is then represented by

$$\mu_m^o = \mu_1^o + RT \ln a_1 \qquad (44)$$

where μ_1^o is the standard chemical potential in solution and a_1 is the activity of surfactant monomer in solution. Putting $a_1 = X_1F_1$, where X_1 is the mol fraction of monomers and if the activity coefficient, f_1, is taken as unity, the standard free energy of micellization per mol of monomer, ΔG_m^o, given by

$$\Delta G_m^o = \mu_m^o - \mu_1^o = RT \ln X_1 \qquad (45)$$

and the c.m.c., X*, may be identified with X_1. For dilute solutions in water at 25°C, the concentration, X* = C*/55.5 mol dm^{-3} and

$$\Delta G_m^o = RT \ln C^* - RT \ln 55.5 \ . \qquad (46)$$

The phase separation model makes no provision for possible variation in the activity of surfactant as the total concentration of surfactant in the solution is increased, and an equation of state analogous to those used to describe adsorbed films is required to refine the model.

Mass Action Law Model

In its simplest form this postulates an equilibrium between monomers, S and a single micellar species of aggregation number, N:

$$N\,S_1 \rightleftharpoons M_N, \qquad K_m = X_N/X_1^N \tag{47}$$

where K_m is the equilibrium constant. The standard free energy of micellization per mol of monomer, ΔG_m^o, is then given by

$$\Delta G_m^o = -(RT/N)\,\ell n\,K_m$$

$$= -(RT/N)\,\ell n\,X_N + RT\,\ell n\,X_1. \tag{48}$$

For large values of N (>50) at total concentrations not greatly exceeding the c.m.c., the first term is negligible and equation (48) reduces to equation (45) of the phase separation model. The relationship between the phase separation and mass action models can be seen from Fig. 6 in which the proportion of an increment in total concentration X_{tot}, which is incorporated into micelles, dX_N/dX_{tot} is shown as a function of the ratio of the total concentration to the c.m.c., X_{tot}/X^*.

It is an assumption inherent in the model that the association of monomers into the micelle is a cooperative process and that aggregates with smaller values of N are unstable.

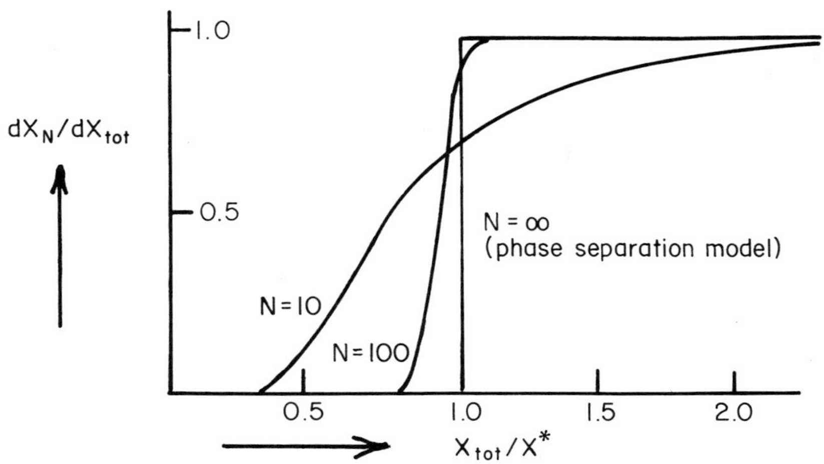

Fig. 6. Distribution of surfactants between monomers and micelles (Lindman and Wennerström, 1980).

This model permits a simple extension to be made to the case of surfactants which are strong electrolytes in dilute solution, but for which the micelles attract a substantial proportion of the counterions into an attached layer so that a uni-valent surfactant with an aggregation number, N, may have an effective charge of p units only, and the degree of dissociation of the micelle is p/N. For such a surfactant, e.g., Na^+S^-, the aggregation process may be represented by

$$NS^- + (N - p)\ Na^+ \rightleftharpoons M^{p-}$$

$$K_m = \frac{X_m}{X_1^N\ X_{Na}^{N-p}} \tag{49}$$

in which it is assumed that activity coefficients are near to unity. A convenient solution (Phillips, 1955) is found at the point of maximum change in gradient of the curve of a colligative property, ϕ, with X_{tot}

$$d^3\phi/dX_{tot}^3 = 0 \tag{50}$$

which leads to the following equation for the free energy of micellization

$$\Delta G_m^o/RT = \frac{\ell n\ 3 + 2\ \ell n\ N}{N} + (1 - \frac{1}{N})\ \ell n(X^* + x) + (1 - \frac{1}{N})\ \ell n\ X^* \tag{51}$$

in which X^* is the c.m.c. and the term, x, is introduced to allow for the effect of added counterions.

The first term in equation (51) is equivalent to the deviations of the curves in Fig. 6 from the discontinuous phase separation graph and becomes negligible for values of $N > 50$. Then in the absence of added non-surface active electrolyte, if $1/N \ll 1$,

$$\Delta G_m^o/RT = (2 - p/N)\ \ell n\ X^* \tag{52}$$

For many ionic surfactants the degree of dissociation of the micelles is found from conductivity measurements to be about 0.2 so that typically the standard free energy, ΔG_m^o, is given by

$$\Delta G_m^o = 1.8\ RT\ \ell n\ X^* \tag{53}$$

and comparison with equation (45) shows why, for similar values of ΔG_m^o, the c.m.c. for ionic surfactants is about two orders of magnitude higher than for nonionic surfactants.

In the presence of an excess of added electrolyte where $x \gg X^*$

$$\Delta G^{o}_{m}/RT = \ln X^{*} + (1 - \frac{p}{N}) \ln x \tag{54}$$

and as x increases the c.m.c. decreases.

For a c.m.c. value, $C^{*} = 8.2 \times 10^{-3}$ mol dm^{-3} ($C_{12}H_{25}SO_4Na^+$ at 25°C), $\Delta G^{o}_{m} = -21.85$ kJ mol^{-1}, and lower values of the c.m.c. would correspond to more negative values of ΔG^{o}_{m}. Ionic surfactants having c.m.c. values in the range 10^{-1} to 10^{-3} mol dm^{-3} in water are likely to exhibit significant deviations from the ideal behaviour outlined above, and a more exact treatment would involve activity coefficients of the ionic species obtained from experimental measurements or by the application of the Debye-Hückel treatment as outlined for adsorption before. There remains the problem of a suitable form for the activity coefficient of a charged micelle for which the approximate forms of the Debye-Hückel theory may not be suitable. It is often assumed that the micellar species behave ideally.

Multiple Equilibrium Model

There is much evidence that the aggregation number of micelles may vary greatly with temperature, especially for nonionic surfactants (Balmbra *et al.*, 1962), and there is also reason to suppose that aggregation numbers may vary with the total concentration of surfactant (Balmbra *et al.*, 1964). An extension of the mass-action model allows for the existence of a range of values of N by postulating a multiple equilibrium characterized by a series of equilibrium constants, K_2, K_3, K_4 ...

$$2\,S_1 \overset{K_2}{\rightleftarrows} S_2;\; S_2 + S_1 \overset{K_3}{\rightleftarrows} S_3;\; \ldots\; S_{N-1} + S_1 \overset{K_n}{\rightleftarrows} S_N \tag{55}$$

These equilibria may also be represented by a series of association reactions of monomers:

$$N\,S_1 \overset{K'_n}{\rightleftarrows} S_N \text{ with } N = 2, 3, 4 \ldots \tag{56}$$

such that

$$K'_n = [S_N]/[S_1]^N \tag{57}$$

and it follows that

$$K'_N = \prod_{i=2}^{N} K_i \quad \text{or} \quad \ln K'_N = \sum_{2}^{N} \ln K_i \,. \tag{58}$$

Associated with these equilibrium constants there will be a series of standard free energy changes which are most conveniently represented by the distinct value for each micellar species of the standard chemical potential, $\mu^{o}_{mic,N}$, per constituent amphiphile molecule (Tanford, 1980). The distribution of surfactant molecules between micelles of different size, N, will then be determined by the variation of $\mu^o_{mic,N}$ with N. The chemical potential per mol of amphiphile in micelles of aggregation number, N, is then given by

$$\mu_{mic,N} = \mu^o_{mic,N} + \frac{RT}{N} \ln \frac{X_N}{N} \tag{59}$$

in which X_N is the mol fraction of amphiphiles in micelles of this size. At equilibrium $\mu_{mic,N}$ must have the same value for all values of N, and must be equal also to the chemical potential of monomer in solution, μ_1 given by

$$\mu_1 = \mu^o_1 + RT \ln X_1 + RT \ln f_1 \tag{60}$$

in which X_1 and f_1 are the mol fraction and activity coefficient of the monomer amphiphile. The standard free energy change of transferring one molecule of amphiphile from the monomer solution to a micelle of aggregation number, N, is then

$$\mu^o_{mic,N} - \mu^o_1 = RT \ln X_1 + RT \ln f_1 - \frac{RT}{N} \ln \frac{X_N}{N} \tag{61}$$

and

$$\ln X_N = -N \frac{\mu^o_{mic,N} - \mu^o_1}{RT} + N \ln X_1 + N \ln f_1 + \ln N \tag{62}$$

The total concentration of amphiphile in micelles, X_{mic}, is given by

$$X_{mic} = \sum_{N=2}^{\infty} X_N \tag{63}$$

and the total amphiphile concentration, X_{tot}, is

$$X_{tot} = X_1 + \sum_{N=2}^{\infty} X_N \tag{64}$$

Equation (62) is an explicit distribution function in terms of micelle size, and a theory to predict the most probable aggregation number, N*, and the sharpness of the distribution function about that value, requires that $\mu^o_{mic,N}$ should have a minimum value for N = N*. The unique value of the aggregation number, N, will then be replaced by a number average value, $\overline{N}_n$, or weight average, $\overline{N}_W$, where

$$\overline{N}_n = \sum_{N=2}^{\infty} X_N \Big/ \sum_{N=2}^{\infty} (X_N/N) \tag{65}$$

and

$$\overline{N}_W = \sum_{N=2}^{\infty} N\ X_N / \sum_{N=2}^{\infty} X_N \ . \qquad (66)$$

For an enclosed volume of any shape, the ratio of surface to volume decreases as the volume increases, so the area of the surface of the core of the micelle per hydrocarbon chain contained in it decreases. Thus, the hydrophobic effect ensures that $\mu^o_{mic,N}$ will decrease sharply as N increases at least until the total volume of the micelle-core attains that of a sphere of radius equal to the extended length of the hydrocarbon chain. This is the cooperative effect which precludes low values of N. Experiment shows that micelles often have aggregation numbers which can be accommodated within a sphere or nearly spherical shape, and that the distribution is quite narrow, with $\overline{N}_n \approx \overline{N}_W \approx \overline{N}^*$. It has therefore been suggested (Tanford, 1980) that as the surface area per amphiphile molecule diminishes, repulsion forces between head-groups cause $\mu^o_{mic,N}$ to increase, and the optimum value, N*, arises from the balance of these opposing forces.

An alternative method of introducing the micelle size distribution into the theory of aggregation is that proposed by Hall and Pethica (1967) who adapted the theory of small system thermodynamics (Hill, 1964) to modify the infinite phase separation model of classical thermodynamics. This theory allows the number of intensive variables to be increased by one, and the micelle size distribution function, F(N), is determined by an equation of state as a dependent of the temperature, T, pressure, p, and the chemical potential of monomer, μ_1.

In the analysis of experimental results, the size distribution is only very rarely available. Light scattering and ultracentrifuge studies give the weight average aggregation number, $\overline{N}_W$, but the number average, $\overline{N}_n$, is very difficult to obtain. However, where measurements have been made for small micelles, it has been found that $\overline{N}_W$ and $\overline{N}_n$ are equal within 2% (Huisman, 1964). Larger micelles, however, exhibit a ratio $\overline{N}_W/\overline{N}_n$ up to 2, indicating very broad distributions.

In the case where $\overline{N}_W \approx \overline{N}_n$, Tanford (1980) rewrites equation (62) in the form

$$\ln X_{mic} = -N^* \frac{\mu^o_{mic} - \mu^o_1}{RT} + N^* \ln X_1 + N^* \ln f_1 + \ln N^* \qquad (67)$$

in which μ^o_{mic} is not an average of $\mu^o_{mic,N}$, but is more negative because X_{mic} is the sum of the terms, X_N.

Putting typical values of N* into equation (67) with $f_1 = 1$ it becomes obvious that while X_1 remains equal to X_{tot} until very nearly the c.m.c., as in the phase separation model, above the c.m.c., X_1 continues to rise with increasing total amount of surfactant, X_{tot}, and the mol fraction of surfactant in micellar form, X_{mic}, begins to increase above zero just below the c.m.c., and then increases almost linearly as further surfactant is added.

Prediction of Critical Micelle Concentration Values

Linear relationships have often been observed between the logarithm of measured c.m.c. values and the number of C atoms, n_c, in the hydrophobic "tail" of the surfactant. Becher (1967) summarizes such results for nonionic surfactants in equation (68)

$$\log C^* = 1.03 - 0.5\, n_c\ . \tag{68}$$

Using equation (46) for the phase separation model, this empirical equation becomes

$$\Delta G_m^o = \mu_{mic}^o - \mu_1^o = 1.85 - 2.85\, n_c \tag{69}$$

with units of kJ mol^{-1}. Similarly, for n-alkyl betaines (Swarbrick and Daruwala, 1970), Tanford (1980) has obtained equation (70) using equation (67) with $f_1 = 1$,

$$\mu_{mic}^o - \mu_1^o = 12.31 - 3.07\, n_c \tag{70}$$

or, using equation (45) and so omitting the size-dependent term,

$$\mu_{mic}^o - \mu_1^o = 10.56 - 2.98\, n_c\ . \tag{71}$$

These relationships strongly resemble those found for the standard free energy of adsorption from water into an oil/water interface, listed in Table 4, giving an average value for W = 3.40, and for adsorption at the air/water interface, W = -2.80 kJ mol^{-1} per CH_2.

In the case of ionic surfactants, in the absence of added electrolyte, a lower value, W = -1.76 is probably the result of an ionic strength effect: the ionic strength is approximately equal to the c.m.c., so that as n_c increases, it partly counteracts the change expected at constant ionic strength.

For nonionic surfactants of the poly(ethylene oxide) monoether type, where the hydrophilic contribution is dependent on the polymer chain length, the contribution to the standard free energy of micellization per ethylene oxide residue is ca. 0.46 kJ mol^{-1} (Becher, 1967), so that an equation similar to equations (69) and (70) holds, but with a much smaller coefficient of n_c.

Energetics and the Size and Shape of Micelles

Above, a qualitative explanation has been given for the co-operative aggregation of surfactant molecules, and for the opposing effect of repulsion between head-groups at the surface of the core of the micelle, which limits the growth of the micelle.

To quantify this model, Tanford (1980) separated the standard free energy of micellization, $\Delta G_m^o = \mu_{mic}^o - \mu_1^o$ into two factors,

$$\mu_{mic}^o - \mu_1^o = U_m^o - \mu_1^o + W_m \tag{72}$$

in which the hydrophobic effect is represented by $U_m^o - \mu_1^o$ negative values leading to association, and W_m is the energy of repulsion between head-groups in which positive values tend to dissociate the micelle.

The micelle is considered to consist of a liquid hydrocarbon core, with some degree of orientation of the chains near the interface with the aqueous medium, surrounded by an hydrophilic sheath. The density of the core being similar to that of liquid alkane, the volume additivity principle (Cohn and Edsall, 1943) gives a value, V nm^3, for the volume of a core containing N hydrocarbon chains of length, n_c'

$$V = N(0.027 + 0.0269\, n_c') \tag{73}$$

There is reason to suppose that the CH_2 group adjacent to the head-group may be exposed to water, so n_c' may be one less than n_c. The length of the hydrocarbon chain, ℓ_{max} nm, cannot exceed

$$\ell_{max} = 0.15 + 0.1265\, n_c \tag{74}$$

and flexing of the chain may lead to real lengths which are shorter, 0.75 or 0.80 ℓ_{max}. This length would be the limit of one dimension of the micelle since it cannot have a void at its centre.

The volume of the core of a micelle is proportional to N and if a spherical shape is assumed, the average surface area per hydrocarbon chain is proportional to $1/N^{\frac{1}{3}}$. Thus, if the hydrocarbon chain length is 0.75 ℓ_{max}, the maximum aggregation number for a surfactant with 12 C atoms is 23, increasing to N = 56 if fully stretched chains are allowed. There is evidence (Mysels and Princen, 1959; Emerson and Holtzer, 1967) that the aggregation number of dodecyl sulphate ranges from 62 to 126 depending upon the amount of added salt, so that Tanford (1980) considers that spherical micelles are rare, and that disc-shaped or rod-shaped models are required to accommodate the observed aggregation numbers.

The hydrophobic associative energy is estimated from the immersion of the hydrocarbon "tail" in the core of the micelle using equations such as (69) or (70), but there remains a hydrocarbon surface area, exposed to water, equal to the average area per surfactant molecule less the cross-sectional area of the chain itself (ca. 0.21 nm^2), so that a shape which minimizes this area is energetically favoured. Figure 7 illustrates the areas calculated by Tanford (1980) for various shapes of micelles at the nearest distance of approach (0.15 nm) to the surface of the core of the micelle. It is evident that disc-shaped or oblate ellipsoidal micelles are energetically favoured. A small allowance C, was made for the restricted freedom of chains near the surface of the core and for penetration of water slightly beyond the first CH_2 group, the effect being considered to be less than that of one CH_2 group. The terminal CH_3 contributes -8.82 kJ mol^{-1}, and $(n_c - 2)$ CH_2 groups contribute $-2.94\ (n_c - 2)$ kJ mol^{-1}, so that if the surface free energy at the hydrocarbon/water interface is assumed to be 10.5 kJ $nm^{-2}mol^{-1}$, the total hydrophobic component, $U^o_m - \mu^o_1$ is given by

$$U^o_m - \mu^o_1 = -8.82 - 2.94\ (n_c - 2) + 10.5\ (\underline{a}_N - 0.21) + C \quad (75)$$

where $\underline{a}_N$ is the surface area per surfactant molecule for a micelle with aggregation number, N. Since $\underline{a}_N$ decreases as N increases, $U^o_N - \mu^o_1$ becomes more negative with increasing micelle size.

The opposing component is the repulsion-energy of the amphiphile head-groups. For ionic surfactants this could be calculated by adapting the Debye-Hückel theory to obtain W_m, the repulsion free energy per head-group in terms of the

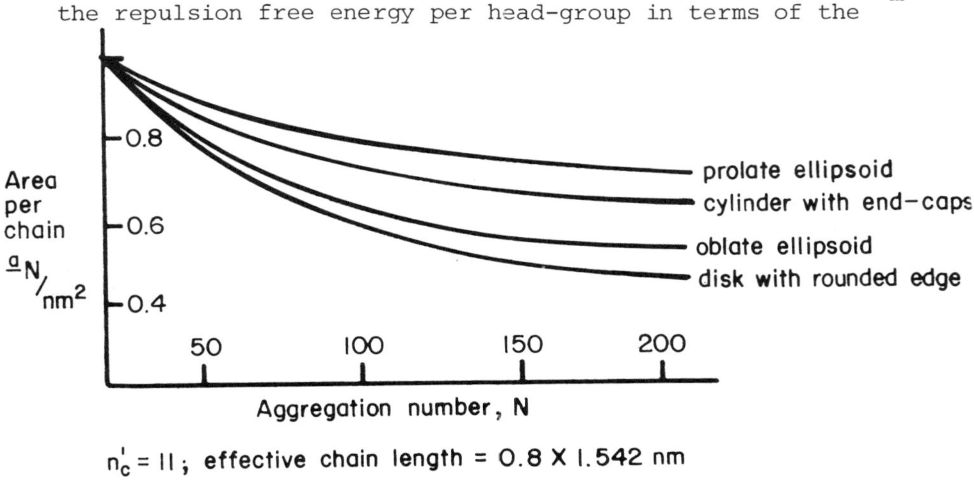

Fig. 7. Areas per surfactant molecule at the surface of the micellar core (Tanford, 1980).

average area per group, a_N. Charged head-groups, however, are largely neutralized by attached counter-ions, the degree of dissociation, p/N, as in equation (52) often being about 0.2. The interaction energy W_m, is therefore more reliably obtained by integrating the equation of state for a monolayer of the surfactant (π vs. a) to obtain the work of compression from which the work of compressing an ideal film ($\pi a = kT$) is subtracted. This procedure can be applied also to evaluate the repulsion energy, arising from steric considerations, of nonionic surfactants.

It is evident that Tanford's theory predicts that disc-shaped micelles (see Fig. 7) would be the most stable. Using equation (75) for the negative associative free energy, and both calculated and experimental estimates of the repulsion energy to give a value of $\ell n\ X_N$ in equation (62) as a function of N, he has calculated the distribution of aggregation numbers as a function of the total surfactant concentration for anionic and cationic surfactants so obtaining the most probable aggregation number N* and also the c.m.c. Very good agreement was obtained between calculated values, N* = 60 and c.m.c. = 6.2×10^{-3} mol dm^{-3} for cetyl trimethylammonium bromide, and the experimental values, N* = 59 and c.m.c. = 6.6×10^{-3} mol cm^{-3}, for example.

This theory of the size and shape of micelles does not predict the formation of prolate ellipsoids or rod-shaped aggregates because their larger surface areas are energetically unfavourable. There is evidence from light-scattering (Balmbra *et al.*, 1962) and from ultracentrifugation (Attwood *et al.*, 1970) that the aggregation number of $C_{12}H_{25}(OCH_2CH_2)_6OH$ increases from ca. 140 at 15°C to ca. 4000 at 45°C with a high asymmetry which accords better with an elongated than a flattened shape. Such surfactants undergo a cloud-point at slightly higher temperature, which is probably associated with the loss of solubility of the poly(ethylene oxide) so that secondary association of nearly spherical micelles may be involved. However, $C_{12}H_{25}(OCH_2CH_2)_8OH$ has an aggregation number of ca. 120 up to high concentration (Corkill and Walker, 1972), so that the size of the hydrophilic group is also important. The effect of the size of the micelle upon the energy of association is expressed in the third, area-dependent term of equation (75), but this uses only the average area per amphiphile molecule a_N, and the restriction of the freedom of chains near the surface of the micellar core was expressed only in the small constant, C. The effect of the shape of the surfactant molecules upon their packing in regions of the micellar surface of different curvature may be introduced by elaborating the term, C, to allow for the effects of the curvature upon the local chemical potential of the surfactant molecule. Clearly, shapes of minimum and uniform curvature (Israelachvili *et al.*, 1976) will be favoured for the accommodation of bulky hydrocarbon chains attached to rather small head-groups. Such pomoid shapes may be expanded into toroidal rings with quite small changes of the average area per amphiphile molecule, $\underline{a}_N$, so that the variation of ΔG_N^o with N may be smaller than that predicted for nearly spherical micelles. The energy penalty in severing a toroid to form a cylindrical rod may not be prohibitive, so that micelles of large N with rod-like asymmetries are energetically possible. The very slight variation of ΔG_N^o with N in such cases, suggests much wider distribution functions of aggregation numbers than with micelles of small N.

Entropy and Enthalpy of Micellization

Since the hydrophobic effect as shown by the solubilities of hydrocarbons in water and the standard free energy of adsorption at air/water and oil/water interfaces described before has been found to result from a small, usually positive, enthalpy of removal of the surfactant from the monomeric dissolved state, and a large positive entropy, it is to be expected that the standard free energy of micellization should be similar. Some of the results of such measurements have been collected by Rosen (1978) showing that the large negative free energies of micellization which becomes increasingly negative as the chain length increases, are made up of a small positive enthalpy which diminishes slightly with an increasing hydrocarbon chain length, and a large positive term, $T\Delta S_m^o$, which becomes more positive as the chain is lengthened. In

TABLE 6

Thermodynamic parameters of micellization of alkyl sulphoxides

Solvent: H_2O *Temperature: 23.6 °C*

Compound	ΔG_m^o/kJ mol^{-1}	ΔH_m^o/kJ mol^{-1}	$T\Delta S_m^o$/kJ mol^{-1}
$C_6H_{13}S(CH_3)O$	-12.0	+10.6	+22.6
$C_7H_{15}S(CH_3)O$	-15.9	+9.2	+25.1
$C_8H_{17}S(CH_3)O$	-18.8	+7.8	+26.4
$C_9H_{19}S(CH_3)O$	-22.0	+7.1	+29.1
$C_{10}H_{21}S(CH_3)O$	-25.5	+5.4	+30.9
$C_{11}H_{23}S(CH_3)O$	-28.7	+3.0	+31.7

the case of the mainly nonionic surfactants listed by Rosen (1978) the positive standard entropy of micellization is clearly the result of the hydrophobic effect, discussed above. An example (Clint and Walker, 1975) is given in Table 6.

Ionic and zwitterionic surfactants are complicated by the entropy associated with the release of water molecules by the surfactant ion upon partial neutralization of the ionic charge by counterions when aggregation occurs. This adds to the positive standard entropy of micellization, but in such cases it is not entirely clear how much is contributed to ΔS^{o}_{m} by solvation effects of the head-group, and how much by water structuring around the hydrocarbon chains.

REFERENCES

Adam, N.K. and Jessop, G. (1926). *Proc. Royal Soc. Ser. A* 110, 423.

Adam, N.K. (1941). "The Physics and Chemistry of Surfaces". 3rd edn, p.28, Oxford. Reprinted 1968, Dover Publications.

Antonow, G. (1970). *J. Chim. Phys.* 5, 364; 372.

Aranow, R.H. and Witten, L. (1960). *J. Phys. Chem.* 64, 1643.

Aranow, R.H. and Witten, L. (1961). *J. Chem. Phys.* 35, 1504.

Arnett, E.M., Kover, W.B. and Carter, J.V. (1969). *J. Amer. Chem. Soc.* 91, 4028.

Attwood, D., Elworthy, P.H. and Kayne, S.B. (1970). *J. Phys. Chem.* 74, 3529.

Aveyard, R. and Haydon, D.A. (1973). "An Introduction to the Principles of Surface Chemistry". Cambridge University Press, Cambridge.

Balmbra, A.R., Clunie, J.S., Corkill, J.M. and Goodman, J.F. (1962). *Trans. Faraday Soc.* 58, 1661.

Balmbra, A.R., Clunie, J.S., Corkill, J.M. and Goodman, J.F. (1964). *Trans. Faraday Soc.* 60, 979.

Becher, P. (1967). *In* "Nonionic Surfactants" (Ed. M.J. Schick) Chapter 15. Dekker, New York.

Betts, J.J. and Pethica, B.A. (1960). *Trans. Faraday Soc.* 56, 1515.

Boucher, E.A., Grinchuk, R.M. and Zettlemoyer, A.C. (1968). *J. Amer. Oil Chem. Soc.* 45, 49.

Clint, J.H., Corkill, J.M., Goodman, J.F. and Tate, J.R. (1968). *J. Colloid Interface Sci.* 28, 522.

Clint, J.H. and Walker, T. (1975). *J. Chem. Soc. Faraday 1* 71, 946.

Cohn, E.J. and Edsall, J.T. (1943). "Proteins, Amino-acids and Peptides". Reinhold, New York.

Corkill, J.M. and Walker, T. (1972). *J. Colloid Interface Sci.* 39, 621.

Cutting, C.L. and Jones, D.C. (1955). *J. Chem. Soc.*, 4067.

David, S.S., Higuchi, T. and Rytting, J.N. (1972). *J. Pharm. Pharmac* 24, supplement. 30 pp.
Davies, J.T. and Rideal, E.K. (1963). "Interfacial Phenomena". Academic Press, London.
De Boer, J.H. (1968). "The Dynamical Character of Adsorption". 2nd edn. Clarendon Press, Oxford.
Eicke, H.F. (1980). *Topics in Current Chemistry* 87, 86. Springer-Verlag, Berlin.
Ekwall, P. (1927). *Acta Acad. Abo.* 4, 1.
Emerson, M.F. and Holtzer, A. (1967). *J. Phys. Chem.* 71, 1898.
Fendler, E.J., Fendler, J.H., Medary, R.T. and El-Seoud, O.A. (1973). *J. Phys. Chem.* 77, 1432.
Frank, H.S. and Evans, M.W. (1945). *J. Chem. Phys.* 13, 507.
Gibbs, J.W. (1928). "The Collected Works of J. Willard Gibbs". 2nd edn. Vol.1, p.218. Longmans-Green, New York.
Gill, S.J., Nichols, N.F. and Wadsö, I. (1976). *J. Chem. Thermodynamics* 8, 445.
Hall, D.G. and Pethica, B.A. (1967). *In* "Nonionic Surfactants" (Ed. M.J. Schick), Chapter 16. Dekker, New York.
Haydon, D.A. and Taylor, F.H. (1960). *Phil. Trans. Royal Soc.* 252, 225.
Haydon, D.A. and Taylor, F.H. (1962). *Trans. Faraday Soc.* 58, 1233.
Hill, T.L. (1964). "Thermodynamics of Small Systems". Benjamin, New York.
Huisman, F. (1964). *Proc. Kon. Ned. Akad. Wetensch. Ser. B* 67, 388; 407.
Israelachvili, J.N., Mitchell, D.J. and Ninham, B.W. (1976). *J. Chem. Soc. Faraday 1* 72, 1525.
Jarvis, N.L. and Zisman, W.A. (1960). *J. Phys. Chem.* 64, 150; 157.
Jones, D.C. and Ottewill, R.H. (1955). *J. Chem. Soc.*, 4076.
Jones, E. and Bury, C.R. (1927). *Phil. Mag.* 4, 841.
Kemball, C. and Rideal, E.K. (1946). *Proc. Royal Soc. Ser. A* 187, 53.
Klevens, H.B. and Davies, J.T. (1957). *Proc. 2nd Int. Congress Surface Activity* 1, 31.
Langmuir, I. (1917). *J. Amer. Chem. Soc.* 39, 1848.
Lindman, B. and Wennerström, H. (1980). *Topics in Current Chemistry* 87, 32. Springer-Verlag, Berlin.
McAuliffe, C. (1966). *J. Phys. Chem.* 70, 1267.
Mitchell, R.W. (1969). Ph.D. thesis, Hull. *Quoted in* Aveyard, R. and Haydon, D.A. (1973). "An Introduction to the Principles of Surface Chemistry". Cambridge University Press, Cambridge.
Mysels, K. and Princen, L. (1959). *J. Phys. Chem.* 63, 1696.
Pappenheimer, J.R., Lepie, N.W. and Wyman, J. (1936). *J. Amer. Chem. Soc.* 58, 1851.

Parsons, R. (1955). *Trans. Faraday Soc.* 51, 1518.
Parsons, R. (1959). *Trudy Chetvertogo Soveshchaniya po Elektrokhimii* 42.
Parsons, R. (1961). *Proc. Royal Soc. Ser. A* 261, 79.
Phillips, J.N. (1955). *Trans. Faraday Soc.* 51, 561.
Posner, A.M., Anderson, J.R. and Alexander, A.E. (1952). *J. Colloid Sci.* 7, 623.
Preston, W.C. (1948). *J. Phys. Colloid Chem.* 52, 84.
Rosen, M.L. (1978). "Surfactants and Interfacial Phenomena", pp.116, 117. Wiley-Interscience, New York.
Ross, S. and Olivier, J.P. (1964). "On Physical Adsorption". Interscience, New York.
Ross, S. and Chen, E.S. (1965). *Ind. Eng. Chem.* 57, 40.
Singleterry, C.R. (1955). *J. Amer. Oil Chem. Soc.* 32, 446.
Smith, R. and Tanford, C. (1973). *Proc. Nat. Acad. Sci., USA* 70, 289.
Swarbrick, J. and Daruwala, J. (1970). *J. Phys. Chem.* 74, 1293.
Tanford, C. (1980). "The Hydrophobic Effect". 2nd edn. Wiley, New York.
Traube, J. (1891). *Ann.* 265, 27.
Ward, A.F.H. and Tordai, L. (1946). *Trans. Faraday Soc.* 42, 399; 408; 413.

PHASE EQUILIBRIA AND MESOPHASES IN SURFACTANT SYSTEMS

R.G. Laughlin

Miami Valley Laboratories,
The Procter and Gamble Company,
Cincinnati, Ohio, USA.

INTRODUCTION

It is the intent of this chapter (1) to review schematically the thermodynamic nonidealities that underlie phase heterogeneity; (2) to state the laws governing phase behaviour and their utility in reading and interpreting phase diagrams; (3) to characterize the distinctive features of the phase behaviour of surfactant systems; and (4) to indicate some aspects of the utility of such information. The emphasis is on two-component surfactant/water systems. Besides being itself important, the phase behaviour of these systems serves as a boundary condition for related systems to which salts or oils are added. These particular ternary systems are important both scientifically and in technology, and will be treated but in lesser detail.

THE THERMODYNAMIC BASIS FOR PHASE SEPARATION

A pure surfactant and pure water, at specified temperature and pressure, each have a characteristic molar Gibbs free energy content relative to some arbitrarily chosen standard state. The dependence of the actual free energy of a molar quantity of hypothetical homogeneous mixtures on mole fraction, and any system displaying partial miscibility, can be depicted as in Fig. 1 provided the standard states are chosen to be the pure components (Bett, 1975). This function is presumably a smooth curve, ABCDE. For surfactant systems it is expected for the most part to lie below the standard state limits, because both the enthalpy of the interaction of water with hydrophilic groups (Laughlin, 1978) and the entropy of

SURFACTANTS
ISBN 0-12-682180-1

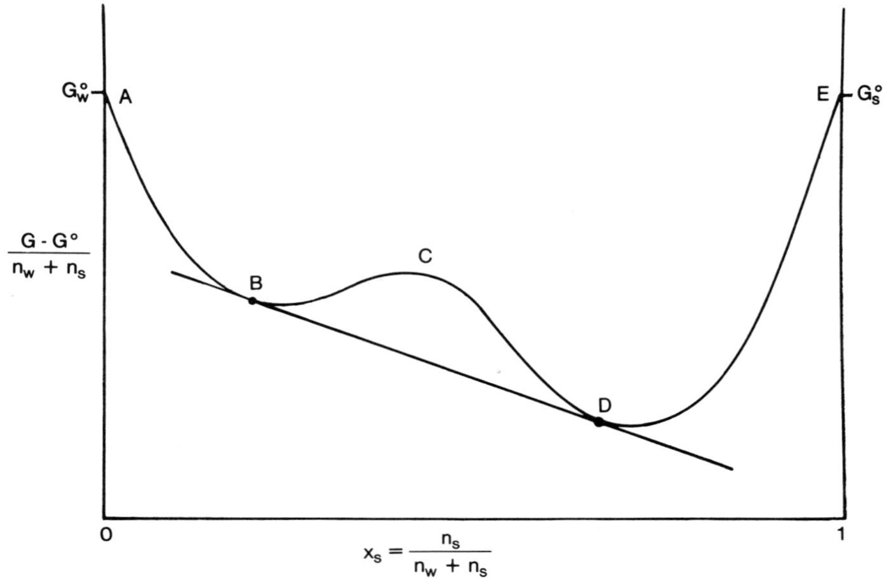

Fig. 1. Free energy vs. composition in mixtures of water and surfactant. The abscissa is mole fraction of surfactant (x_s); n_s and n_w = the number of moles of surfactant and water, respectively. G^o is the standard state free energy of each component, and the ordinate is the difference in free energy between an actual mixture and the standard state free energy, per mole of mixture.

mixing term tend to favour miscibility.

Over certain composition ranges, e.g., A-B and D-E, the homogeneous mixture constitutes the lowest free energy state of the system. Within these ranges we say that the surfactant and water "are miscible", "are soluble in one another", or that "the system is one-phase".

Between B and D, however, the homogeneous mixture is <u>not</u> the lowest free energy state of the system - even though its free energy may lie below that of the pure components (Prigogine, 1954; Bett, 1975). Here we say that a "miscibility gap exists" or that "the system is two-phase". At any composition between B and D, such as C, a momentarily homogeneous system is either unstable or metastable (Prigogine, 1954) and will, at equilibrium, become heterogeneous. That is, it will spontaneously rearrange itself, by countercurrent mass flow across interfaces, into two different phases. The space within each new phase is macroscopically homogeneous. For all values of C these phases are B, of higher water content than C, and D, of higher surfactant content. The phases B and D are described as "coexisting" or "conjugate" phases.

The coexisting phases B and D correspond to those unique points on the molar free energy curve such that a line drawn between them is tangential to the curve at both points. The molar chemical potential of the surfactant, μ_S, is the first derivative of the total free energy of the system with respect to the number of moles of surfactant, n_S, at constant temperature and pressure:

$$\mu_s = \left(\frac{\partial G}{\partial n_s}\right)_{T,P,\ n_w} \quad (1)$$

The chemical potential of the water, μ_w, is similarly defined:

$$\mu_w = \left(\frac{\partial G}{\partial n_w}\right)_{T,P,\ n_s} \quad (2)$$

As equilibrium, both of these potentials (as well as temperature and pressure) must be the same in both phases. Where composition is stated as mole fraction, as in Fig. 1, the slope of the curve at any point corresponds to the difference in chemical potentials, $\mu_S - \mu_W$, at that point. That is, at B, the slope = $\mu_S^B - \mu_W^B$ and at D the slope = $\mu_S^D - \mu_W^D$. Since at equilibrium $\mu_W^B = \mu_W^D$ and $\mu_S^B = \mu_S^D$, it must follow that the slope at points B and D must be the same as well.

Thus, Fig. 1 graphically depicts the general form of a free energy/composition function which dictates that at all compositions between B and D separation into two phases will occur at equilibrium. Figure 1 further depicts how the chemical potential of both surfactant and water can be identical (and thus the system be at equilibrium) in two phases that differ both in composition and in molar free energy.

These concepts are taken directly from general texts dealing broadly with chemical thermodynamics. It is important to recognize that they apply equally as well to surfactant systems as to simpler molecules, irrespective of the greatly enhanced activity at surfaces and exaggerated association in solution that surfactants display. There are no known exceptions to this generalization.

THE DIMENSIONS OF PHASE INFORMATION

A phase diagram is a graphical portrayal of phase information, which answers three basic questions: (1) how many phases exist?; (2) what is the qualitative nature of each phase?; (3) what is the composition of each phase? From a properly constructed and sufficiently complete diagram one can efficiently answer by inspection, or derive by calculation, the answers to each of these three questions for any coordinates of composition and temperature within the diagram. Further, from such diagrams other kinds of information may be deduced. The qualitative and quantitative definition of solubility and its dependence on temperature, the variance of the system, or (within multiphase regions) the proportion of the system in each of the coexisting phases are examples of derived information.

It is worth noting that the micellar regions of surfactant/water systems where the lipophilic group of the surfactant is not unusually large, such as sodium dodecylsulphate or cetyltrimethylammonium bromide, have been shown beyond reasonable doubt (Mukerjee, 1958; Wennerström, 1979; Danielsson, 1981) to be single phases and "homogeneous" in the macroscopic phase sense. In less well-studied systems, such as polar lipids in water, the long-lived stability of colloidal dispersions of one phase in another and the physical similarity of these dispersions to true micellar solutions render the decision as to "how many phases?" a non-trivial experimental problem that is yet to be definitively resolved.

LAWS GOVERNING HETEROGENEOUS EQUILIBRIA: THE GIBBS PHASE RULE

When the number of components, c, is specified, then information as to the number of phases present at equilibrium, p, determines the "variance" or degrees of freedom, f, of any system - including surfactant systems. Around 1875, Gibbs (1928) demonstrated that the relationship between these parameters must be (within clearly stated boundary conditions)

$$f = c - p + 2 \qquad (3)$$

The boundary conditions are that the influence of electrical fields and gravity can be neglected, that pressure is uniform in all directions, and that surface energies are small in comparison with bulk energies. The phase rule has proven to be exceedingly useful in guiding both the determination of phase information and its proper use.

For nonionic surfactants or oils, the number of components equals the number of different molecules present. Each homologue, isomer, or ethoxylogue counts. For inorganic salts or ionic surfactants, which are electrically neutral sets of formally charged molecules, each set is a component. For acidic or basic surfactants (of either class) that undergo significant but variable hydrolysis, the hydrolysis product must also be counted as a component (Ekwall, 1975; Laughlin, 1981). For example, phase studies of a pure soap in pure water yields data that seemingly violate the phase rule - unless fatty acid formed by the hydrolysis reaction is reckoned with:

$$RCO_2^-, Na^+ + H_2O \rightleftharpoons RCO_2H + Na^+, OH^-$$

This has been shown to be a general phenomenon for any anionic surfactant whose basicity resembles that of soaps (Laughlin, unreported). On the other hand, the amine salts (R_3NH^+, Cl^-), which hydrolyse much less strongly in water, seem to behave as single components (Broome, 1951).

The "variance" is the number of intensive thermodynamic variables (typically pressure, temperature, or chemical potentials) that must be specified to precisely determine the phase information (Tamas, 1970; Ricci, 1951; Findley, 1938). In surfactant/water systems ($c = 2$) that are one-phase ($p = 1$), the variance $f = 3$. If $p = 1$, the applied pressure must exceed the vapour pressure of the system so that no gas phase exists. If pressure is fixed, above this limit, $f = 2$.

Figure 2 is the phase diagram of the $C_{10}E_5$*/W system. Referring back to Fig. 1, the molar free energy along the <u>isotherm</u> (a line of constant temperature) at 65°C should behave according to the hypothetical profile ABCDE of Fig. 1. If we look at any composition on the isotherm between D and E (or the very narrow composition range between A and B), we can see that a small variation in both x_S and temperature is possible without qualitatively changing the phase information; the system remains one liquid phase and <u>both</u> x_S and temperature must be specified to define the system. We say, at this point, that $f = 2$ (pressure will from here on be specified as described above).

Suppose, however, we consider a point on the isotherm between B and D. As noted earlier, if we vary x_S by a small amount along this isotherm <u>absolutely no change occurs in the phases present</u>. But, if we vary temperature along the <u>isopleth</u> (a line of constant composition but variable temperature, which is vertical in our representation), e.g., at $x_S = 0.02$, the phases in equilibrium <u>do</u> change (the compositions at B' and D' differ slightly). Thus, within this two-phase region the variance $f = 1$, one less than in the one-phase region. For any given system an increase in the number of phases requires a corresponding reduction in the variance, and vice versa.

As one decreases temperature within the miscibility gap of Fig. 2 the composition differences between coexisting phases

---

*$C_{10}E_5 = C_{10}H_{21}(OCH_2CH_2)_5OH$ = 3,6,9,12,15-pentaoxapentacosan-1-ol, and
$C_{10}E_4 = C_{10}H_{21}(OCH_2CH_2)_4OH$ = 3,6,9,12-tetraoxadocosan-1-ol.

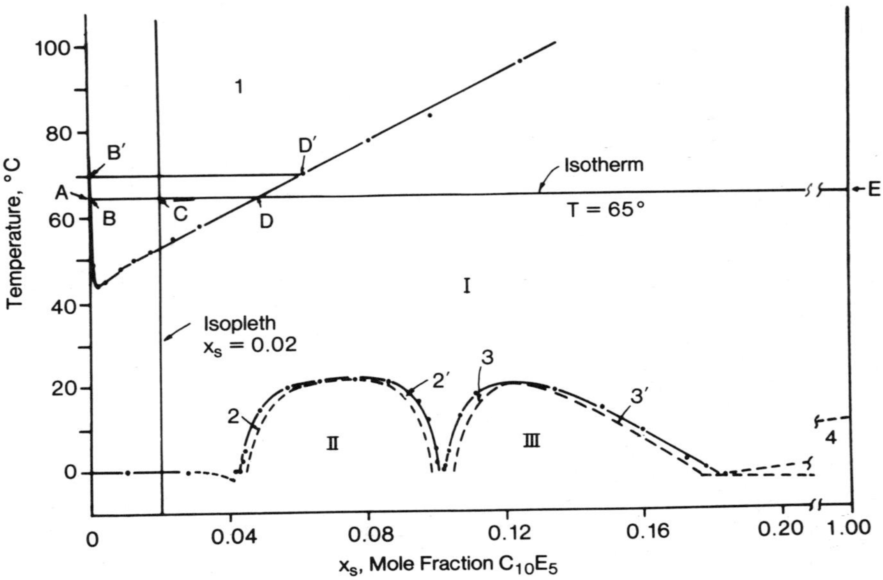

Fig. 2. Phase diagram of the $C_{10}E_5$-water system, composition in mole fraction units. I, II, and III are one-phase liquid, hexagonal liquid crystal, and lamellar liquid crystal regions, respectively. Regions marked with arabic numerals are two-phase regions. The isotherm at 65°C, the isopleth at x_s = 0.02, and a tie-line (B'-D') within the miscibility gap 1 at 70°C are also marked on the diagram.

are seen to diminish. At a unique point this difference vanishes; this is the critical point. Near critical points distinctive behaviour in both the coexisting bulk phases and the interface between them is observed; the experimental and theoretical investigation of critical phenomena have long been and still remain an area of active research (Roozeboom, 1901; Griffiths, 1970).

At a three-phase line (e.g., at T_u, the upper limit of the Krafft boundary, Fig. 3; Laughlin, 1978), the compositions of all three phases are determined and the system is invariant. Varying x_s along such a line has no influence whatever on the compositions of the three coexisting phases, while changing the temperature causes one phase to disappear. Thus, when three phases coexist the variance f = 0.

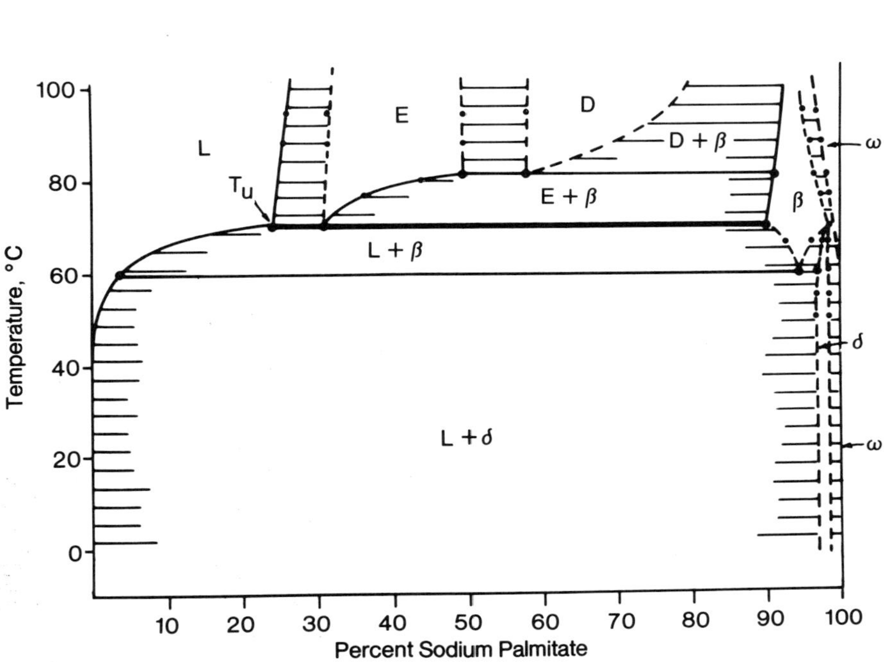

Fig. 3. Phase diagram of the sodium palmitate-water system below 100°C. L = micellar solution; ω, δ, and β are crystalline phases. T_u is the upper limit of the Krafft (crystal solubility) boundary. Near the upper limit of the E + β region two previously unrecognized liquid crystalline phases exist according to recent calorimetric (Madelmont, 1976) and optical (Rendall, 1983) data.

Increasing the number of components to three (adding salt or oil to a surfactant/water system) increases the variance accordingly. For a one-phase region f = 3; temperature and two potentials or compositions must all be specified. For a two-phase region f = 2. That is, temperature and any chemical potential within either phase must be defined; this automatically specifies the nature of the conjugate phase. For a three-phase region f = 1; temperature alone is sufficient to define the system. Not until p = 4 does f = 0. This situation would exist if for example the pressure over a system of three condensed phases were adjusted to its vapour pressure - thus creating a fourth gas phase.

The primary aspect of phase information to which the phase rule speaks is "number of phases". The qualitative nature of the phases and the magnitude of their composition or the properties of the phases are irrelevant. Compositions and properties (e.g., viscosities) do significantly influence rates of equilibration and the selection of appropriate experimental methods for determining phase information.

READING PHASE DIAGRAMS AND CALCULATING PHASE RATIOS

Two-component Diagrams

The arbitrary convention used herein places temperature on the ordinate and composition on the abscissa, with water to the left and surfactant to the right. The left and right borders represent the individual components at different temperatures. Phase information for these components such as m.p., b.p., etc., is indicated on these borders and represents boundary conditions to which the phase behaviour of mixtures must conform.

Each point within the diagram has the coordinates (x_S, T), and corresponds experimentally to adjusting the temperature of a given composition, x_S, to a specified value, T. Experimentally defined lines drawn within the diagram, termed "phase boundaries", divide the diagram into regions. At every point within a region or along a given boundary the number of phases and their qualitative nature are similar. For example, in Fig. 2, regions I, II, and III are all one-phase regions. In I, the system is an isotropic liquid or solution. In II it is a hexagonal liquid crystal, and in III, a lamellar liquid crystal (see later). Regions, 1, 2, 2', 3, 3', and 4 are all two-phase regions. In region 1 two immiscible liquids (both lying in region I) coexist; in 2 and 2', hexagonal liquid crystal (II) and isotropic liquid (I) coexist; in 3 and 3', lamellar liquid crystal (III) and isotropic liquid (I) coexist; and in 4, dry crystalline surfactant coexists with isotropic liquid (I). (It is assumed, for the sake of simplicity, that no crystal hydrates exist.) The dashed boundaries are known to exist but are not ordinarily defined quantitatively; only their approximate location is known.

Tie-lines and the Lever Rule

At any temperature within any miscibility gap in Fig. 2, horizontal lines may be drawn that terminate at the phase boundaries. The line B-D is one such line, termed a "tie-line". Tie-lines are lines that connect coexisting phases. Tie-lines within two-component diagrams must be horizontal since (by definition) phases in equilibrium must also be at the same temperature.

Obviously, gross composition (C) may be smoothly varied over the entire length of a tie-line. While the compositions of the coexisting phases do not vary along the tie-line, the proportions of each do vary. They may be calculated using the lever rule - a strict relationship between compositions and phase ratios that is dictated by conservation of mass.

We may isolate the tie-line B-D diagrammatically as follows:

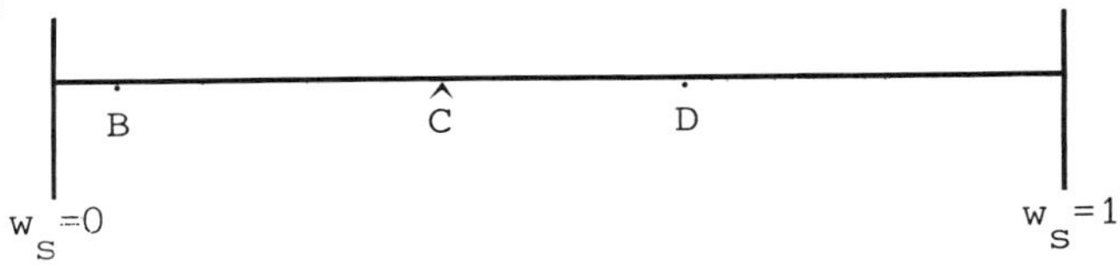

The dimensional unit of the tie-line is composition. Previously composition has been stated in mole fraction units, to accommodate the conventional definition of chemical potential. Both theoretically and experimentally we are often interested in the mass fraction of each phase, so from here on we will shift from mole fraction (x) to mass fraction (w) (or weight per cent) composition units.

The lever rule states that, at C,

$$f^B = \frac{D - C}{D - B} \tag{4}$$

and

$$f^D = \frac{C - B}{D - B} \tag{5}$$

where f^B and f^D denote the fraction of the system in phases B and D, respectively. Thus, the fraction of each phase is proportional to the length of the "opposite" lever arm. As C moves to the left, closer to B, the proportion of phase B increases. When C = B, $f^B = 1$ and the entire system is the B phase. As C moves closer to D, the fraction of D increases similarly.

The lever rule may be applied to any tie-line to calculate the proportion of the system in each phase but is never used to calculate proportions within three-phase lines. A latent heat is always associated with them, and the proportion of each phase is determined by gross composition and the heat content of the system. As a mixture of L and β phases is heated past the 70°C transition in Fig. 3, either L or β is converted to E plus the other phase - to a degree determined by the isothermal heat input into the system.

It is clear from this analysis that passing phase transitions along an isopleth creates a driving force for the formation of new interfaces and for countercurrent transport of material across interfaces. The consequences of this predictable phenomenon on physical and colloidal behaviour have been largely neglected.

Three-component Diagrams

By far the simplest and best way to depict three-component phase information (implicitly at constant pressure) when an undistorted overall view of the system is desired is with a right triangular prism (Fig. 4). Temperature is scaled along the length of the prism, and an isotherm through the prism is

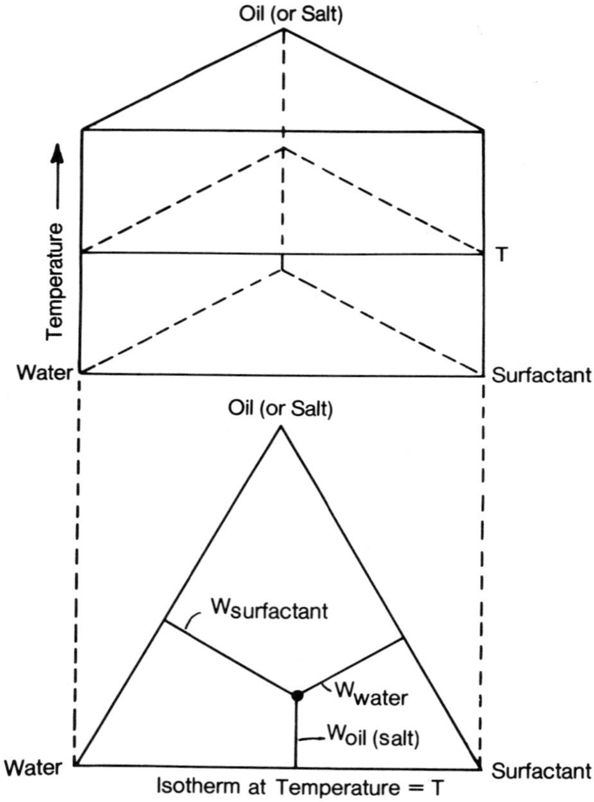

Fig. 4. Right triangular prism used to depict phase information within a three-component system. The front face of the prism depicts the surfactant-water diagram. An isotherm at T is a triangular plane within the prism. Within the isotherm, the weight fractions, w, of each component are given by the lengths of the corresponding perpendiculars (shown).

an equilateral triangle.

There are times when it is useful to distort this presentation to expand the scale along which one component is presented. In this circumstance the right triangular presentation is particularly useful. In the sodium palmitate - sodium chloride - water diagram (Figs 5 and 7), the latter presentation depicts clearly phase behaviour that is highly compressed and difficult to perceive in the equilateral triangular presentation (Fig. 5).

Suppose we add oil to a surfactant/water system and select an isotherm from the triangular prism (Fig. 4). At any point within the triangle the weight fraction of the surfactant, w_S, is proportional to the length of the perpendicular dropped

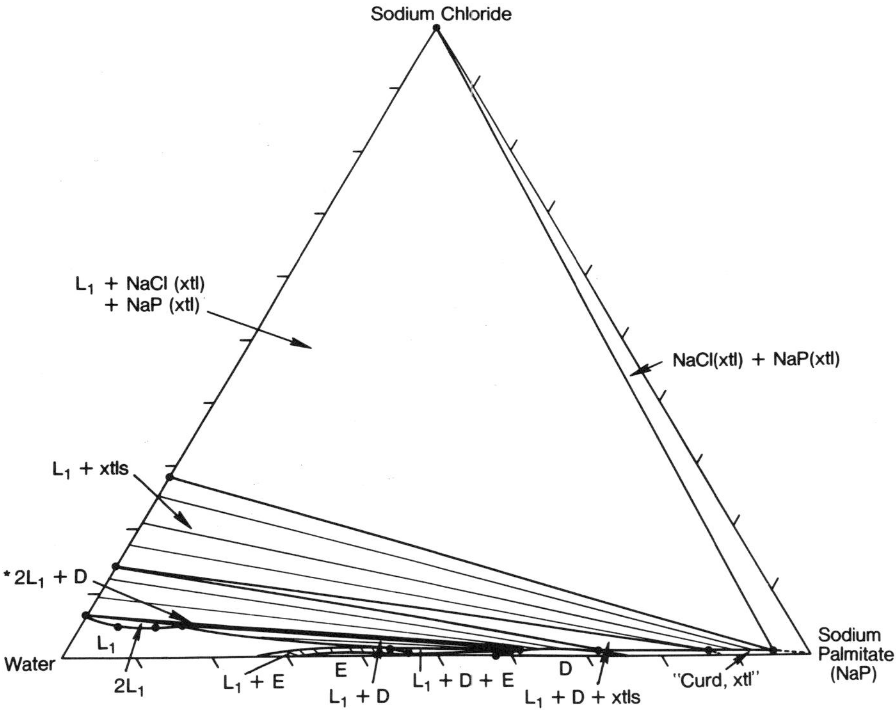

Fig. 5. The sodium palmitate - sodium chloride - water system at 90°C. Composite diagram compiled by R.D. Vold (1937) from vapour pressure, dilatometry, and X-ray data incorporating isothermal phase data of McBain. L = liquid, D = lamellar, and E = hexagonal phases. The ($2L_1$ + D) three-phase triangle, marked with the asterisk, is the "soap boiler's triangle" (see Fig. 7).

to the opposite (oil - water) edge (Taylor, 1930). The fractions of the other two components are defined similarly. The long-recognized beauty of the equilateral triangle is that the sum of the lengths of these three perpendiculars equals the altitude of the triangle. Thus, if the point is moved to a corner, two perpendiculars vanish and the third becomes the altitude; i.e., the fraction of that component at the corner becomes unity. If the point is moved to any side, the fraction of one component becomes zero and we now have a two-component system. With a little practice, one can subjectively sense at a glance the proportion of three components at any point in a triangular diagram.

While the isotherm becomes a plane in three components, the isopleth remains a line. Every isopleth is parallel to the axis of the prism and passes through each isotherm at the same coordinates (by definition).

A two-phase region along a two-component isotherm ordinarily extends a finite distance into the triangular diagram (Ekwall, 1975). The orientation of tie-lines within such regions cannot be specified *a priori*, except in the two-component limit. Unless interrupted by another phase boundary the length of these tie-lines must also vanish at some point on this boundary. This is termed the plait point; it is essentially an isothermal critical point.

Provided a tie-line is experimentally defined, the lever rule may be applied (in principle) within two-phase regions of triangular diagrams to calculate phase ratios exactly as in two-component diagrams (Masing, 1944; Palatnik, 1964). Performing the calculation analytically is far more complex; in fact, lever rule calculations within three-component diagrams have been handled by direct measurements from graphs heretofore (Tamas, 1970). Analytical expressions that permit precise lever rule calculations in three-component systems, under certain restrictions, have recently been developed (Laughlin, unreported).

Three-phase regions are depicted as irregular triangles within the triangular isotherm. To calculate proportions of each phase at any point within this "tie-triangle", lever-lines are drawn from each vertex through the coordinates of the three-phase mixture to the opposite edge. The lever arm ratio is determined within each line, and from the additional condition that the sum of the three-phase fractions must equal unity the phase ratios may be determined (Laughlin, unreported). While exceedingly laborious by hand, these calculations are easily handled using modern computers.

EXPERIMENTAL METHODS FOR DETERMINING PHASE INFORMATION

In order to locate a phase boundary one must cross it, either by changing composition or by changing temperature. Thus, all known methods for determining phase information fall into two broad classes: isothermal or isoplethal (Hill, 1923). These methods will now be considered, with emphasis on the special problems that surfactant systems present.

Isothermal Methods

In these methods, multiphase regions are located by preliminary exploration of composition. Then compositions within these regions are equilibrated, the individual coexisting phases sampled and their compositions (and qualitative natures) determined by appropriate methods. The classical phase studies of nonsurfactant systems (Roozeboom, 1904; Schreinemakers, 1911) and of surfactant systems (Ekwall, 1975) were largely performed by this method. In recent times it has been significantly refined with the use of modern instrumentation (Lang, 1980; Vorob'eva, 1967).

The qualitative identification of ordered phases rests ultimately on X-ray data (Fontell, 1974). Ordinarily, texture observed using the polarized light microscope will both distinguish crystal from liquid crystal and classify the liquid crystal. The pioneering microscopy of Friedel (1922) described the microscopic textures of liquid crystals, and the correlations between these textures and X-ray-based structure were established by Rosevear (1954).

Where applicable, isothermal methods provide the most precise phase information. Unfortunately, relatively large samples, good analytical methods and a lot of work are required. Also there are many boundaries and systems to which this classical method is not applicable. Rapid and clean separation of phases must occur, and therefore both phases must ordinarily be mobile liquids or liquid crystals. Coalescence of the dispersed conjugate phase droplets must be fast - but this may not be true when interfacial tensions are low or the solubility is small, as with polar lipid/water systems. A new principle for carrying out isothermal phase studies which both accelerates data collection and sidesteps these problems, the "diffusive interfacial transport" method, is currently being developed (Laughlin, unreported).

Isoplethal Methods

The alternative to crossing phase boundaries isothermally is to cross them isoplethally - i.e., by varying temperature at constant composition. This old method (Alexejew, 1882) is very widely used due to its experimental simplicity, and is especially useful for delineating the range of the isotropic solution phase. Depending upon the amount of sample and time invested it can either yield very precise data (Hill, 1923) or be used as a quick screening method for locating solubility boundaries on small samples (Laughlin, 1976). When the sample is observed by the polarized light microscope during the heating/cooling process, it can be used to locate inner boundaries as well as solubility boundaries (Balmbra, 1962).

Ordinarily, the best way to establish that equilibrium has been attained using this method is to approach the boundary both on heating and on cooling and observe the same boundary temperature. However, the inherently slow formation of crystals, from either solutions or liquid crystals, ordinarily precludes determination of the Krafft boundary by cooling. It is reliably determined only during a heating cycle, after which the sample must be chilled sufficiently below the boundary to again induce nucleation and crystal growth. Fortunately, small fractions of surfactant crystals are easily detected - especially when the system is observed between crossed polarizers.

The kinetics of separation (and ordinarily the coalescence) of liquid crystals from solution is invariably fast; readily apparent supercooling past a liquid crystal-separation boundary is unknown (McBain, 1933; Laughlin, 1976). When the liquid crystal is birefringent, visual detection between crossed polarizers provides very sensitive detection by isoplethal studies of the boundary that defines the composition of the coexisting liquid phase. Delineating the range of existence of the liquid crystal itself is exceedingly difficult and rarely, if ever, accomplished.

The cubic liquid crystals also separate from fluid solutions very rapidly, but no method for accurately locating their solubility boundary presently exists. Loss of birefringence is observed with isoplethal microscopic techniques when birefringent liquid crystals are transformed to cubic phases, but distinguishing this phase from fluid isotropic phases by this method alone is impossible. Dramatic changes in viscosity and the acoustic properties of the system have been used to perceive the appearance of the cubic phase (Lutton, 1965; Laughlin, 1976).

Isoplethal methods have significant inherent limitations. First, the proportion of the coexisting phase precisely at a phase boundary is zero. Thus, one must pass the true temperature of the boundary by a finite temperature increment in order to produce a detectable amount of the coexisting phase. For this reason nearly vertical boundaries or boundaries connected to long tie-lines are not easily located. Furthermore, the data are sensitive to the kinetics of phase separation and coalescence, and to the ability of the experimentalist (or the instrumentation) to detect the appearance of the coexisting phase.

Secondly, critical opalescence in the one-phase region near the critical point of a miscibility gap is very difficult to distinguish subjectively from separation of small amounts of a coexisting phase. Significant errors stemming from this problem have been documented (Lang, 1980). For this reason, "cloud point" data are always to some degree subjective and imprecise.

Finally, stirring may induce long-range and long-lived shear birefringence within systems that, at rest, are optically isotropic. This phenomenon is ordinarily observed below an upper temperature limit at compositions close to liquid crystal solubility boundaries. It is of course dependent on the level of input of shear energy and is not an equilibrium property of the system. Cubic phases often (perhaps always) display shear birefringence. Shear birefringence must exist for other liquid crystals but would be difficult to detect.

SURFACTANT PHASE BEHAVIOUR

It was emphasized earlier that surfactant systems obey the same thermodynamic laws that govern phase equilibria in general. While hardly a surprising result, these laws were in the main established in classical studies of compounds that are neither as grossly nonideal in solution, nor as active at interfaces, as are surfactants, and deserved re-examination in surfactant systems.

Next, emphasis will be given to the distinctive aspects of surfactant phase behaviour. These are: (1) the unusually rich variety of states of matter that can exist; (2) the distortion, in shape or position within the diagram, of phase boundaries that superficially resemble those also found in non-surfactant systems; and (3) the complexity of the phase behaviour that results.

States of Matter in Surfactant Systems

Crystals The most highly ordered state of matter is the crystal, and pure, dry, surfactants easily crystallize. Surfactant crystals have to date always been found to be lamellar in structure - qualitatively resembling crystalline amphiphilic molecules in general (Larsson, 1976). A head-to-head and tail-to-tail arrangement of molecules exists, so that planes of either polar functional groups or nonpolar terminal methyl groups alternate within the structure. The repeating structural unit normal to the planes in this arrangement is the bilayer. The molecules are oriented more or less perpendicular to the structural planes but the angle with respect to the planes varies within different crystalline phases, giving rise to the rich polymorphism characteristic of these structures. Optically, such crystals may be either uni- or bi-axial and their X-ray patterns display sets of high-angle lines. Among these is a sharp Bragg spacing that is determined by the separation between chains. The conformational structure of n-alkyl chains is ordinarily all-trans.

Water is commonly found in crystalline surfactant phases that coexist with either the isotropic solution phase or a liquid crystal (Fig. 3; Clunie, 1969). Its presence allows participation of water in the crystal structure in a way that is compatible with its chemical potential in the coexisting phase. While crystal hydrates are often unique compositions, it has been suggested that hydrated soap crystals span ranges of composition. These may be regarded as solid solutions, and are not restricted to surfactant systems (Cox, 1971). Details as to this aspect of surfactant phase equilibria are exceedingly sparse.

Crystal energies are always strongly influenced by molecular symmetry. Terminally substituted straight chain surfactants with compact, highly polar hydrophilic groups (e.g. soaps) form the highest melting crystals. Deviations from this linear structure, i.e., by branching, reduce crystal stability.

The precipitation of crystals from mixtures of many different molecules is always difficult and surfactants are no exception. Commercial polyoxyethylene nonionics, surfactants derived from petroleum alkylates of benzene or phenol, and homologous mixtures derived from natural fatty acids all crystallize with difficulty relative to the pure individual components, for this reason.

Mesophases or liquid crystals Lying between the high degree of order of the single crystal and the smeared-out order of the liquid state are the liquid crystals (or mesomorphic phases, or mesophases). These are distinctive states of matter that display, within each individual state, both order and disorder. In each liquid crystal, liquid-like disorder exists in one or more dimensions while crystal-like order exists in others. Because of this hybrid structure liquid crystals are non-rigid fluids, with rheological parameters falling between those of crystals and of liquids. At the same time they display one or more sharp X-ray diffraction lines - a crystal-like characteristic.

Liquid crystals may be conveniently divided into two large classes - thermotropic and lyotropic. Thermotropic liquid crystals are formed by the partial disordering of crystals of elongated dipolar molecules or mixtures of such molecules (Gray, 1962). The disorder is induced by elevating thermal, electrical, or magnetic energies beyond a critical level. Lyotropic (meaning solvent-induced) liquid crystals are formed by the action of solvent on selected amphiphilic molecules. This is the major and characteristic class of liquid crystals that surfactants form; indeed, the existence of lyotropic liquid crystal regions has been used to infer association colloid behaviour (Laughlin, 1978). By inspection of phase information, surfactants may be distinguished from nonsurfactant amphiphiles. Excepting soaps, pure dry surfactants rarely display (according to existing data) thermotropic liquid crystal phases; but then no other systems have been studied to the degree that have been soaps.

There are, in theory, a great many possible liquid crystal states. Purely on mathematical symmetry grounds Hermann (1931; Mabis, 1962) predicted the existence of 18 different liquid crystals, many of which have been actually seen in various systems. Some are found only in chemically complex multicomponent systems (Rosevear, 1954). Fortunately for the sake of simplicity, only three major types of liquid crystals have been found to date in two-component surfactant/water systems. Furthermore, these same types dominate the scene in surfactant/oil/water and surfactant/salt/water systems. These are:

i) The hexagonal phase - a highly viscous class of liquid crystals whose structure is based on a close-packed array of cylindrical elements of indefinite length. There are two sub-classes (following Ekwall's (1975) symbolism):

   1) an oil-core (normal, E, middle) hexagonal phase typical of most surfactant/water systems) where the lipophilic groups form the core of each cylinder and the hydrophilic groups lie on the surface; and
   2) a water-core (inverted, F) hexagonal phase, where the core of the cylinder is water and the lipophilic groups fill the space between the cylinders.

ii) The lamellar (neat, D) liquid crystal phase. Except for the fact that the chains have a disordered conformational structure, this class of liquid crystals structurally closely resembles the typical crystalline state of surfactants (see above). The typical lamellar phase is optically uniaxial. The level of order may vary both continuously and discontinuously. For example, the dipalmitoyl phosphatidylcholine/water system is now known to display at least four different lamellar regions. The highest temperature region is the familiar lamellar liquid crystal. The "gel" phases (Chen, 1980) that exist at lower temperatures appear to display levels of order higher than the familiar lamellar phase but less than the true crystals, thus compounding the complexity of the phase behaviour of such systems.

(iii) The cubic (viscous isotropic, I) phases. These are a presently ill-defined collection of clear, optically isotropic liquid crystal phases that are often extraordinarily viscous. They are highly varied structurally, and may be formed from:

1) spherical, micelle-sized aggregates arrayed in one of the cubic (primitive, body-centered, or face-centered) lattices (Clunie, 1965; Balmbra, 1969);
2) cylindrical structural elements formed into a complex isotropic lattice from two interwoven anisotropic lattices (Luzzatti, 1968); or
3) bilayer elements, arrayed in a bicontinuous structure (Lindblom, 1979).

Thus, it appears at this time that the cubic phases may be assembled from structural elements of any of the other major classes. They may be found to either side of the hexagonal phase in surfactant/water systems (Clunie, 1965).

Cubic phases are the most difficult to detect and characterize and the least well-known liquid crystals. They are nevertheless very common and have been observed within all five hydrophilic group sub-classes.

These, then, are the major liquid crystalline states encountered in surfactant systems. It is worth re-emphasizing that they are mostly lyotropic phases; i.e., their existence hinges on the water - surfactant interaction. Unlike crystals or (most) crystal hydrates, they span regions (not lines) in phase diagrams.

The same liquid crystalline states found within surfactant/water systems are also found within surfactant/salt/water systems. Adding an oil to a surfactant/water systen often allows that system to form a more complete set of liquid crystal states than is possible without the oil.

Liquids and solutions The condensed state of matter with the lowest level of order is the liquid or solution. In surfactant/water systems simple molecular solutions always border the water ordinate, but as surfactant composition is increased these become, ordinarily, micellar solutions. This transition, while sudden, is smooth and has been shown <u>not</u> to be a thermodynamic discontinuity (Lindman, 1980).

Starting from the surfactant ordinate at temperatures above

or near the surfactant's m.p., one finds that water is invariably miscible with the liquid surfactant to a finite degree and depresses the crystal m.p. Sometimes the surfactant and water are infinitely miscible along an isotherm (Fig. 2). If one is not constrained to a single temperature, most surfactants display a continuous band of isotropic solution between pure water and pure liquid surfactant.

Thus, it is useful to view the liquid or solution phase as the backdrop upon which the more highly ordered phases are superimposed. This view is compatible with the philosophical notion (Ekwall, 1975; Laughlin, 1978) that the micellar structure in solution is the definitive characteristic of surfactant behaviour: the liquid crystal appears as a consequence of stripping water from (or, equivalently, reducing water activity in) the micellar system.

Micellar solutions are not unstructured, but the structural elements within them are randomly arrayed and have molecular dimensions. Some of the dimensions of micellar structures are often similar to structural dimensions within coexisting liquid crystalline phases. The addition of salt does not fundamentally alter this picture (Fig. 5), but the addition of oil deserves further comment.

When an amphiphilic oil that itself is not a surfactant (usually a fatty acid or alcohol) is added to a soluble surfactant/water system, two distinctive isotropic solution regions often develop (Fig. 6). One, termed L_1 by Ekwall (1975), is simply an oil-containing extension of the micellar solution region and is tied to the water corner. Extending from the oil corner, however, is another isotropic solution region that, while superficially similar to L_1, is distinguished from the latter by its structure and properties. This region was termed L_2 by Ekwall, and is the "microemulsion" phase that has received so much attention for so many years. Shulman (Hoar, 1943; Shinoda, 1975) dramatically elevated the level of interest in this phase by showing that its range could be significantly enlarged by adding a fourth amphiphilic component (e.g., a C_4-C_6 alcohol). He also showed how one could construct the phase in a manner such that equilibrium was rapidly attained.

These, then, are the most widely encountered phases that surfactant systems display. Our next concern is the shape of the phase boundaries that define their range of existence.

Boundary Shapes in Surfactant Systems

Given the multiplicity of phases that surfactant systems display and the gross nonideality of their solutions, it is hardly surprising (in retrospect) that the boundaries that

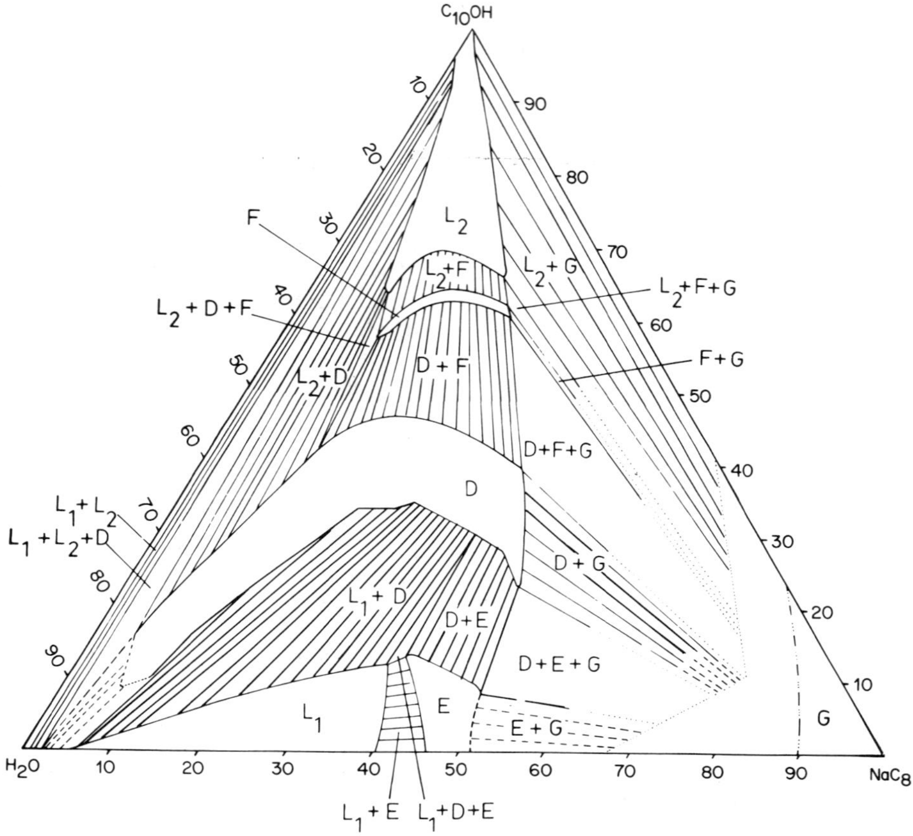

Fig. 6. The sodium octanoate-decanol-water diagram of Ekwall et al. *(1975), as recently revised by Friman* et al. *(1982). L = liquid, E = hexagonal liquid crystal, D = lamellar liquid crystal, F = inverted hexagonal liquid crystal.*

delineate the range of existence of the various phases assume distinctive shapes.

The Krafft boundary It was first documented by Krafft (1899) that soap crystals displayed an unusual dependence of solubility on temperature. At low temperatures the solubility is low and the solubility boundary displays a very steep slope (low temperature coefficient of solubility). The solubility then suddenly increases and the crystal solubility boundary becomes nearly horizontal - signifying a very large temperature coefficient of solubility. Inorganic salts display comparatively smaller temperature coefficients of solubility and much less dramatic changes in this parameter, even when the nature of the coexisting crystal phase changes.

The coexisting phase below the Krafft boundary is always (by definition) a surfactant crystal - usually a crystal hydrate. The slope of the boundary within the region of the plateau varies substantially, but must always be positive. With ionic surfactants, the width of the plateau in the temperature dimension is 10-15°C while with nonionics it is much smaller. At its upper limit (T_u, Fig. 3), the Krafft boundary intersects another kind of solubility boundary (usually a liquid crystal, sometimes a liquid) at unique coordinates that lie at a terminus of an invariant three-phase line. T_u is one of the important limiting phase boundary parameters in surfactant diagrams, for below T_u liquid crystals do not exist at any composition.

The "Krafft point" (Shinoda, 1967), often cited in the literature, is the point of intersection of the trajectory of c.m.c. within the solution phase (above the Krafft boundary) with the Krafft boundary. As such it is not a thermodynamic discontinuity. When this intersection occurs, interestingly, it means that micellar solutions do not exist at temperatures below this point of intersection. Below this temperature phase separation (of crystals) occurs at compositions below the hypothetical c.m.c. - which now exists within a metastable solution phase.

The composition at T_u may vary widely. If the phase coexisting with the dilute solution above T_u is very insoluble, usually due to the existence of either very weak hydrophilic groups (Lutton, 1965) or very large lipophilic groups (Kunleda, 1978; Franses, 1980), the tie-lines to the coexisting phase become very long and the plateau region of the Krafft boundary will vanish. The definition of T_u remains unchanged; it is still the temperature at which crystalline surfactant coexists with isotropic solution and liquid crystal. Polar lipids, dilong chain quaternary ammonium surfactants, monoglycerides, and primary amines all display this kind of behaviour.

Mesophase or liquid crystal boundaries Like true liquid phases, lyotropic liquid crystals always exist over a range of composition and temperature. Typically the region they occupy is bullet- or dome-shaped. Sometimes the dome is tilted to one side and, in at least one case (dodecylammonium chloride), it is detached as an island (Broome, 1951). Distortion of the shape also occurs in the $C_{10}E_4$-like systems, where the liquid crystal and miscibility gap regions interact (Lang, 1980).

So far as is presently known, all phase transitions in surfactant systems are first-order transitions. As a result, liquid crystal regions are always separated from coexisting phases by two-phase regions. In many cases these two-phase regions are very thin (Clunie, 1969).

Liquid crystals invariably "melt", above a specified temperature limit (T_m), because they are highly ordered relative to the liquid state. Although hard proof does not exist (because of experimental difficulties), it seems probable that the inner boundary denoting the limits of the liquid crystal region is tangential to the outer boundary, denoting the limits of the coexisting liquid phase. The phase situation may thus resemble that at the azeotropic point in vapour/ liquid equilibria (Taylor, 1930).

Three-component systems Liquid crystal regions within three-component triangular isotherms often assume one of two basic shapes. They may simply extend inward from their range of existence along the two-component border but vanish beyond a specific level of the third component. The result, agains, is a dome-like shape.

The lamellar (D) and inverted hexagonal (F) phases in surfactant/oil/water systems often assume an angular or kidney-like shape (Ekwall, 1975). Sometimes the D region touches the surfactant/water border, but often (with short lipophilic group surfactants) it exists only within the interior - i.e., does not exist in the surfactant/water system. At specific points, each of these regions touch invariant three-phase triangles, but along most of their periphery they border two-phase regions of various kinds.

Miscibility gaps The term "miscibility gap" can be broadly used to denote any two- or three-phase region, but the gap in miscibility within a liquid phase region has special significance. It is very common for solutions of single bond and semipolar surfactants to "cloud" on heating due to separation of a second liquid phase. The locus of these "cloud points", versus composition defines the position of this miscibility gap, which is alternatively termed the "consolute" boundary (Laughlin, 1978).

Actually, surfactant systems (like non-surfactant systems display two classes of consolute boundaries: the upper and the lower consolute boundary. The former displays an upper critical point, above which the boundary does not exist, while the latter displays a lower critical point. The lower consolute boundary is far more commonly encountered - the upper consolute boundary being found so far only in selected zwitterionic surfactants (Laughlin, 1978). It was shown only recently that the lower consolute boundary is actually part of a closed loop of coexistence, in representative single bond and semipolar surfactants (Lang, 1980).

The upper and lower consolute boundaries are <u>not</u> similar in shape. The upper consolute boundary observed in zwitterionic surfactant systems has a symmetrical shape superficially resembling that observed in partially miscible solvent systems, except that it is centred towards the water side of the diagram (Laughlin, 1978).

The lower consolute boundary is severely distorted, by comparison, in that the critical composition lies very close to the water ordinate. The arm lying towards the water ordinate is thus necessarily almost vertical, while the other arm displays a shallow slope until the temperature is well above the critical temperature. The small semipolar groups (especially the sulphone diimine) display more compact lower consolute boundaries than do the large polyfunctional polyoxyethylene structures (Laughlin, 1978).

It is a curious fact that we speak of the shape of the one-phase liquid crystal region, but refer to the shape of the "gap" within the one-phase liquid region - a complementary concept.

Within three-component surfactant/oil/water systems, the L_1 and L_2 regions themselves have characteristic shapes (Ekwall, 1975). Like the liquid crystal, the L_1 region simply penetrates into the interior a finite distance, developing a whale-like form (Fig. 6). The L_2 (microemulsion) region <u>does</u> have a more or less characteristic shape. As in Fig. 7, it often displays a roughly triangular form that is convex towards the oil apex along the interior side. The shape of this side roughly parallels the shape of the lamellar (D) (or hexagonal (F)) liquid crystal region with which it coexists.

UTILITY OF PHASE INFORMATION

Why all this concern with surfactant phase behaviour? The answer is that phase information is possibly the most widely useful kind of physical data that exists for surfactant, or other, systems. There are several reasons for this.

1) First, the trajectory of any physical process in which composition and/or temperature are varied, within a defined system, may be charted on that system's phase diagram. This superposition depicts, at a glance the phase situation at any point and discontinuities in it during the process.

   For example, heating or cooling any system of fixed composition corresponds to rising or falling along an isopleth. The evaporation of solvent from a solution corresponds to moving to the right; dilution of the system is depicted by moving to the left. Mixing any two

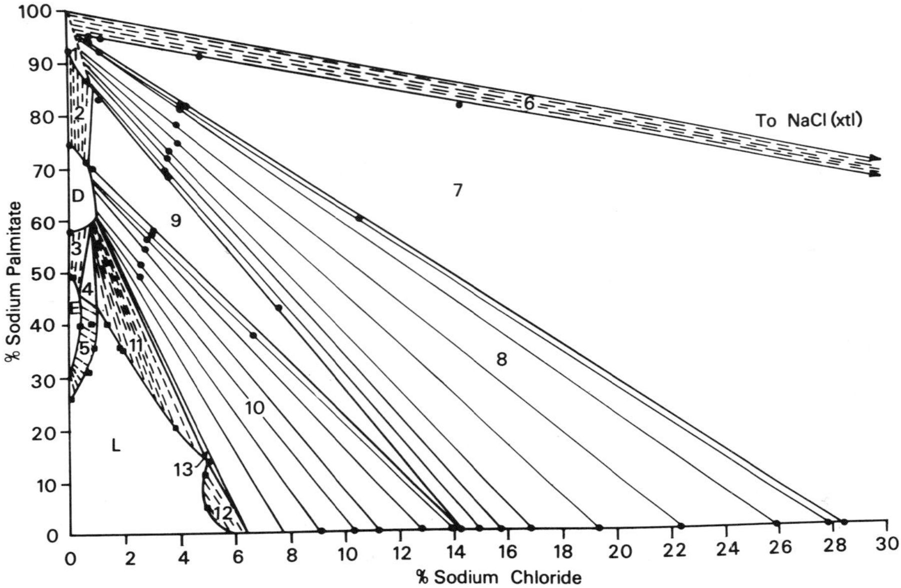

Fig. 7. The sodium palmitate - sodium chloride - water diagram at 90°C. Same data as in Fig. 5, but with the sodium chloride axis greatly expanded and presented in right triangular format to facilitate visualization of the phenomena underlying commercial soap boiling. The diagram does not accurately depict the complex phase behaviour of the crystalline phases (region 1). The goal of the soap boiler was to finish inside the thin triangle 13 (2L + D). D, E, and L regions were previously defined. Regions 2, 3, 5, 6, 8, 10, 11, and 12 are two-phase regions. Regions 4, 7, 9, and 13 are three-phase regions.

compositions produces exclusively those compositions that lie on a straight line between them. Process engineers can utilize phase information to design a process; e.g., to circumvent entry into an unmanageably difficult liquid crystalline phase.

2) The proper interpretation of physical studies of any system for which erroneous phase behaviour is assumed is jeopardized in some degree. Where chemical potentials play a role the assumption of a finite variance where none in fact exists, or vice versa, can lead one astray. If, for example, a surfactant is soluble within a certain composition range then the vapour pressure and osmotic

properties will vary over this range. If insoluble, they will not.

3) Phase heterogeneity underlies most separation processes. In the art of soap-making the soap boiler first salted out ("pitched") the kettle by adding salt to the saponified fat then "smoothed" by adding water. His goal (Fig. 7) was eventually to enter a three-phase triangle (neat - nigre - lye, $2L_1 + D$, or lamellar liquid crystal - micellar soap solution - concentrated electrolyte solution, in modern terminology). The lye phase extracted the bulk of the unwanted electrolyte, and the micellar soap solution (nigre phase) extracted most of the coloured impurities. This left the greatly refined soap in the liquid crystal phase. The hot neat soap phase could be (and still is) cooled into a crystal/solution region to form framed soap bars.

The three-phase triangle that was the soap boiler's target is small in area and rather thin, as it happens (Figs 5 and 7). Adding too much water in the "smoothing" operation could easily generate the middle (hexagonal) liquid crystal - a difficult situation and costly mistake. The pioneering phase studies of McBain (1926), and later Ferguson (1932) and others, transformed this process from an art to an engineerable science. Ironically, not long after this occurred soap boiling on a large scale was being "phased out".

Another application of phase information to surfactant systems lies in the refining of fats. Polar lipid impurities are removed from fats by first precipitating the liquid crystal by addition of water. Rydhag (1979) has recently clarified the phase behaviour underlying this process, which again involves entering a three-phase triangle (oil - water - liquid crystal). The liquid crystal phase can be removed by filtration after it is precipitated.

4) Phase information provides a complete (both qualitative and quantitative) description of solubility. In surfactant systems, given the numerous and varied states of matter that are possible, both the numerical value of solubility and the qualitative nature and composition of the coexisting phase will change with temperature. This sort of complexity is most easily depicted and communicated by the use of phase diagrams. Highly complex behaviour, as in the $C_{10}E_4$/water or monoglyceride/water systems, virtually demands this approach.

5) The utility of surfactants is on occasion limited by inadequate solubility. An example of this is the detergency of sodium tallow alkyl sulphate (predominantly a

C_{18} surfactant mixture). At 60°C this surfactant performs well as a detergent while at 25°C it does not. At 60°C its solubility boundary lies well above while at 25°C it lies well below the use concentration; the detergency at 25°C is thus solubility-limited. Whenever a solubility boundary lies well below an intended use concentration, the lack of solubility can possibly impair the effectiveness of the material for the intended use.

6) Phase behaviour often dictates the selection of suitable components for use in a product. It is important in soap bars, for example, to have both a soluble and an insoluble soap. Milling can alter the phase situation, and thus the lathering properties and utility of the bar, by changing texture, erosion rates, etc. Components of liquid detergents, which are ordinarily concentrated solutions, must obviously be selected with both phase behaviour and utility in mind.

7) Bulk phase compositions determine the tensions between coexisting phases (Widom, 1981). Bulk phases are far more amenable to direct chemical analysis than surface "phases", so the control of interfacial properties via control of bulk compositions is possible, in principle. Utilizing the extensive Swedish phase data Friberg (1978, 1979; Larsson, 1976) has suggested correlations between both foam and emulsion stability with the qualitative nature of the phase behaviour of the system. For example, foam stability is exceptionally good when the system has two phases, one being a small amount of a lamellar liquid crystal. Conversely, foam stability is exceptionally poor in neighbouring two-phase regions where the conjugate phase is the L_2 liquid. A triumph of this approach was clarification of how a fatty alcohol could function, at different compositions, as <u>both</u> a foam stabilizer and foam killer.

8) Finally, phase behaviour has proven to be correlatable with molecular structure (Laughlin, 1978). In spite of the inherent complexity and gross nonideality of surfactant systems, perturbations in hydrophilic or lipophilic group structural features alter phase boundary locations in surprisingly predictable ways. As mentioned earlier, phase behaviour has also been used to distinguish nonsurfactant amphiphilic molecules from surfactants. The existence of predictable relationships between phase behaviour and structure is potentially very significant, for where the need for a specific kind of phase behaviour can be stated (Laughlin, 1981) then molecules can be efficiently selected or designed to fill this need. This, in the long run, could be the most important aspect of

phase information of all.

REFERENCES

Alexejew, W. (1882). *J. Prakt. Chem.* 25, 518.
Balmbra, R.R., Clunie, J.S., Corkill, J.M. and Goodman, J.F. (1962). *Trans. Faraday Soc.* 58, 1661.
Balmbra, R.R., Clunie, J.S. and Goodman, J.F. (1969). *Nature* 222, 1159.
Bett, K.E., Rowlinson, J.S. and Saville, G. (1975). "Thermodynamics for Chemical Engineers". The MIT Press, Cambridge, Mass., USA.
Broome, F.K., Hoerr, C.W. and Harwood, H.J. (1951). *J. Am. Chem. Soc.* 73, 3350.
Chen, S.C., Sturtevant, J.M. and Gaffney, B.J. (1980). *Proc. Natl Acad. Sci.* 77, 5060.
Clunie, J.S., Corkill, J.M. and Goodman, J.F. (1965). *Proc. Royal Soc. A* 285, 520.
Clunie, J.S., Goodman, J.F. and Symons, P.C. (1969). *Trans. Faraday Soc.* 65, 287.
Cox, J.S.G., Woodward, G.D. and McCrone, W.C. (1971). *J. Pharm. Sci.* 60, 1458.
Danielson, I. and Lindman, B. (1981). *Colloids and Surfaces* 3, 391.
Ekwall, P. (1975). *In* "Advances in Liquid Crystals" (Ed. G.H. Brown), Vol.1, pp.1-142. Academic Press, New York.
Ferguson, R.H. and Richardson, A.S. (1932). *Ind. Eng. Chem.* 24, 1329.
Findley, A. (1938). "The Phase Rule and Its Applications". 8th edn. Longmans, London.
Fontell, K. (1974). *In* "Liquid Crystals and Plastic Crystals" (Eds Gray and Windsor), Vol.2, pp.80-109. Ellis Horwood Ltd, Chichester, Wiley.
Franses, E.I., Puig, J.E., Talmon, Y., Miller, W.G., Scriven, L.E. and Davis, H.T. (1980). *J. Phys. Chem.* 84, 1547.
Friberg, S. (1978). *In* "Advances in Liquid Crystals" (Ed. G.H. Brown), Vol.3, pp.149-165. Academic Press, New York.
Friberg, S. (1979). *J. Soc. Cos. Chemists* 30, 309.
Friedel, G. (1922). *Ann. de physique* 18, 300.
Friman, R., Danielsson, I. and Stenius, P. (1982). *J. Colloid Interface Sci.* 86, 501.
Gibbs, J.W. (1928). *In* "The Collected Works of J. Willard Gibbs" (Ed. W.R. Longley). Longmans, Green & Co., New York.
Gray, G.W. (1962). "Molecular Structure and the Properties of Liquid Crystals". Academic Press, New York.
Griffiths, R.B. and Wheeler, J.C. (1970). *Phys. Rev. A* 2, 1047.
Hermann, C. (1931). *Z. Krist.* 79, 186.
Hill, A.E. (1923). *J. Am. Chem. Soc.* 45, 1143.

Hoar, T.P. and Schulman, J.H. (1943). *Nature* 152, 102.
Krafft, F. (1899). *Ber. der Deut. Chem. Gesell.* 32, 1596.
Kunleda, H. and Shinoda, K. (1978). *J. Phys. Chem.* 82, 1710.
Lang, J.C. and Morgan, R.D. (1980). *J. Chem. Phys.* 73, 5849.
Larsson, K. (1976). *In* "Food Emulsions" (Ed. S. Friberg), pp.39-66. Marcel Dekker, New York.
Laughlin, R.G. (1976). *J. Colloid Interface Sci.* 55, 239.
Laughlin, R.G. (1978). *In* "Advances in Liquid Crystals" (Ed. G.H. Brown), Vol.3, pp.41-148. Academic Press, New York.
Laughlin, R.G. (1981). *J. Soc. Cosmet. Chemists* 32, 371.
Lindblom, G., Larsson, K., Johanssen, L., Fontell, K. and Forsen, S. (1979). *J. Am. Chem. Soc.* 101, 5465.
Lindman, B. and Wennerström, H. (1980). *In* "Fortschritte der chemischen Forschung" (Ed. F.L. Boschke), pp.1-83. Springer-Verlag, Berlin.
Lutton, E.S. (1965). *J. Am. Oil Chemists Soc.* 42, 1068.
Luzzati, V., Tardieu, A., Gulik-Krzywicki, T., Rivas, E. and Reiss-Husson, F. (1968). *Nature* 220, 485.
Mabis, A.J. (1962). *Acta Crystallogr.* 15, 1152.
Madelmont, C. and Peron, R. (1976). *Colloid and Polymer Sci.* 254, 581.
Masing, G. (1944). "Ternary Systems". Reinhold Publishing Corp., New York.
McBain, J.W. (1926). *In* "Colloid Chemistry" (Ed. J. Alexander) Vol.I, pp.137-164. The Chemical Catalog Co. Inc., New York.
McBain, J.W. and Field, M.C. (1933). *J. Am. Chem. Soc.* 55, 4776.
Mukerjee, P., Mysels, K.J. and Dulin, C.I. (1958). *J. Am. Chem. Soc.* 62, 1390.
Palatnik, L.S. and Landau, A.I. (1964). "Phase Equilibria in Multicomponent Systems" (Ed. J. Joffee), pp.40-48. Holt, Rinehart and Winston, Inc., New York.
Prigogine, I. and Defay, R. (1954). "Chemical Thermodynamics" (Ed. D.H. Everett), pp.242-243, 208-209. Longmans, Green & Co. Ltd, London.
Rendall, K., Tiddy, G.J.T. and Trevethan, M.A. (1983). *J. Chem. Soc. Faraday Trans. 1* 79, 637.
Ricci, J.E. (1951). "The Phase Rule and Heterogeneous Equilibrium". D. Van Nostrand Co., Inc., New York.
Roozeboom, H.W.B. (1901). "Die Heterogene Gleichgewichte", Vol.I. F. Vieweg u. Sohn, Braunschweig.
Roozeboom, H.W.B. (1904). "Die Heterogene Gleichgewichte", Vol.II. F. Vieweg u. Sohn, Braunschweig.
Rosevaer, F.B. (1954). *J. Am. Oil Chemists Soc.* 31, 628.
Rydhag, L. (1979). *Fette, Seifen, Anstrichm.* 81, 168.
Schreinemakers, F.A.H. (1911). *In* "Die Heterogene Gleichgewichte" (Ed. H.W.B. Roozeboom), Vol.III. F. Vieweg u. Sohn, Braunschweig.

Shinoda, K. (1967). *In* "Solvent Properties of Surfactant Solutions" (Ed. K. Shinoda), pp.12-16. Marcel Dekker, New York.
Shinoda, K. and Friberg, S. (1975). *Adv. Colloid Interface Sci.* 4, 281.
Tamas, F. and Pal, I. (1970). "Phase Equilibria Spatial Diagrams". Iliffe Books, Butterworth, London.
Taylor, H.S. (1930). "A Treatise on Physical Chemistry". D. Van Nostrand Co., Inc., New York.
Vold, R.D. (1937). Unpublished report.
Vorob'eva, A.I. and Karapet'yants, M.Kh. (1967). *Russian J. Phys. Chem.* 41, 1061.
Wennerström, H. and Lindman, B. (1979). *Physics Reports* 52, 1.
Widom, B. (1981). *Faraday Symp. Chem. Soc.*, No.16.

STRUCTURAL ASPECTS OF SURFACTANT MICELLAR SYSTEMS

B. Lindman

Physical Chemistry 1, Chemical Centre, Post Office Box 740, S-220 07 Lund, Sweden.

INTRODUCTION

Surfactants are amphiphilic substances which have their hydrophilic and hydrophobic parts well separated. Adsorption at interfaces is one way of avoiding the entropically unfavourable contact between water and the hydrophobic part (generally one or two alkyl chains) while retaining the aqueous hydrophilic part contact. Self-association into structures with marked hydrophobe-hydrophile separation is another way of achieving this. Surfactant self-association is strongly cooperative and starts generally with the formation of roughly spherical micelles around a fairly well-defined concentration, the critical micelle concentration (c.m.c.). With varying composition in the system or with the addition of a third component, there may be micellar growth and a phase change into various liquid crystalline phases or other solution phases.

In considering micellization and, in particular, phase diagrams, one encounters a very complex pattern with large changes in stability ranges of different phases following subtle changes in chemical structure and temperature. This situation arises mainly from the quite small energy differences between different phases. Considerable progress in the understanding of surfactant micellization and self-association in general has been made recently (Tanford, 1980; Mittal, 1977; 1979; Mittal and Fendler, 1982; Mittal and Lindman, 1983; Israelachvili *et al.*, 1980; Mitchell and Ninham, 1981; Gunnarsson *et al.*, 1980; Jönsson, 1981; Wennerström and Lindman, 1979; Lindman *et al.*, 1980; Kahlweit and Teubner, 1980). In particular the importance of simple geometrical

SURFACTANTS
ISBN 0-12-682180-1

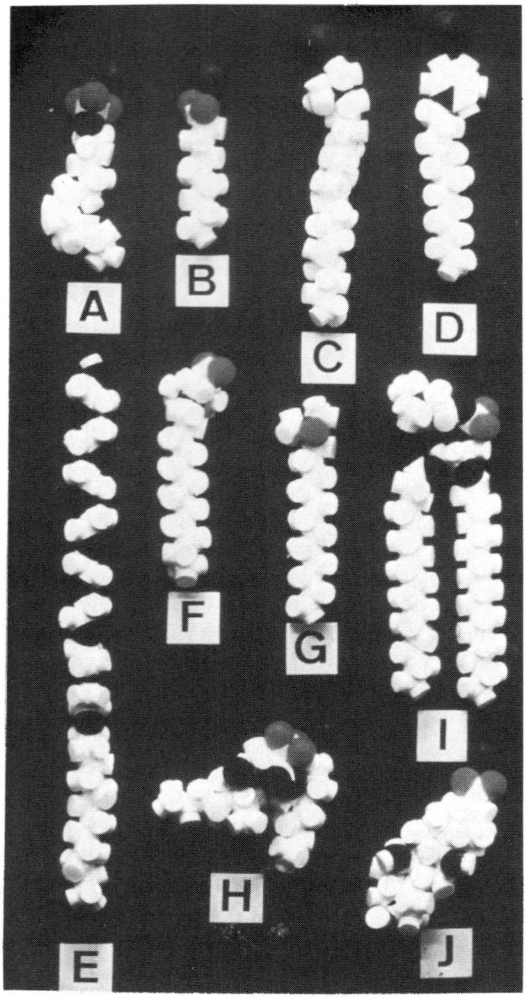

Fig. 1. Space-filling models of the following amphiphile compounds:

a. sodium n-dodecylsulfate (SDS)
b. potassium n-octanoate
c. n-hexadecyltrimethylammonium bromide (CTAB)
d. n-dodecylpyridinium chloride
e. n-dodecyloctaethyleneglycolmonoether ($C_{12}E_8$)
f. N-n-dodecyl-N,N-dimethylbetaine
g. dimethyl-n-dodecylamineoxide
h. sodium di-2-ethylhexylsulfosuccinate (Aerosol OT)
i. dipalmitoyllecithin
j. sodium cholate.

(From Wennerström and Lindman, 1979.)

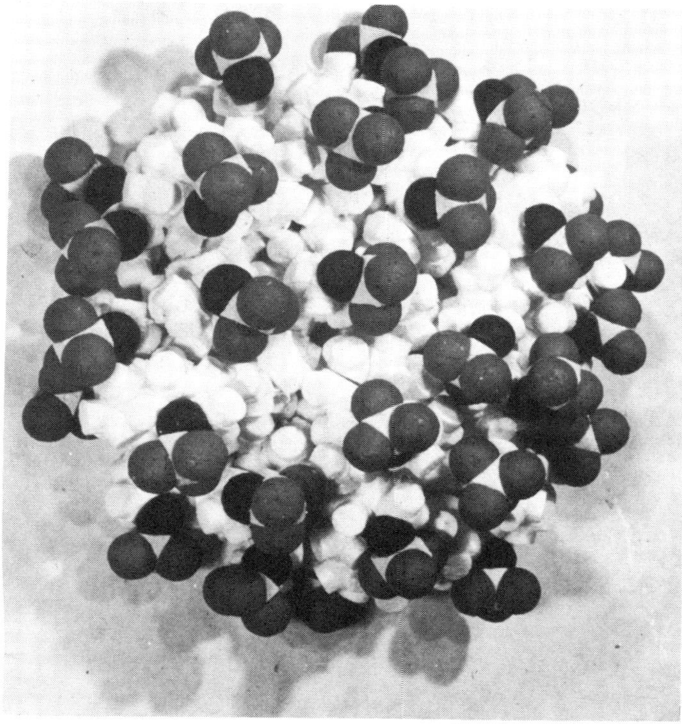

Fig. 2. Space-filling model of a SDS micelle. (From Wennerström and Lindman, 1979.)

features and electrostatic effects have been rationalized. Space-filling models are useful in illustrating possibilities and limitations in the packing of amphiphile molecules. Different classes of amphiphiles are illustrated in Fig. 1. For a single alkyl chain surfactant like sodium dodecylsulfate (SDS) it is possible to pack the molecules in micelles with a rather distinct separation between the hydrophobic parts and the aqueous medium (Fig. 2). The area per polar group in a spherical micelle is rather large so there is inevitably rather extensive hydrocarbon - water contact at the surface. Space-filling models of this type have been made also for several other aggregate geometries characteristic of different phases and serve as a useful aid in the design and evaluation of experiments. These models emphasize the equilibrium structural state but it is important to stress that micelles are very dynamic entities characterized by high intramicellar mobility, by frequent monomer exits and entrances and by a relatively short life-time.

It is the purpose of this Chapter to review broadly structural and dynamic features of surfactant micelles. For more comprehensive treatises the reader is referred to the above-mentioned works which also give a more complete bibliography than can be included here. Special emphasis in this Chapter will be on recent progress in the field.

MICELLE SIZE AND SHAPE

The self-association of an amphiphile occurs in a stepwise manner with one monomer added to the aggregate at a time. For a long-chain amphiphile, the association is strongly cooperative up to a certain micelle size where counteracting factors become increasingly important. Typically the micelles have a closely spherical shape in a rather wide concentration range above the c.m.c. Often there is no great change in shape until at the surfactant solubility limit, where a liquid crystalline phase normally separates out. In other cases, there is a marked transition to large micelles at higher concentrations.

While these general aspects of micellar size and shape have been described in the literature for a long time, their closer investigation and definite and detailed establishment are of relatively recent date. By scattering techniques, like classical light scattering, one monitors the hydrophobic-hydrophilic interface and thus obtains information on the volume occupied by the surfactant molecules, and so on the radius of the "dry" micelle. In quasi-elastic light scattering studies one may, from the determined mutual diffusion coefficient, deduce the hydrodynamic radius of the micelle. This "wet" micelle radius is also obtained from the micelle self-diffusion coefficient. A problem with all these techniques is that in order to obtain the micelle radius one must know the intermicellar interactions, or in some way eliminate them. Very strong micelle - micelle repulsions occur with ionic surfactant micelles and it is common practice then to add high electrolyte concentrations to reduce almost completely the electrostatic interactions. This, of course, reduces the range of conditions under which micelle size is obtainable.

The micellar radius deduced from different types of experiments corresponds closely in a large number of cases to the length of the extended amphiphile molecule. This, of course, shall not be taken to imply that the amphiphile molecules are extended. Rather, considering the packing limitations, only one or a few are in a fully extended state. The difference between the "wet" and "dry" micellar radii is rather small demonstrating that the micelles have no long-range perturbing effect on the solvent.

Most methods, like those mentioned above, provide information on some micellar size but it is also very important to

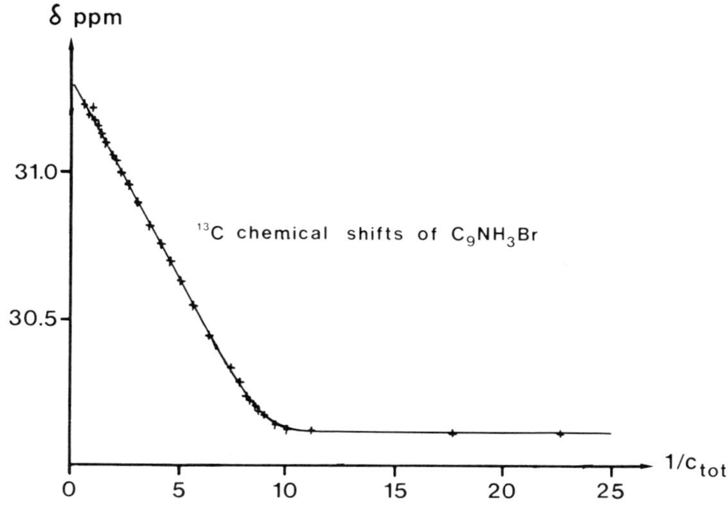

Fig. 3. ^{13}*C NMR chemical shifts of carbon 4 (numbering starting from the polar head) in nonylammonium bromide as a function of the inverse surfactant concentration. A positive shift is down-field. Solid line calculated for an assumed micelle aggregation number of 33. (After Persson* et al., *1979.)*

know the micelle aggregation number. Translating the micelle size into the number of monomers in the micelle is by no means easy and requires assumptions, the validity of which it is difficult to test experimentally. The aggregation number (in some average form) of a micelle can be obtained in different ways. Firstly, the number of micelles may be "counted" and, with the additional information about the amount of surfactant in micelles (see below), the micelle aggregation number is obtained; this forms the basis of the very powerful fluorescence quenching method (Turro and Yekta, 1978; Almgren and Swarup, 1982; 1983a; 1983b), which gives ca. 70 monomers in a SDS micelle. Secondly, a spectroscopic or other property of the surfactant which is distinctly different in the monomeric and micellar states may be monitored as a function of concentration. There are several possibilities but the ^{13}C NMR chemical shift is probably the most generally and conveniently studied parameter. This is exemplified in Fig. 3 where data are compared with a simple mass action law model, $nA \rightleftharpoons A_n$. (It would be necessary in addition to account for the electrostatic effects.) It might further be added that multi-field

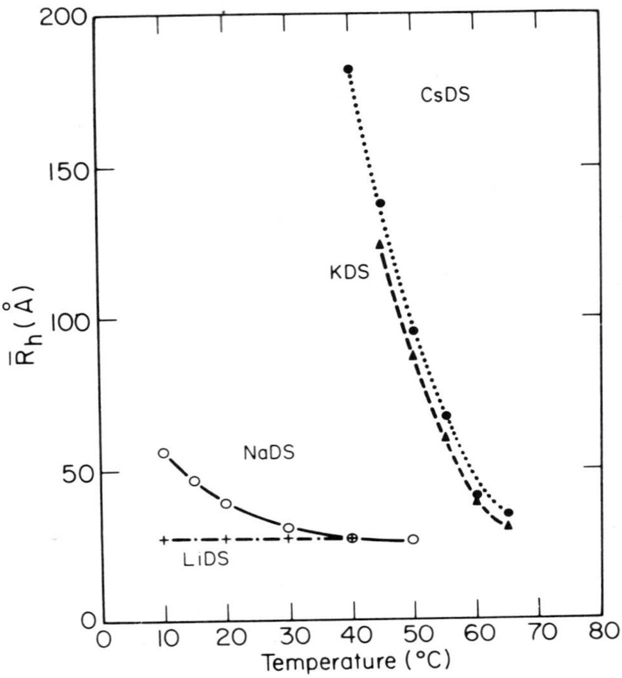

Fig. 4. The sizes of alkali dodecylsulphate micelles decrease with increasing temperature and are sensitive to the counter-ion. The hydrodynamic radius from quasi-elastic light scattering studies for LiDS (20 g per 1 + 1 M LiCl), NaDS (20 g per 1 + 0.45 M NaCl), KDS (5 g per 1 + 0.45 M KCl) and CsCl (5 g per 1 + 0.45 M CsCl). (From Missel et al., *1982.)*

NMR relaxation (see below) may give a micelle radius independent of intermicellar interactions.

From the directly determined aggregation numbers, the radius of a minimal sphere may be calculated using surfactant molecular volumes. In general the radii measurements obtained by this method are in close agreement with those obtained from direct measurements as described above. A direct implication of this is that there is a very distinct separation of the alkyl chains and water with essentially no water penetration into the micelles.

The most obvious evidence of micellar growth is probably the dramatic increase in viscosity with increasing concentration which is observed in several surfactant solutions. It is

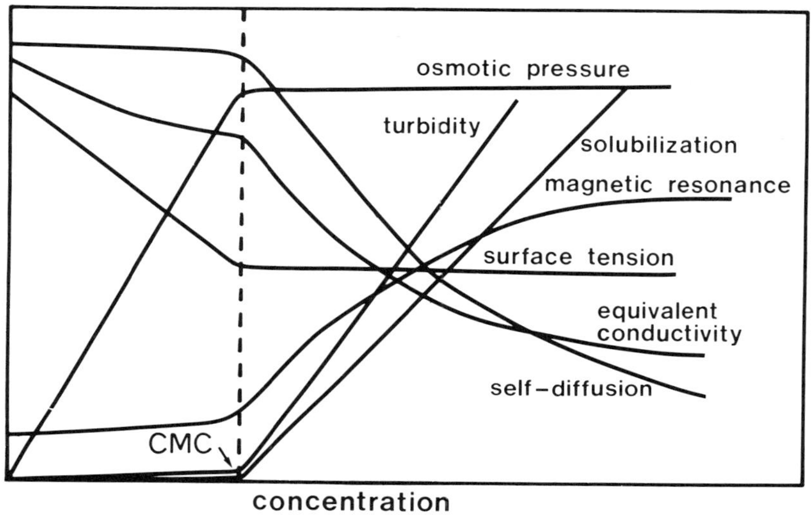

Fig. 5. Changes in the concentration dependences of a wide range of physico-chemical quantities around the critical micelle concentration. (From Lindman et al., *1980.)*

now well established that for common ionic single-chain surfactant systems, micellar growth leads to long rod-like aggregates. Micellar growth is very sensitive to system features, i.e. the energy difference between spheres and rods is small. Micellar growth is favoured by decreasing the temperature, adding electrolyte and lengthening the surfactant chain length and is, furthermore, very sensitive to the nature of the counterion. The sphere-to-rod transition of alkali alkyl sulfate micelles has been documented in detail by Mazer *et al.*, 1976; 1977; Missel *et al.*, 1982; 1983; Young *et al.*, 1978. From the quasi-elastic light scattering investigations (exemplified in Fig. 4) a thermodynamic theory of micelle growth was presented. The analysis of the QLS data is supported by calorimetric studies which provide direct information on the thermodynamic parameters (Mazer and Olofsson, 1982). The enthalpy of micelle formation is -2.6 kJ/mol for SDS in water but the process becomes much more exothermic in the presence of NaCl; this is due to the decrease in the partial molar enthalpy of micellized SDS with added NaCl. The enthalpy change for transferring molecules from spherical to rod-like micelles is weakly exothermic.

For cationic surfactant systems, like hexadecyltrimethylammonium and hexadecylpyridinium halides, there may be a dramatic micellar growth in the absence of added electrolyte. Growth increases very strongly with decreasing temperature and with increasing counterion size (Cl^-, Br^-, I^-).

THE ASSOCIATION PROCESS

From the variation of various physico-chemical properties with surfactant concentration one may obtain information about the activities and concentrations of free and micellized surfactant (Shinoda *et al.*, 1963). It has long been established that there are quite abrupt changes in the concentration dependence of a large number of properties at a particular concentration (Fig. 5); this led to the c.m.c. concept. At low concentrations, many parameters show that there is no appreciable aggregation of the surfactant while above the c.m.c. an extensive aggregation to large aggregates starts. The c.m.c. concept has an exact meaning within the so-called phase separation model of micelle formation. This model considers micelle formation as analogous to a phase separation and the c.m.c. is then the saturation concentration of the amphiphile in the monomeric state whereas the micelles constitute the separated pseudophase. While it is now recognized that the phase separation model only provides a rather approximate description (see below), the c.m.c. concept is extensively used as a very informative characterization of the self-association of surfactants. Mukerjee and Mysels (1971) produced a most valuable compilation of c.m.c. data which, in addition to providing an extensive tabulation of c.m.c., critically discusses procedures and techniques for c.m.c. determination. Table 1 provides selected examples of c.m.c. values.

The variation of the c.m.c. with chemical and physical parameters provides a good insight into the nature of surfactant self-association. The c.m.c. decreases strongly as the hydrophobic part of the surfactant is increased; the decrease is logarithmic being more rapid for nonionic than for ionic surfactants. Fluorocarbon chain surfactants have much lower c.m.c. than hydrocarbon ones; with partially fluorinated surfactants, the c.m.c. may be higher than in either of these cases. With respect to the polar head-group, charge is the decisive factor, nonionics giving much lower c.m.c. than ionics. The difference between different ionics (for example, cationics versus anionics) is relatively small. Similarly the c.m.c. varies little between different monovalent inorganic counterions. Divalent counterions, as well as partially hydrophobic counterions, give much lower c.m.c. (by a factor of ca. 4). Neither temperature nor pressure has

TABLE 1

Examples of critical micelle concentrations at 25°C (mainly from Mukerjee and Mysels, 1971).

Surfactant	c.m.c.*
Dodecylammonium chloride	1.47×10^{-2} M
Dodecyltrimethylammonium chloride	2.03×10^{-2} M
Decyltrimethylammonium bromide	6.5×10^{-2} m
Dodecyltrimethylammonium bromide	1.56×10^{-2} m
Hexadecyltrimethylammonium bromide	9.2×10^{-4} m
Dodecylpyridinium chloride	1.47×10^{-2} M
Sodium tetradecylsulfate	2.1×10^{-3} M
Sodium dodecylsulfate	8.3×10^{-3} M
Sodium decylsulfate	3.3×10^{-2} M
Sodium octylsulfate	1.33×10^{-1} M
Sodium octanoate	4×10^{-1} m
Sodium nonanoate	2.1×10^{-1} m
Sodium decanoate	1.09×10^{-1} m
Sodium undecanoate	5.6×10^{-2} m
Sodium dodecanoate	2.78×10^{-2} m
Sodium p-octylbenzenesulfonate	1.47×10^{-2} m
Sodium p-dodecylbenzenesulfonate	1.20×10^{-3} m
Dimethyldodecylamineoxide	2.1×10^{-3} M
$CH_3(CH_2)_9(OCH_2CH_2)_6OH$	9×10^{-4} M
$CH_3(CH_2)_9(OCH_2CH_2)_9OH$	1.3×10^{-3} M
$CH_3(CH_2)_{11}(OCH_2CH_2)_6OH$	8.7×10^{-5} M
$CH_3(CH_2)_7C_6H_4(CH_2CH_2O)_6OH$	2.05×10^{-4} M
Potassium perfluorooctanoate	2.88×10^{-2} m

* In moles/dm^3 (M) or moles/kg H_2O (m).

any large effect on the c.m.c. which often displays a minimum around room temperature. The addition of electrolyte strongly reduces the c.m.c. for ionics but has little effect for non-ionics. The addition of compounds with hydrophobic groups has a general effect of reducing the c.m.c.

The phase separation model predicts a sharp change in physico-chemical properties around the c.m.c. while a closer analysis of experimental data reveals a smooth transition (more so the shorter the alkyl chain) and, furthermore, the monomer concentration and activity are not constant above the c.m.c. A number of alternative models have, therefore, been proposed. In the <u>mass action law model</u>, it is assumed that a single micellar species is in equilibrium with the monomers,

$$nA_1 \rightleftharpoons A_n \tag{1}$$

$$[A_n] / [A_1]^n = K \tag{2}$$

and

$$\text{c.m.c.} \cong K^{1/(n-1)}. \tag{3}$$

The assumption of a single aggregation number is a simplification but this approach illustrates well that the larger the micelle the more cooperative is the association and the closer one is to a phase separation behaviour. As already noted, n can be obtained from the smooth variation of a property around the c.m.c.

In reality micellar size is not restricted to a single aggregation number but there is a micellar size distribution. This can be analyzed using the multiple equilibrium model which can be best formulated as a stepwise aggregation

$$\begin{aligned} A_1 + A_1 &\rightleftharpoons A_2 \\ A_2 + A_1 &\rightleftharpoons A_3 \\ &\vdots \\ A_{i-1} + A_1 &\rightleftharpoons A_i \\ &\vdots \end{aligned} \tag{4}$$

As noted in particular in the analysis of kinetic data (Aniansson and Wall, 1974; Aniansson *et al.*, 1976; Kahlweit and Teubner, 1980) there are aggregates over a wide range of aggregation numbers, from dimers and well beyond the most stable micelles. However, for typical surfactants with not too high c.m.c., the size distribution curve has a very deep minimum, the least stable aggregates being present in concentrations many orders of magnitude below those of the most abundant micelles. Furthermore, for cases with predominantly spherical micelles the polydispersity is low and there is thus a particularly preferred micellar size. In cases where micellar growth occurs the polydispersity is large and a wide range of micelle sizes occur.

With respect to the thermodynamics of micellization it is clear from the weak temperature dependence of the c.m.c. that the enthalpy of micelle formation (ΔH) is close to zero. Direct calorimetric studies (Mazer and Olofsson, 1982) have investigated this more directly and have also demonstrated that there is quite a large decrease in the partial molar heat capacity, C_p, on micellization. The low ΔH value implies that there is a large entropy increase which is the main driving force for the surfactant self-association. On the whole, with good approximation, the thermodynamics of micelle formation has the same characteristics as the transfer of aqueous hydrocarbons to a liquid hydrocarbon phase.

Fast kinetic investigations of micellar solutions showed two relaxation times on widely different time scales (Aniansson and Wall, 1974; Aniansson *et al.*, 1976; Kahlweit and Teubner, 1980). This can be understood on the basis of the above-mentioned deep minimum in the size distribution curve while assuming that the elementary process that governs the dynamic behaviour is the equilibrium between a monomer and a micellar aggregate, i.e. $A_1 + A_{n-1} \rightleftharpoons A_n$. After a sudden temperature change in a T-jump experiment, the aggregate size distribution will be somewhat different from the equilibrium state. Under the given assumptions Aniansson and Wall (1974) could demonstrate that the return to equilibrium can be approximated as a two-step process. Firstly, there is a rapid process representing a redistribution of monomers to a quasi-equilibrium at a fixed total number of micelles. Secondly, there is a much slower process involving a change in the number of micelles to attain true equilibrium. During this process, aggregates at the minimum in the size distribution curve are involved. Since these have a very low probability the process is typically very slow. The kinetic studies have provided insight into the mechanism of micelle formation and dissolution in terms of the entering or leaving of one monomer at a time. They have also given information on the width of the micelle size distribution; as an example the standard deviation of the micelle size distribution is ca. 13 for SDS micelles (Aniansson *et al* ., 1976). For this system, the average monomer life-time in a micelle is 6×10^{-6} seconds while the micelle life-time is in the range $40\text{-}70 \times 10^{-3}$ seconds (Aniansson *et al.*, 1976; Wall, personal communication). (Data refer to low micelle concentrations.)

IONIC SURFACTANTS: COUNTERION BINDING AND ELECTROSTATIC EFFECTS

In the discussion so far we have not considered the effects of charged polar head-groups and of counterion binding to the micelles. One possibility of accounting for the counterions is to include them in the equilibria, e.g. with counterions M,

$$mM + nA_1 \rightleftharpoons M_m A_n \tag{5}$$

If M is a monovalent ion, the micelle will have a net charge of n-m. Counterion binding in micellar systems has been extensively investigated by a number of experimental techniques and we will briefly summarize the main findings here (Wennerström and Lindman, 1979). From these results it is found that a description in terms of chemical equilibrium is unsuitable. The reason is that such a model presupposes short-range forces while the electrostatic forces governing the counterion distribution are of a long-range nature. Counterion binding does not involve specific complex formation with sites on the micelle but rather a nonspecific continuous counterion distribution. Then, in principle, one would expect different experimental approaches to give different results. A general electrostatic model (Gunnarsson *et al.*, 1980; Jönsson, 1981; Jönsson and Wennerström, 1981) of ionic interactions in surfactant systems which accounts well for various observations has recently been developed and will be briefly reviewed below.

According to the phase separation model, the surfactant monomer activity stays constant above the c.m.c. while, according to the mass action law model, it increases gradually. It was observed in investigations using surfactant specific electrodes that the monomer activity decreases markedly above the c.m.c. (Cutler *et al.*, 1978) and in kinetic investigations that there is a corresponding decrease in monomer concentration (Aniansson *et al.*, 1976). We will illustrate these and some other general features of ionic surfactant micelles by self-diffusion studies.

The principles of self-diffusion studies of surfactant systems are simple. The micelles, being relatively large entities, diffuse slowly. Therefore, any molecule confined to the micelles (due to a low solubility in the intermicellar solution) will also have a very low self-diffusion coefficient. Molecules free in the intermicellar medium, on the other hand, will diffuse as free molecules and hence be characterized by large self-diffusion coefficients. A molecule which is distributed between the micelles and the intermicellar solution will have a self-diffusion coefficient between these two extremes. From the value of the self-diffusion coefficient the distribution can be quantified according to the equation,

$$D = P_{free} D_{free} + P_{mic} D_{mic} \tag{6}$$

Here P_{free} and P_{mic} are the fractions of free and micellar molecules or ions (surfactant, counterion, etc.). D_{free} is the free ion or molecule self-diffusion coefficient, while D_{mic} is the self-diffusion coefficient of the micelles.

We exemplify these studies by the case of sodium p-octylbenzenesulfonate (Lindman *et al.*, 1982) which has a c.m.c. of ca. 13×10^{-3}mol/kg. As can be inferred from Fig. 6, water diffusion is rapid and the same applies to ions which have the same sign of the charge as the surfactant. These species clearly diffuse for a quite small or insignificant amount with the micelles. The self-diffusion coefficient of

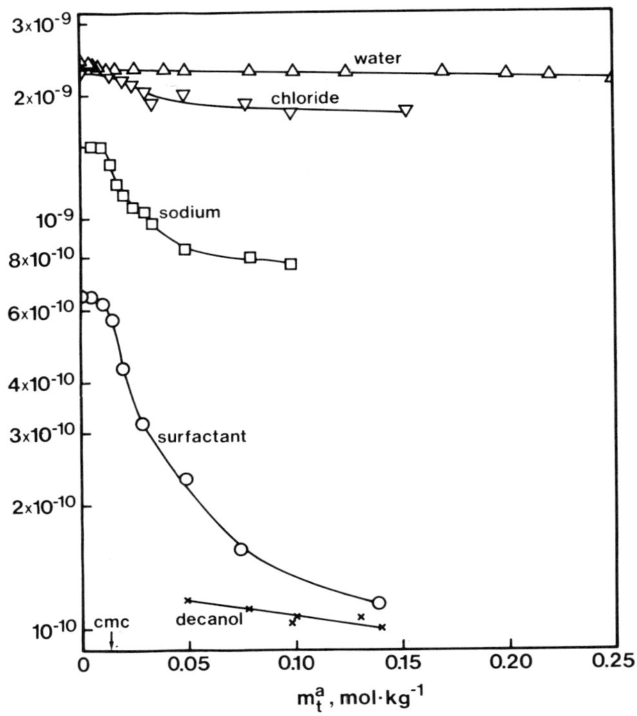

Fig. 6. Self-diffusion behaviour of solutions of sodium p-octylbenzenesulfonate. Self-diffusion coefficients (33°C) of surfactant ions, counterions and water molecules as well as of small amounts of added decanol and chloride are given as a function of surfactant concentration. (From Lindman et al., *1981.)*

solubilized decanol is low as would be expected from its very low aqueous solubility. Counterion and surfactant diffusion fall in between these cases and are, above all, characterized by a strong concentration dependence. There is clearly a marked distribution of these ions between the micelles and the intermicellar solution and a distribution that changes considerably with the surfactant concentration. Using the procedure indicated above, the concentrations of free and micellized surfactant and counterion may be quantified. This is illustrated in Fig. 7 which also shows the degree of counterion binding, β, that is the ratio of counterions and surfactant ions in the micelles. We should especially note in

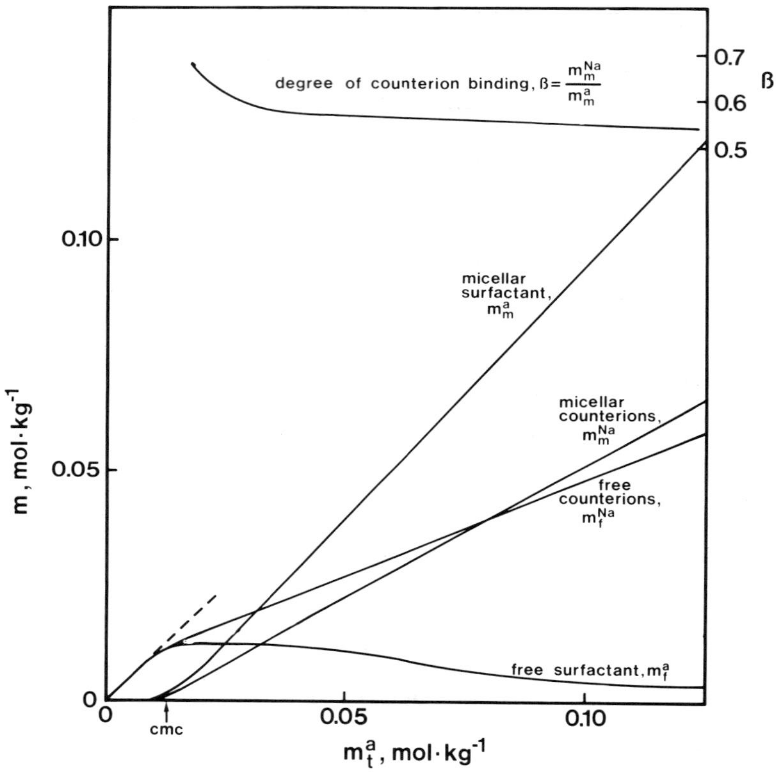

Fig. 7. Concentrations of free and micellized surfactant ions and counterions for solutions of sodium p-octylbenzenesulfonate. Additionally, β, the degree of counterion binding is given. (From Lindman et al., *1982.)*

Fig. 7 the marked decrease in free surfactant concentration above the c.m.c. as well as the insensitivity of β to concentration changes.

The approximate constancy of β with very large concentration changes is a rather general observation for both micelles and other ionic surfactant phases. Significant observations for simple inorganic counterions are (Wennerström and Lindman, 1979; Lindman and Wennerström, 1982; Wennerström *et al.*, 1980).

a) For most cases, β lies in the range 0.5 to 0.8.
b) Different experimental approaches give the same general

picture of counterion binding. The different experimental techniques can be classified into thermodynamic (e.g. ion activity measurements), transport (such as self-diffusion) and spectroscopic (NMR etc.) measurements (Gunnarsson *et al.*, 1980).

c) Ion specificity is small among monovalent ions for the c.m.c. and for β, but may be very important for micelle size and shape and for phase equilibria.
d) Ion competition effects may be marked. For example, divalent ions replace monovalent ions effectively and there may also be marked ion competition effects among monovalent ions.
e) β is approximately independent of surfactant concentration, temperature and alkyl chain length.
f) β increases slightly with aggregate shape from spheres over rods to lamellae.
g) β is only slightly affected by added electrolyte.

Several of these observations are difficult to reconcile with conventional chemical equilibrium, e.g., a mass action law model involving the counterions. One would, for example, expect β to change with temperature and added electrolyte. Furthermore, one would expect the counterions to be dissociated on diluting the system. The pattern resembles, on the other hand, the polyelectrolyte behaviour observed for linear polymers with a high charge density. Here the so-called <u>ion condensation model</u>, in which counterion binding occurs to an extent that reduces the residual charge density to a certain critical value that is independent of temperature and added electrolyte, has been found to rationalize a number of experimental observations (Manning, 1978). A limitation of the ion condensation model is that it treats the polymer as a line charge and does not take into account the finite radius and moreover it is restricted to a single geometry. These problems are overcome using another electrostatic model based on the Poisson-Boltzmann equation. Such models, considering the amphiphile aggregates as uniformly charged and the counterions as point charges, have identified an ion condensation behaviour for different aggregate geometries and have predicted counterion association in agreement with experimental results (Gunnarsson *et al.*, 1980; Jönsson and Wennerström, 1981).

A very useful model of surfactant systems in general, based on the Poisson-Boltzmann equation, has been developed by Jönsson and Wennerström (1981). We will briefly consider it for surfactant micellization. In these calculations, the system is considered to be divided into subsystems, the geometries of which are given by the phase symmetry (the so-called cell model). In the case of spherical micelles one

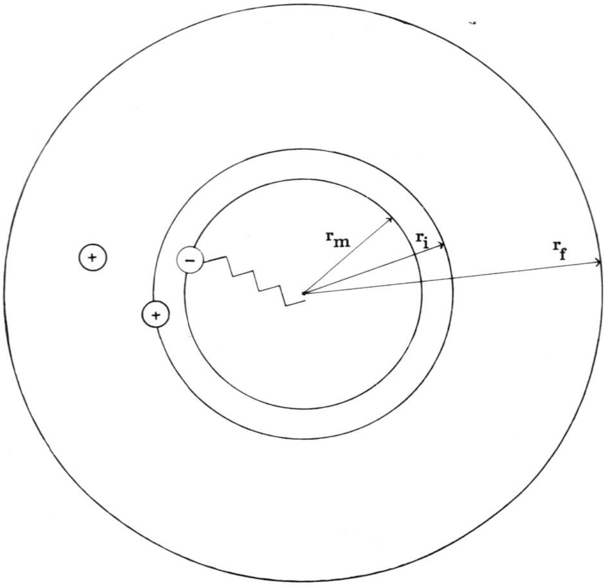

Fig. 8. The model for a spherical micelle used in the electrostatic approach to micellization (Gunnarsson et al., *1980). r_m is the micellar radius, r_i the radius for closest approach of counterions and r_f the radius of a sphere representing the volume per micelle. (From Lindman* et al., *1980.)*

considers then a spherical cell as drawn in Fig. 8, i.e. the spherical micelle is in the centre of a sphere with a concentration dependent radius. The contributions to the free energy of one subsystem are considered to be: a hydrophobic interaction term which is the difference in chemical potential for a nonpolar solute between a hydrocarbon solvent and water; an interfacial energy which is proportional to the area between the aggregates and the aqueous medium; an entropy term for the ideal mixing of the components in the aqueous region; an electrostatic free energy; and a van der Waals' attraction energy between the aggregates.

Because of the high micellar charge, there will be an accumulation of counterions near the micelle while the free surfactant ions will be repelled. The (monovalent) counterion distribution is given by,

$$m^i(r) = m_o^i \exp\{-e\Psi(r)/kT\} \tag{7}$$

as a function of r, the distance from the centre of the cell, and the electrostatic potential, Ψ. m_o^i is the counterion concentration at the surface of the cell. The ion distribution apparently characterizes the ion binding which is denoted by a single parameter, β. The counterion concentration is continuously decaying as one moves away from the micelle and the separation into free and bound ions is ambiguous. In self-diffusion studies one may crudely consider those ions to have diffused with the micelle which are attracted by more than the thermal energy, kT. In agreement with experiment, theory predicts β to be approximately independent of concentration both for spherical, cylindrical and planar surfaces. (This is also true for other measures of the counterion binding.)

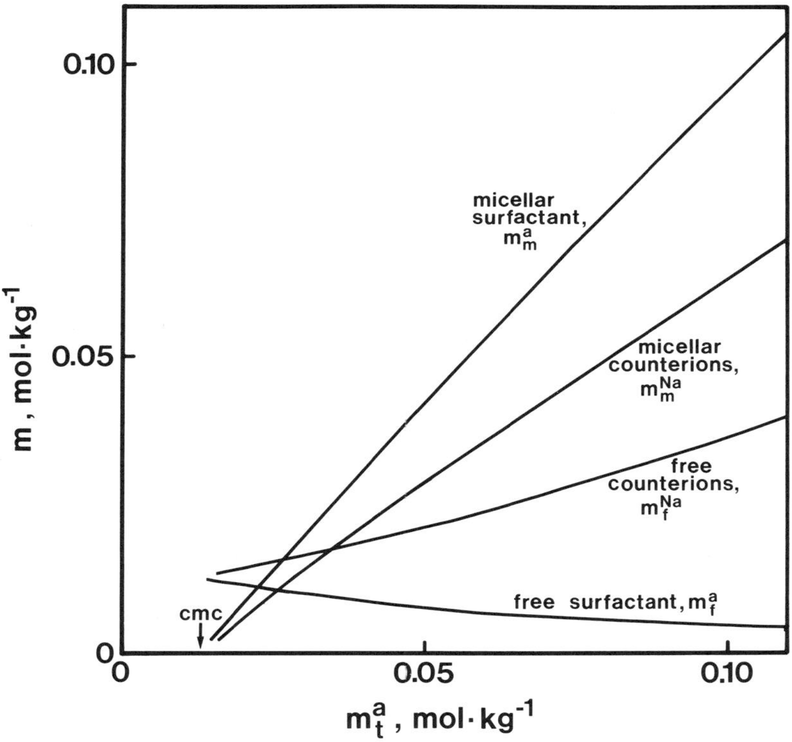

Fig. 9. Calculated concentrations of free and micellized surfactant ions and counterions as obtained for sodium p-octylbenzenesulfonate from the theoretical model (Gunnarsson et al., *1980) using the Poisson-Boltzmann equation for the electrostatic effects. (From Lindman* et al., *1982.)*

The nonuniform ion distribution influences the free energy of the solution and thus also the surfactant chemical potential and activity. Due to the partial counterion dissociation from the micelles, the uneven counterion distribution is levelled out partially as the total surfactant concentration increases. The enhanced free counterion concentration stabilizes the micellar state and thus decreases the free surfactant activity and concentration. The changes in the various activities and concentrations may be quantified using the electrostatic model and an example is given in Fig. 9. In Fig. 10, it is demonstrated that also for a case with low cooperativity in the association (c.m.c. ca. 0.4 m), the electrostatic model may quantitatively rationalize the

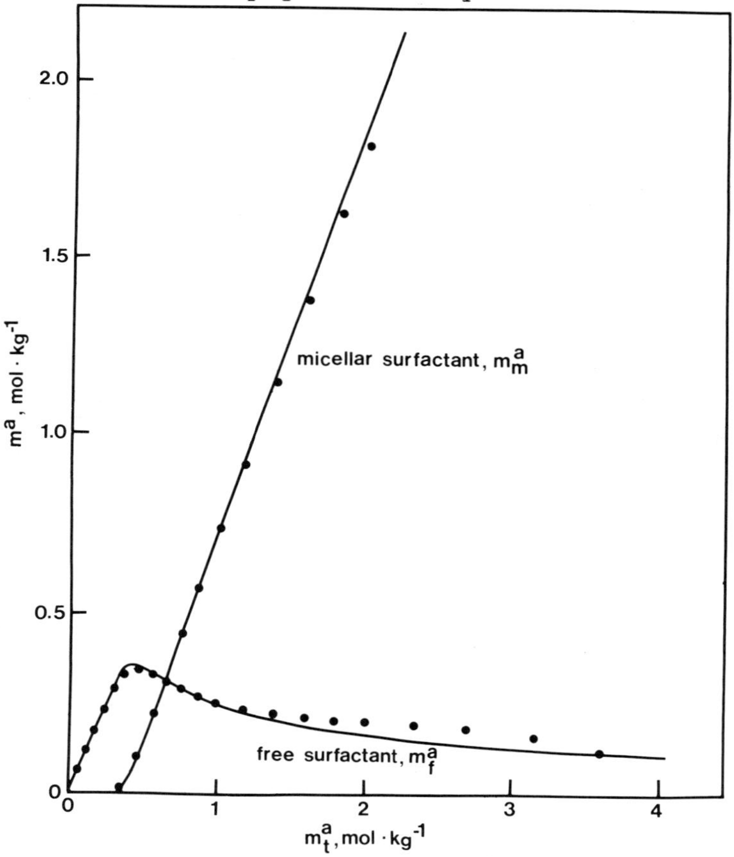

Fig. 10. Experimental (circles; from self-diffusion) and theoretical (curves; Gunnarsson et al., *1980) concentrations of micellized and free surfactant for sodium octanoate. (From Lindman* et al., *1983.)*

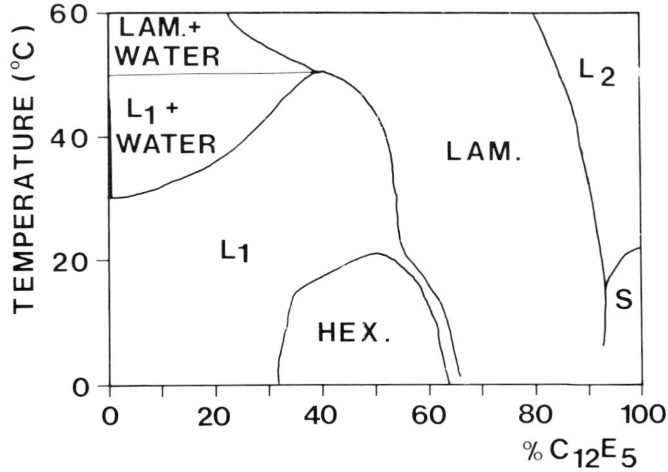

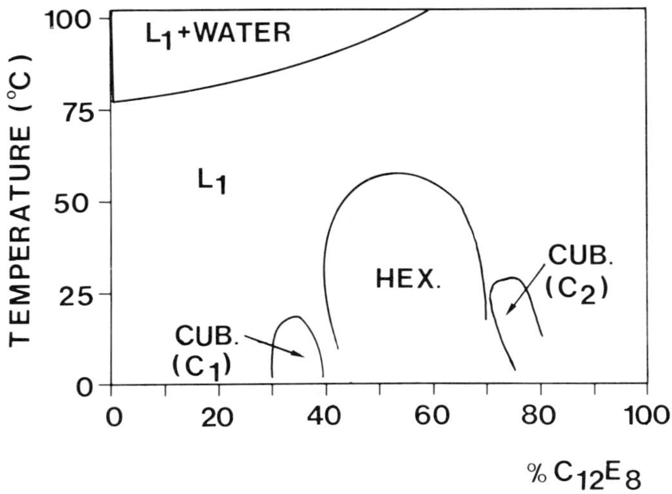

Fig. 11. Schematic phase diagrams of the systems $C_{12}E_5$-H_2O (Harusawa et al., *1974) and $C_{12}E_8$-H_2O (Shinoda, 1970) modified from the original references. Hexagonal, lamellar and cubic liquid crystalline phases are denoted by hex., lam., and cub., respectively, and L_1 and L_2 denote isotropic solutions. S denotes solid surfactant. (From Nilsson* et al., *1983.)*

concentrations of free and micellized surfactant.

NONIONIC SURFACTANT MICELLES

The electrostatic repulsions strongly destabilize ionic surfactant micelles, especially in the absence of added electrolyte. For nonionics, this effect is absent and cooperativity in association is much greater as manifested *inter alia* in the much lower c.m.c. This difference also appears in the feasibility of closer packing of the head-groups as is required in the transformation to other geometries (micellar growth). Much interest at present concerns nonionics of the poly(ethylene oxide) alkylether type (here abbreviated C_nE_x, n being the number of carbons in a n-alkyl chain and x the number of ethylene oxide groups).

The phase behaviour of these nonionic surfactant - water systems is very sensitive to temperature changes as illustrated in Fig. 11. As can be seen a dilute solution of C_nE_x may, on heating, separate into two isotropic solutions and an important consideration is the aggregate structure in these phases. A great deal of work, using a range of experimental techniques, is concerned with the aggregate shape and size just below the minimum in the phase separation curve. One such approach is the strong light scattering by the solution which is referred to as the clouding phenomenon. Here we briefly review some self-diffusion work but reference should be made also to other work (Staples and Tiddy, 1978; Corti and Degiorgio, 1981; Zulauf and Rosenbusch, 1983; Kjellander, 1982; Mitchell *et al.*, 1983; Cebula and Ottewill, 1982; Triolo *et al.*, 1982) which is partly at variance with ours.

From water self-diffusion, one observes strong hydration of the ethylene oxide groups. This hydration is reduced as the water - to - ethylene oxide molar ratio is reduced and as temperature is increased. On the other hand, hydration seems to be quite insensitive to the particular system, aggregate size and shape, etc. (Nilsson and Lindman, 1983). Simply from the head-group area - to - alkyl chain volume ratio one expects micellar growth to be easy for small x/n ratios in C_nE_x while the minimal spherical shape should be favoured if x is large. Furthermore, a micellar growth and shape change sphere → rod → lamellae should be facilitated by a temperature increase. These and other expected phenomena are nicely confirmed as far as phase behaviour is concerned and there is good reason to believe that they should apply also within the isotropic phase regions. Support for this is given by the self-diffusion work (Nilsson *et al.*, 1983). Thus $C_{12}E_5$ micelles grow both with increasing temperature and with increasing concentration (Fig. 12). For $C_{12}E_8$, there is also micellar growth but to a

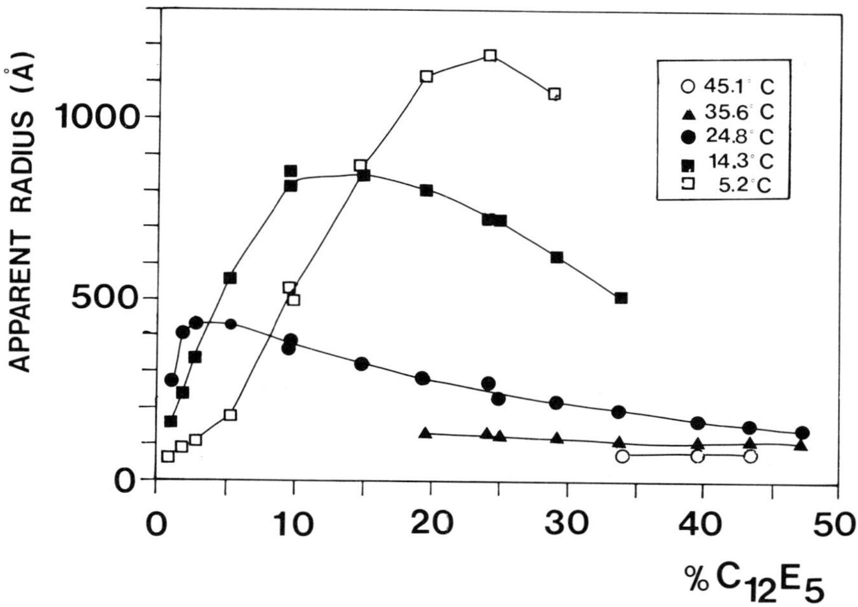

Fig. 12. The apparent radius of $C_{12}E_5$ micelles deduced from NMR self-diffusion data (Nilsson et al., 1983).

much smaller extent. From proton NMR relaxation, it can be deduced that the order of the surfactant molecules is reduced with increasing temperature. Over wide concentration ranges in the L_1 phase, rod $C_{12}E_5$ micelles are certainly formed on micellar growth. Very close to the cloud point some other shape may not be excluded. The cloud point seems generally to lie just above the upper stability limit of the hexagonal phase, which would point to a rod → lamellar transition in this range. To explain the phase separation in general, it seems necessary to assume formation of anisometric aggregates (Kjellander, 1982; Nilsson *et al.*, 1983). The preferred aggregate shape may be an intermediate between a disc and a rod, i.e. the aggregates close to the cloud point may not have a preferred shape and there may be large fluctuations in size and shape. As regards micelle shape, rods rather than discs are supported by the obstruction effect in water self-diffusion over wide ranges in the L_1 phase. In a high temperature isotropic phase discs seem, on the other hand, to be the preferred shape (Nilsson and Lindman, to be published).

Much interest is devoted to the establishing of critical effects in the vicinity of the cloud point and Corti and Degiorgio (1975; 1980; 1981) in particular have done some useful work on this aspect. These authors demonstrate that the correlation range of concentration fluctuations as studied by light scattering diverges on approach of the cloud point. Furthermore, it is found that usual models of critical phenomena for binary mixtures may not be appropriate for describing critical micelle solutions. It seems particularly important to separate effects due to critical phenomena from those associated with micelle growth.

INTERNAL STRUCTURE AND DYNAMICS OF MICELLES

A picture of a micelle such as that in Fig. 2 has several structural implications which can be directly tested by experiment. Thus the model suggests a sharp division between hydrophobic and hydrophilic groups with water molecules only in the vicinity of the head-groups, a substantial coiling of the alkyl chains which may be similar to that in a liquid hydrocarbon and a disordered location (mainly laterally) of the polar head-groups (and of the counterions). There are also a number of important questions on dynamics, such as: Dynamics of the alkyl chains? Lateral diffusion of surfactant ions and of counterions? Life-times of counterions and water molecules on the micelle surface?

With respect to micelle hydration and water molecule distribution, several experimental approaches are ambiguous but in particular Cabane has devised powerful methods for this problem using NMR relaxation with paramagnetic counterions (Cabane, 1981) and neutron scattering (Cabane *et al.*, 1983). From studies on SDS micelles, it is concluded that the water/hydrocarbon contact is limited to a very thin layer comparable in thickness to the dimensions of a CH_2 group. Such a sharp hydrophilic/hydrophobic interface is also the starting point in theoretical treatments of surfactant systems and is further demonstrated *inter alia* in water self-diffusion (Lindman *et al.*, 1982) and ^{17}O NMR relaxation studies (Halle and Carlström, 1981). The latter also demonstrate that water molecules are very mobile when bound to the micelles. They rotate almost isotropically at a rate which is two or three times slower than in pure water and exchange rapidly with the intermicellar solution (life-times in the micelles of the order of 10^{-8} seconds).

Monovalent counterions also exchange very rapidly with

life-times which often lie in the range 10^{-8} to 10^{-9} seconds. Furthermore, the counterions appear to be laterally very mobile and not to be associated with (single) specific groups on the micelles. The association with the micelle as a whole rather than with individual groups is, of course, strongly supported by the success of electrostatic models treating a micelle as a uniformly charged sphere (see above).

It is generally agreed that the micelle interior qualitatively resembles a liquid hydrocarbon "droplet" but it has not been easy to quantify alkyl chain orientation and flexibility. For the liquid crystalline phases, static NMR parameters provide quantitative information on alkyl chain orientation via the so-called order parameters and for micelles, Cabane (1981) has used ^{1}H and ^{13}C NMR relaxation to demonstrate a disordered alkyl chain packing. We will here briefly introduce a multi-field ^{13}C relaxation method capable of providing information on both the amplitudes and dynamics of alkyl chain motion in micelles.

A general but important observation in NMR studies of associated surfactant systems is that of rather long relaxation times combined with a marked magnetic field dependence. This shows that motions on widely different time scales contribute to relaxation. To account for this, Wennerström (Wennerström *et al.*, 1974; Halle and Wennerström, 1981) has developed a two-step model of relaxation. The model assumes that part of the interaction is modulated by rapid local motions which do not give complete isotropy but reduce the interactions to a level which can be characterized by an order parameter. This is defined as the degree of orientation of a molecular vector with respect to the micelle surface. The remainder of the interaction is modulated by slow motions in the range 10^{-7} to 10^{-9} seconds like micelle rotation and surfactant lateral diffusion along the micelle surface. Using Wennerström's theory it is possible to analyse multi-field relaxation data to provide three parameters, i.e. one correlation time, τ_c^f, of the fast local motion; one, τ_c^s, of the slow overall motion; and one order parameter, S (Wennerström *et al.*, 1979; Walderhaug *et al.*, 1983). τ_c^f and S are obtained (normally) for all carbons along the alkyl chain. Information obtained is exemplified for sodium p-octylbenzenesulfonate micelles in Fig. 13. It can be inferred that the order parameters are in the range 0.05-0.2 and similar in magnitude to those of liquid crystalline phases. For a two-component surfactant - water system, which offers a better comparison, a quantitative agreement between micellar and liquid crystal order parameters was found. The fast motion is more rapid as one moves away from the polar head but occurs throughout in times of the order of 10^{-11} seconds. This is not

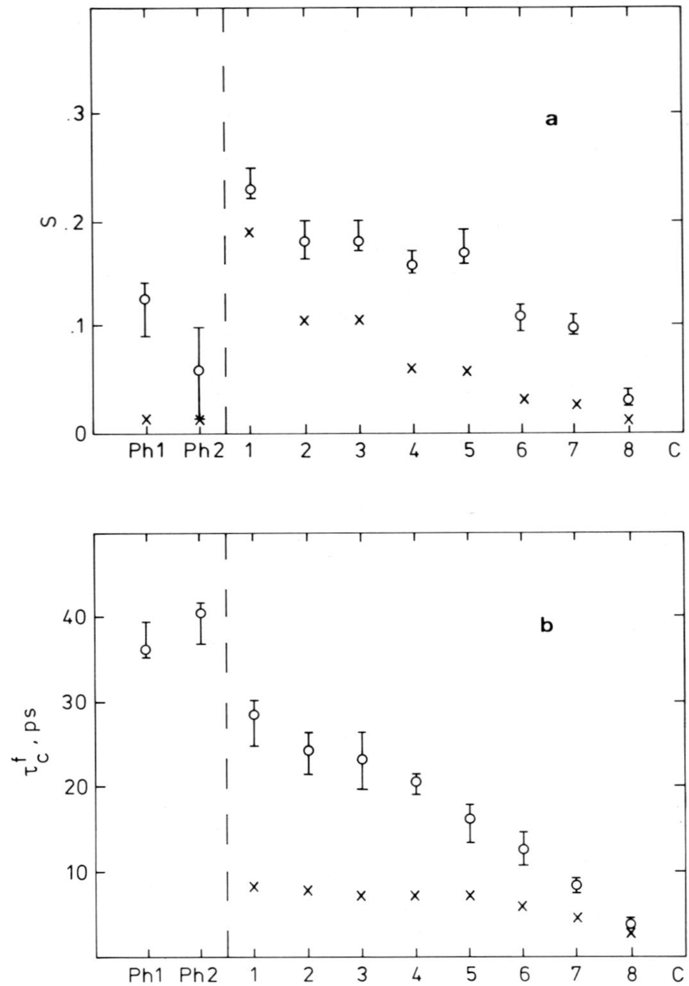

*Fig. 13*a. *The order parameter for micelles of sodium p-octylbenzenesulfonate at 42°C (O). Numbering of alkyl chain carbons starts from the polar head; Ph 1 and Ph 2 denote phenyl carbons. X denotes corresponding order parameters (30°C) for a lamellar phase sample with pentanol as the third component. (From Walderhaug* et al., *1983.)*

*Fig. 13*b. *The fast motion correlation time (42°C) as a function of chain position for sodium p-octylbenzenesulfonate micelles. X gives corresponding data for octylbenzene in acetone at 30°C. (From Walderhaug* et al., *1983.)*

far from what is observed for pure liquid hydrocarbons of similar size or for hydrocarbons in other solvents. It is clearly possible to quantify intramicellar dynamics to demonstrate that the model of a liquid hydrocarbon droplet is not far from reality.

REFERENCES

Almgren, M. and Swarup, S. (1982). *J. Phys. Chem.* 86, 4212.

Almgren, M. and Swarup, S. (1983a). *J. Colloid Interface Sci.* 91, 256.

Almgren, M. and Swarup, S. (1983b). *J. Phys. Chem.* 87, 876.

Aniansson, E.A.G. and Wall, S.N. (1974). *J. Phys. Chem.* 78, 1024.

Aniansson, E.A.G., Wall, S.N., Almgren, M., Hoffmann, H., Kielmann, I., Ulbricht, W., Zana, R., Lang, J. and Tondre, C. (1976). *J. Phys. Chem.* 80, 905.

Cabane, B. (1981). *J. Physique, Paris* 42, 847.

Cabane, B., Duplessix, R. and Zemb, T. (1983). *In* Mittal and Lindman (1983) in press.

Cebula, D.J. and Ottewill, R.H. (1982). *Colloid and Polymer Sci.* 260, 1118.

Corti, M. and Degiorgio, V. (1975). *Optics Communication* 14, 358.

Corti, M. and Degiorgio, V. (1980). *Phys. Rev. Lett.* 45, 1045.

Corti, M. and Degiorgio, V. (1981). *J. Phys. Chem.* 85, 1442.

Cutler, S.G., Meares, P. and Hall, D.G. (1978). *J. Chem. Soc. Faraday 1* 74, 1758.

Gunnarsson, G., Jönsson, B. and Wennerström, H. (1980). *J. Phys. Chem.* 84, 3114.

Halle, B. and Carlström, G. (1981). *J. Phys. Chem.* 85, 2142.

Halle, B. and Wennerström, H. (1981). *J. Chem. Phys.* 75, 1928.

Harusawa, F., Nakamura, S. and Mitsui, T. (1974). *Colloid and Polymer Sci.* 252, 613.

Israelachvili, J.N., Marcelja, S. and Horn, R.G. (1980). *Quart. Rev. Biophys.* 13, 121.

Jönsson, B. (1981). "The Thermodynamics of Ionic Amphiphile Water Systems, A Theoretical Analysis". Thesis, Lund.

Jönsson, B. and Wennerström, H. (1981). *J. Colloid Interface Sci.* 80, 482.

Kahlweit, M. and Teubner, M. (1980). *Adv. Colloid Interface Sci.* 13, 1.

Kjellander, R. (1982). *J. Chem. Soc. Faraday 2* 78, 2025.

Lindman, B. and Wennerstrom, H. (1982). *In* Mittal and Fendler (1982), p.3.

Lindman, B., Wennerström, H. and Eicke, H.F. (1980). "Micelles", Springer-Verlag, Heidelberg.

Lindman, B., Puyal, M.C., Kamenka, N., Brun, B. and Gunnarsson, G. (1982). *J. Phys. Chem.* 86, 1702.
Lindman, B., Kamenka, N., Puyal, M.C., Brun, B. and Jönsson, B. (1983). *J. Phys. Chem.* In press.
Manning, G.S. (1978). *Quart. Rev. Biophys.* 11, 179.
Mazer, N.A. and Olofsson, G. (1982). *J. Phys. Chem.* 86, 4584.
Mazer, N.A., Benedek, G.B. and Carey, M.C. (1976). *J. Phys. Chem.* 80, 1075.
Mazer, N.A., Carey, M.C. and Benedek, G.B. (1977). *In* Mittal (1977), p.359.
Missel, P.J., Mazer, N.A., Carey, M.C. and Benedek, G.B. (1982). *In* Mittal and Fendler (1982), p.373.
Missel, P.J., Mazer, N.A., Benedek, G.B. and Carey, M.C. (1983). *J. Phys. Chem.* 87, 1264.
Mitchell, D.J. and Ninham, B.W. (1981). *J. Chem. Soc. Faraday 2* 77, 601.
Mitchell, D.J., Tiddy, G.J.T., Waring, L., Bostock, T. and McDonald, M.P. (1983). *J. Chem. Soc. Faraday 1* 79, 975.
Mittal, K.L. (Ed.) (1977). "Micellization, Solubilization and Microemulsions". Plenum Press, New York. Vols 1 and 2.
Mittal, K.L. (Ed.) (1979). "Solution Chemistry of Surfactants". Plenum Press, New York. Vols 1 and 2.
Mittal, K.L. and Fendler, E.J. (Eds) (1982). "Solution Behaviour of Surfactants". Plenum Press, New York. Vols 1 and 2.
Mittal, K.L. and Lindman, B. (Eds) (1983). "Surfactants in Solution". Plenum Press, New York. Vols 1-3.
Mukerjee, P. and Mysels, K.J. (1971). "Critical Micelle Concentrations of Aqueous Surfactant Systems". NSRDA-NBS-36. U.S. Government Printing Office, Washington D.C.
Nilsson, P.-G. and Lindman, B. (1983). *J. Phys. Chem.* In press.
Nilsson, P.-G., Wennerström, H. and Lindman, B. (1983). *J. Phys. Chem.* 87, 1377.
Persson, B.O., Drakenberg, T. and Lindman, B. (1979). *J. Phys. Chem.* 83, 3011.
Shinoda, K. (1970). *J. Colloid Interface Sci.* 34, 278.
Shinoda, K., Nakagawa, T., Tamamushi, B.-I. and Isemura, T. (1963). "Colloidal Surfactants, some Physico-chemical Properties". Academic Press, New York.
Staples, E.J. and Tiddy, G.J.T. (1978). *J. Chem. Soc. Faraday 1* 74, 2530.
Tanford, C. (1980). "The Hydrophobic Effect: Formation of Micelles and Biological Membranes". Wiley, New York. 2nd edn.
Triolo, R., Magid, L.J., Johnson, J.S. Jr and Child, H.R. (1982). *J. Phys. Chem.* 86, 3689.
Turro, N.J. and Yekta, A. (1978). *J. Am. Chem. Soc.* 100, 5951.

Walderhaug, N., Söderman, O. and Stilbs, P. (1983). To be published.
Wennerström, H. and Lindman, B. (1979). *Phys. Rep.* 52, 1.
Wennerström, H., Lindblom, G. and Lindman, B. (1974). *Chem. Scr.* 6, 97.
Wennerström, H., Lindman, B., Söderman, O., Drakenberg, T. and Rosenholm, J.B. (1979). *J. Am. Chem. Soc.* 101, 6860.
Wennerström, H., Lindman, B., Engström, S., Söderman, O., Lindblom, G. and Tiddy, G.J.T. (1980). *In* "Magnetic Resonance in Colloid and Interface Science" (Eds J.P. Fraissard and H.A. Resing), p.609. D. Reidel Publishing Company, Dordrecht, Netherlands.
Young, C.Y., Missel, P.J., Mazer, N.A., Benedek, G.B. and Carey, M.F. (1978). *J. Phys. Chem.* 82, 1375.
Zulauf, M. and Rosenbusch, J.P. (1983). *J. Phys. Chem.* 87, 856.

MICROEMULSIONS

J.T.G. Overbeek, P.L. de Bruyn and F. Verhoeckx

Van 't Hoff Laboratory, University of Utrecht, Padualaan 8, 3584 CH Utrecht, The Netherlands.

SUMMARY

Microemulsions are apparently homogeneous mixtures of water and oil with large amounts of surfactants. Microemulsions form spontaneously, are thermodynamically stable and contain particles that are so small that turbidity is low. The surfactants create an extremely low interfacial tension between oil and water, which promotes emulsification. In many, but not all cases, microemulsions can be regarded as rather monodisperse droplets of water-in-oil or oil-in-water. They may be in equilibrium with excess oil, excess water or both. Phase equilibria in these multicomponent systems are quite complicated. Often only two-dimensional cross-sections through the complete phase diagrams are used. Droplet size in microemulsions appears to be limited and this may be explained by the effect of curvature on the interfacial tension.

Most of this chapter deals with the combination of ionic surfactants and simple nonionic cosurfactants such as pentanol, but attention is also given to microemulsions stabilized by nonionic polyethylene oxide derivatives. Application of microemulsion systems in enhanced oil recovery and other applications are mentioned.

INTRODUCTION

Oil and water do not mix, but if some soap or another micelle forming surfactant is added, a small amount of oil is solubilized (Vold and Vold, 1983a) in the micelles and after vigorous shaking or stirring turbid emulsions are formed (Vold and Vold, 1983b). These emulsions are never really

SURFACTANTS
ISBN 0-12-682180-1

stable. The free energy contained in the interfacial area between the droplets and the continuous phase tends to destroy this interface and thus emulsions coarsen by Ostwald ripening (growth of large droplets at the expense of smaller ones) and coalescence.

In 1943 Hoar and Schulman described transparent or translucent systems, formed spontaneously when oil and water were mixed with relatively large amounts of an ionic surfactant combined with a cosurfactant, e.g. a medium size alcohol. These systems are dispersions of very small drops (radius of the order of 10 nm) of water-in-oil (W/O) or of oil-in-water (O/W). Since for small particles light scattering is proportional to the square of the volume of the scattering particle and obeys the equation (Rayleigh, 1871; Kerker, 1969)

$$I(\text{scatt}) = I_0 f n V^2 \tag{1}$$

these systems scatter little light and are not turbid. In equation (1) I(scatt) is the intensity of the scattered light, I_0 that of the incident light, n the number of particles in the scattering volume, V the volume of one particle and f a factor which takes into account the refractive indices of the particles and their surroundings, the angle between incident and scattered light, the planes of polarization, and the lack of ideality of the dispersion.

These systems have been called microemulsions (Schulman *et al.*, 1959; Prince, 1977a). They differ from ordinary emulsions not only in their lack of turbidity but, more essentially, in being thermodynamically, i.e. indefinitely, stable. Because continuous transitions occur from molecular solutions via swollen micelles (or inverse micelles) to microemulsions, the transition from swollen micelles to microemulsions can only be located arbitrarily at a radius of the order of 3 nm.

In recent years microemulsions have attracted a great deal of attention because, if oil and water can be made to mix spontaneously, there may be hope of enhanced oil recovery by water flooding of dried up oil wells (Bansal and Shah, 1977).

REASON FOR COMBINING SURFACTANT AND COSURFACTANT

Spontaneous formation of an emulsion with decrease of the free energy can only be expected if the interfacial tension is so low that the remaining free energy of the interface can be overcompensated by the entropy of dispersion of the droplets in the medium. Surfactants do lower the interfacial tension, but in most cases the c.m.c. or the limit of solubility is reached before the interfacial tension is close to zero. Addition of a second surfactant of a completely

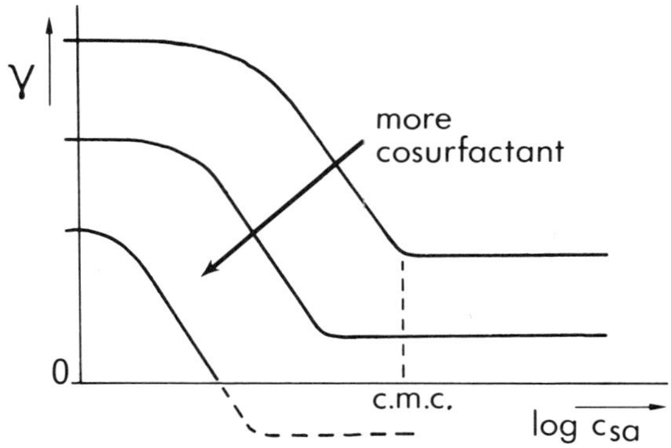

Fig. 1. Interfacial tension, γ, plotted against the logarithm of the surfactant concentration for three different concentrations of cosurfactant. The dotted part, below γ = 0, cannot be realized.

different nature then lowers the interfacial tension further and very small, even transiently negative values, may be reached, as illustrated in Fig. 1.

The well-known Gibbs adsorption equation for a two-component system (Adamson, 1967)

$$\left(\frac{\partial\gamma}{\partial\mu_{\text{solute}}}\right)_{p,T} = -\left(\frac{\partial n_{\text{solute}}}{\partial A}\right)_{p,T,n_1} = -\Gamma^1_{\text{solute}} \qquad (2)$$

can be extended to multicomponent systems to yield

$$d\gamma = -\sum_i \Gamma_i d\mu_i \simeq -\sum_i \Gamma_i RT d\,\ell n\, c_i. \qquad (3)$$

In these equations γ is the surface tension or interfacial tension, μ_i is the chemical potential of component i, c_i its concentration, n_i its amount, Γ_i the amount of i adsorbed per unit area (= the surface excess), and A the surface or interfacial area. Component 1 is the solvent. Equation (3) shows that the addition to the system of a new surfactant which is positively adsorbed will always cause a further lowering of γ.

If γ were to become negative, as suggested by the dotted line in Fig. 1, the interface would expand, adsorb surfactants, until their concentration is lowered enough to make γ positive again.

In some cases single surfactants will lower γ far enough for microemulsion formation to become possible. Examples are: AOT (sodium diethylhexylsulphosuccinate) and many nonionic

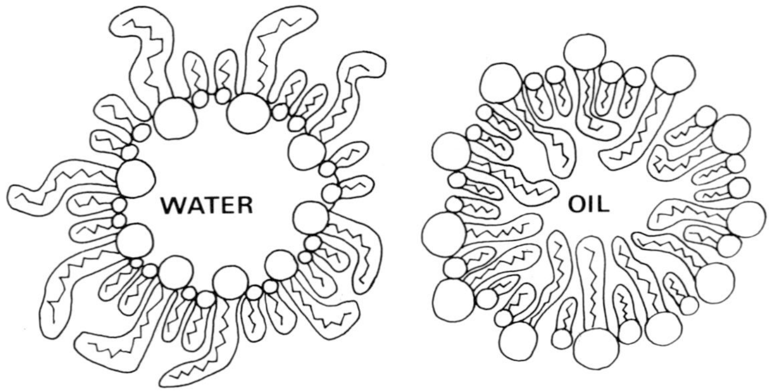

Fig. 2. Schematic pictures of a W/O and an O/W droplet with a long surfactant and a short cosurfactant. It should be clear that the polar heads are more crowded in W/O whereas the hydrocarbon tails are more crowded in O/W.

surfactants near the temperature where they change from being water-soluble to being oil-soluble.

MODEL CONSIDERATIONS

As a typical example of a microemulsion system we consider a combination of sodium dodecylsulphate (SDS), the surfactant ("sa"), pentanol, the cosurfactant ("co"), water with or without NaCl ("w") and a hydrocarbon, e.g. n-hexane, cyclohexane, hexadecane or toluene, the oil ("o").

To begin with, we assume that a microemulsion contains spherical droplets of uniform size. Fig. 2 gives a schematic picture of W/O and O/W droplets. The crowding of surfactant and cosurfactant in the surface as shown corresponds to reality.

In a microemulsion the concentration of sa in the aqueous phase is at or below the c.m.c. Most of the sa is therefore at the interface. The cosurfactant will be distributed more evenly between the oil phase and the interface, with a small but finite concentration in the water phase. The surface excesses of surfactant and cosurfactant can be found from measurements of the interfacial tension between macroscopic water and oil phases and application of the Gibbs adsorption equation (2,3). In the concentration ranges occurring in microemulsions the adsorptions show saturation behaviour in the sense that $\partial\gamma/\partial \ell n c_i$ is constant (Van Voorst Vader, 1960),

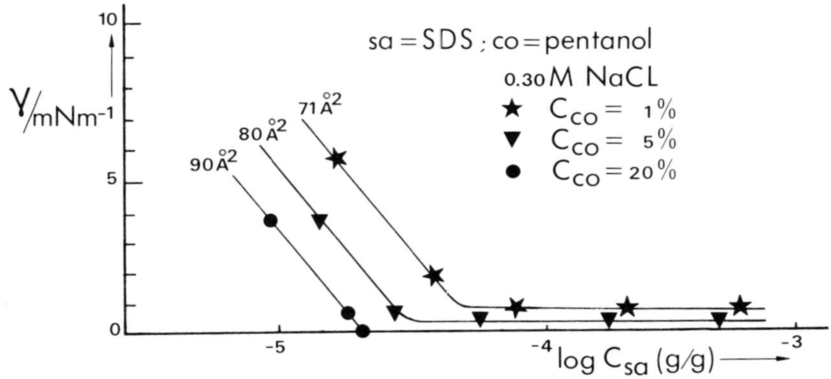

Fig. 3. Interfacial tensions between 0.30 M aqueous NaCl and cyclohexane, containing SDS in the aqueous phase and pentanol, mostly in the oil phase. From the slopes of the lines below the c.m.c. the area per SDS molecule is found, using equation (4). At 20% pentanol any excess of SDS above $c_{sa} = 2 \times 10^{-5}$ *g/g was taken up in the oil phase.*

but the value of Γ_{sa} depends on c_{co}, and Γ_{co} depends on c_{sa}. Fig. 3 shows surface excesses, Γ_{sa}, as from

$$\Gamma_{sa} = - \; [\partial\gamma/(RT\partial \ln c_{sa})]_{c_{co}} \tag{4}$$

in the presence of excess NaCl. These values may be assumed to equal those in the microemulsion, which cannot be easily determined. Depending on c_{co} and c_{NaCl} the area per SDS molecule is found to vary between 70 Å^2 and 110 Å^2. Under conditions where W/O microemulsicns form the area per pentanol molecule is about 30 Å^2. These values mean that every 90 Å^2 of interface contains about one SDS molecule plus three pentanol molecules.

The model now becomes much more precise. In a W/O emulsion, the total volume, V, of the droplets is equal to the volume of water + salt, the total interfacial area, A, is equal to

$$A = n_{sa}/\Gamma_{sa} \simeq n_{sa} \times N_{Avog} \times 90 \text{ Å}^2 \tag{5}$$

where n_{sa} is the amount of surfactant. With volume and area known, the radius of the droplets, a, is then

$$a = 3 \times \frac{4}{3}\pi a^3/4\pi a^2 = 3V/A. \tag{6}$$

With $V = n_{\text{water}} \times 18\ \text{cm}^3\text{mol}^{-1}$ and $A = n_{\text{sa}} \times N_{\text{Avog}} \times 90\ \text{Å}^2$ equation (6) transforms into

$$a = (n_{\text{water}}/n_{\text{sa}})\,\text{Å}. \tag{7}$$

These surprisingly simple expressions give quite good values for the droplet radius. Other methods for the determination of the droplet size are: Rayleigh and dynamic light scattering (Vrij *et al.*, 1978), X-ray diffraction (Schulman and Riley, 1948), small angle neutron diffraction (M. Lagües *et al.*, 1978; Cebula *et al.*, 1982). These various methods lead to slightly different values for a, since equations (6) and (7) give the water radius, whereas the other methods also include a smaller or larger part of the adsorbed sa and co.

For O/W microemulsions, usually formed at lower concentrations of co, analogous equations are valid. The number of cosurfactant molecules per sa in the interface is smaller, perhaps 2 rather than 3.

Of course the actual numbers given in this section will be different if other components are chosen.

PHASE EQUILIBRIA

If the spontaneous formation of a large interface would be the only essential requirement for microemulsion formation, there would be no preference for O/W over W/O or the reverse.

However, the entropy of mixing of the droplets with the continuous phase would favour the phase with the smaller volume to form the droplets since, at a given interfacial area, the number of droplets is inversely proportional to the square of their total volume and a larger number of droplets leads to a larger entropy of mixing.

Moreover, experiment shows that microemulsions of either type may become saturated with respect to the droplets, long before a volume fraction of one half is reached. If more of the internal liquid is presented, such microemulsions form an equilibrium with the excess aqueous (for W/O emulsions) or oily (for O/W emulsions) phase and even three-phase equilibria, water-microemulsion-oil, exist although in that case our droplet model is suspect.

The understanding of such phase equilibria is an essential test of our understanding of microemulsions. Before proceeding further we must first consider whether microemulsions are to be treated as one- or two-phase systems.

ARE MICROEMULSIONS ONE- OR TWO-PHASE SYSTEMS?

A phase may be defined as a region of space filled homogeneously with matter. But it depends on the fineness of the grid that we use to determine homogeneity whether a solution, a colloidal suspension or a microemulsion are homogeneous or heterogeneous. This point of view seems to give difficulties with the phase rule. If we change arbitrarily the number of phases by one, then the phase rule tells us that we also change the number of degrees of freedom by one. And that cannot be correct.

The problem is already an old one and has been reviewed by Hermans (1952). The phase rule can be written:

$$f = n - p + 2 \tag{8}$$

where f is the number of degrees of freedom, n the number of components, and p the number of phases of the system. The number 2 refers to the independent variables temperature and pressure. One may consider $(n-p)$ in equation (8) as the number of concentrations that can be varied independently. In suspensions or emulsions one deals not only with the concentrations of the low molecular weight components but one must also consider the concentration of the particles, c_p. In a coarse suspension or emulsion this, however, is a "negligible variable" (Tolman, 1913) but in a microemulsion it affects the thermodynamic properties of the system, e.g. the chemical potential of the suspending liquid.

There are therefore two legitimate, equivalent but different ways to describe a microemulsion. It is a one-phase system, obeying equation (8) with $(n-1)$ independently variable concentrations; or, it is a two-phase system with $(n-2)$ independent concentrations plus an independently variable concentration of droplets, and thus again $(n-1)$ degrees of freedom.

The one-phase point of view is often the more obvious choice, especially in considering equilibria with other phases, but in calculating properties of microemulsions from their composition the two-phase approach is often preferable.

PHASE DIAGRAMS

Phase equilibria in microemulsions are best described with more or less schematic phase diagrams. Such phase diagrams are bound to be complicated, as we have to deal with four or five components and because the two-component water-surfactant phase diagram is already often complicated. Even at constant temperature and pressure, a four-component system can only be represented in three dimensions, e.g. in a tetrahedron. For simplicity we shall start with some three-component and pseudo-three-component diagrams.

Figure 4 shows schematically a phase diagram of a system containing only water (+ salt), sa, co, but no oil (Friberg, 1977; Danielsson *et al.*, 1982). The cosurfactant fulfils the

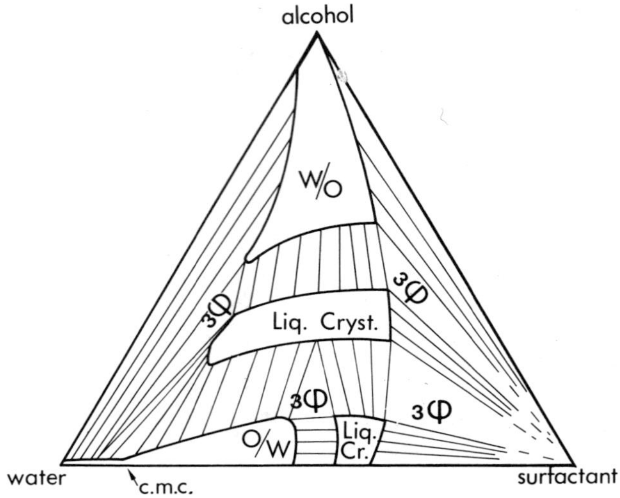

Fig. 4. Schematic three-component phase diagram for a water, surfactant, cosurfactant combination. It shows four one-phase regions, four three-phase (3φ) regions and a number of two-phase regions with tie-lines. The surfactant corner has been left vague. There may be hydrates, and/or further mesomorphic phases.

double role of co and oil. The diagram contains four one-phase regions, first a W/O microemulsion growing out from the region of true solubility of water in the alcohol. At the left-hand border (high water-surfactant ratio, large droplets, equation (7)) this phase is in equilibrium with a dilute sa solution in water, at the right-hand border with solid surfactant, and at the lower border (high droplet concentration) with a birefringent mesomorphic phase (assumed to be lamellar), itself the second one-phase region. An (inverted) hexagonal phase between the lamellar phase and the W/O region has been left out for simplicity.

A third phase is the isotropic O/W microemulsion, extending from the water corner as a molecular solution below the c.m.c. and with alcohol solubilized in the surfactant micelles above it.

The fourth one-phase region is again mesomorphic presumably containing hexagonally-packed cylinders, and growing out of the "middle soap" region in the water-surfactant phase diagram.

Between the various one-phase regions, two-phase regions

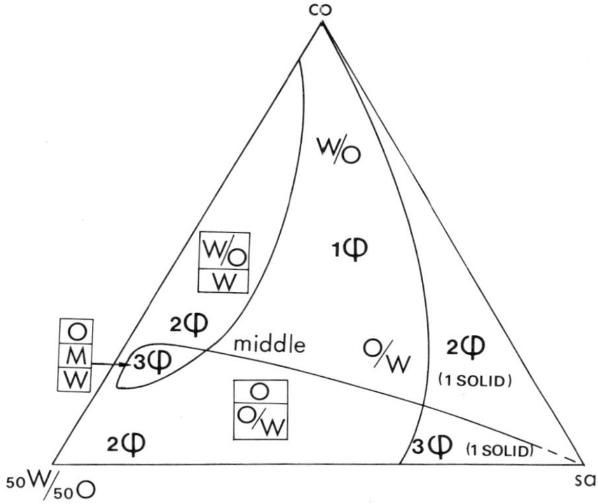

Fig. 5. Schematic cross-section through the complete three-dimensional phase diagram, with equal quantities of oil and water in one corner and co and sa in the other corners. In the two- and three-phase regions the kind of equilibria are indicated, but tie-lines cannot be shown since they are not in the plane of the figure. Middle refers to middle-phase microemulsion. For further details see text.

occur where the phases in equilibrium are connected by tie-lines, and triangular three-phase (3ϕ) regions, having zero degrees of freedom at constant p and T in a three-component system.

Addition of a small amount of oil, miscible with the cosurfactant, but not with water and the surfactant, changes the phase diagram only slightly, but addition of more oil leads to more fundamental changes as illustrated in Fig. 5 in which the boundaries are shown of one-phase, two-phase and three-phase regions for a system containing equal volumes of oil and water. Tie-lines are not shown, since they are not in the plane of this figure.

Near the co corner the changes as compared to Fig. 4 remain small, but the O/W microemulsion near the water-sa axis is now not in equilibrium with the lamellar phase, but with a non-colloidal oil + cosurfactant phase. If co is added to such a two-phase equilibrium at fairly high sa concentration all oil is taken up and a one-phase microemulsion appears.

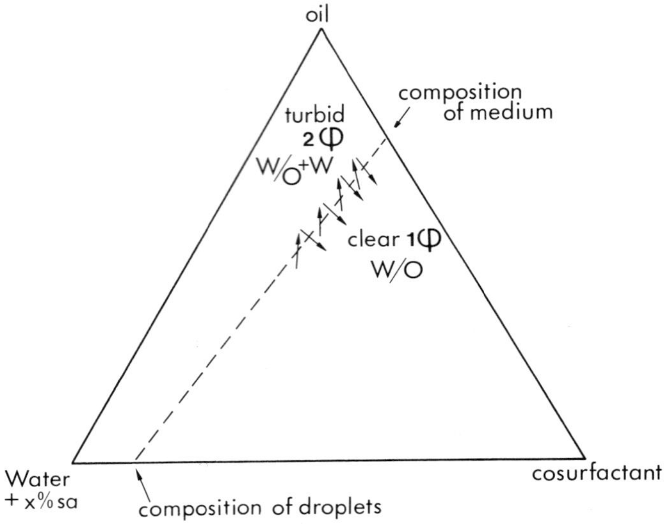

Fig. 6. Illustrates the titration procedure, which allows dilution of the microemulsion without changing the composition of the droplets.

But addition of co at low sa concentration may lead to separation of an excess aqueous phase before all oil is taken up in the microemulsion. A three-phase system is formed, containing a microemulsion that cannot be clearly identified as W/O or O/W and that presumably is similar to the lamellar phase swollen with oil or to a still more irregular intertwining of aqueous and oily regions. We have called this a middle-phase microemulsion.

The interfacial tensions between the three phases are very low, between 0.1 and 0.0001 mNm^{-1} or even lower than that. Further addition of co to the three-phase system makes the oil phase disappear and leaves a W/O microemulsion in equilibrium with a dilute aqueous sa solution. In the large one-phase region continuous transitions from O/W to middle-phase to W/O microemulsions are found.

In Fig. 6 still a further way of looking at the four-component system is given. Here a fixed water-surfactant ratio is shown in one corner, the other corners being co and oil. This figure also illustrates the titration procedure (Bowcott and Schulman, 1955), by which the composition of the droplets and of the continuous medium can be determined. A microemulsion (W/O in this case) is just at the border between the one-phase

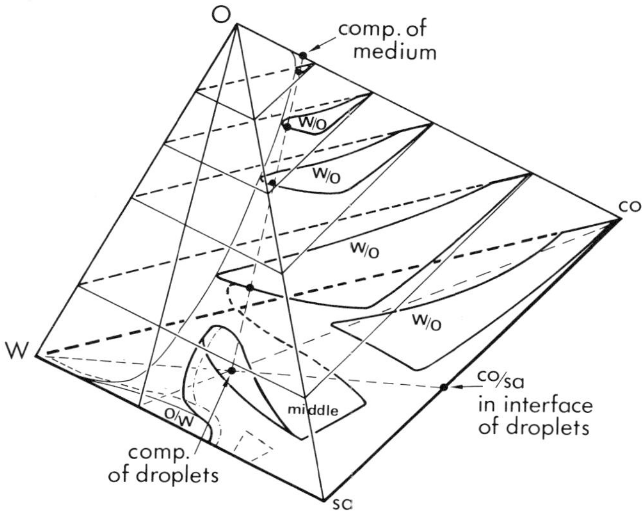

Fig. 7. Schematic four-component phase diagram. The merging of the W/O, O/W and middle-type one-phase regions is indicated. A titration plane as given in Fig. 6 is drawn in.

and the two-phase region. Oil and cosurfactant can be added alternately while just straddling the borderline, on one side of which the system is transparent, on the other side markedly turbid. The ratio of oil and co added in this process gives the composition of the continuous medium. Extrapolation of the borderline thus found to the w/sa-co axis gives the composition of the droplet and also, since the cosurfactant is hardly water-soluble and most of the sa is in the interface, the ratio of sa to co in the interface.

By using this titration procedure one can dilute a microemulsion without changing the droplet composition. It is assumed implicitly that this composition is hardly affected by the droplet concentration.

Finally a very schematic phase diagram in a four-component tetrahedron is shown in Fig. 7. On addition of oil to the w, sa, co system, the one-phase W/O region expands, changes shape, retracts again and disappears at high oil contents. The one-phase O/W region disappears already at modest oil contents from the o, w, sa triangle, but shifts to the inside of the tetrahedron, and may form a continuous microemulsion region with the middle and the W/O microemulsion. In this tetrahedron a titration triangle as given in Fig. 6 is drawn showing a dilution curve in its three-dimensional location.

It hardly needs to be mentioned that given the freedom to change surfactant, cosurfactant, oil and salt content large variations in the phase diagrams may occur. Nevertheless Figs 4 to 7 have their utility as a qualitative mnemonic device.

THERMODYNAMICS OF MICROEMULSIONS

Considering the complicated phase behaviour of microemulsions, we are left with a number of questions and few answers. A low interfacial tension O/W appears to be essential for the formation of a microemulsion, but we do not understand why O/W, W/O or still another structure is formed nor how the phase equilibria are controlled.

These and other questions may find their answers in thermodynamics. We therefore choose a model, minimize its calculated free energy and check whether the model explains typical phase equilibria.

Such programmes have been carried out by several authors of whom we mention Adamson (1969), Ruckenstein *et al.* (1975, 1979, 1980), Reiss (1975), Wagner (1976), Overbeek (1978), Miller and Neogi (1980), Mukherjee *et al.* (1983), Mitchell and Ninham (1981). None of the results so far obtained is completely satisfactory. We shall therefore give a rather simplified treatment of the W/O case, just to show which factors are involved. We use the model of spherical drops of equal size, large enough to make the description, as a drop of constant composition with a surface layer of adsorbed surfactants, acceptable. We "prepare" the emulsion in a number of steps and for each step calculate the Helmholtz energy, F, involved. The Helmholtz energy is chosen because the pressure inside the drops is higher by the Laplace pressure, $2\gamma/a$, than the pressure in the continuous medium.

First step Prepare the oil phase in its final concentration

$$F_1 = \sum_i n_i'\mu_i' - p_1V_1 \tag{9}$$

where n_i' and μ_i' are the amount and chemical potential of oil and cosurfactant in the continuous phase, without droplets being mixed in. p_1 is the atmospheric pressure, and V_1 the volume of the oil phase.

Second step Prepare the aqueous phase in its final concentration

$$F_2 = \sum_i n_i'\mu_i' - p_1V_2$$

where i are now water, surfactant and salt and V_2 is the volume of the water phase.

Third step Form the water phase into droplets, close packed in the oil phase, and add all the adsorbed material.

$$F_3 = \gamma A + \Gamma_{sa} A(\mu'_{sa} + \frac{2\gamma}{a} \overline{V}_{sa}) + \sum_i \Gamma_i A\mu'_i \tag{11}$$

where i refers to cosurfactant and oil. The oil must be negatively adsorbed in order to keep the volume of the adsorption layer zero, in accordance with the assumption of a Gibbs dividing plane. It is assumed in equation (11) that the Gibbs plane, the surface of tension in this case (see Hill, 1968), lies close enough to the surface, where $\Gamma_{water} = 0$.

Fourth step We allow the close packed emulsion to expand to its final concentration, thus generating the Helmholtz energy of mixing

$$F_4 = n_{dr} RT \, f(\phi) \tag{12}$$

where n_{dr} is the amount of droplets (in moles) and $f(\phi)$ is a function of the volume fraction, ϕ, of the droplets, most simply written ($\ln \phi - \ln 0.74$) and, more accurately, using the Percus-Yevick-Carnahan Starling (1969) approximation for hard spheres as

$$f(\phi) = \ln \phi + \phi \frac{4-3\phi}{(1-\phi)^2} - \ln 0.74 - 0.74 \frac{4 - 3 \times 0.74}{(1-0.74)^2}$$

$$= \ln \phi + \phi \frac{4-3\phi}{(1-\phi)^2} - 19.25. \tag{13}$$

The constant term, -19.25, deviates from the one given earlier (Overbeek, 1978), because now the free energy of mixing of droplets and medium is assumed to be zero at close packing, rather than deriving the standard chemical potential of the droplets from an assumed behaviour at high dilution.

Combining equations (9) to (13) gives the Helmholtz energy, F of the complete emulsion. The F is minimized with respect to a change in the interfacial area, A. This involves transfer of adsorbed components to or from the interface, thereby changing the bulk concentrations and thus γ. The result is then that

$$\gamma = -\text{const.} \times \frac{1}{a^2} \times g(\phi) \tag{14}$$

where $g(\phi)$ is similar to but not identical with $f(\phi)$. Since the radius, a, has to fit both equations (6) and (14), and γ is the most easily varied quantity in these equations, the correct value of γ is obtained by adaptation of c_{sa}. So far we have used the two-phase approach with n_{dr} (equation (12)) as the extra variable. We now switch over to the one-phase approach by calculating μ_w, μ_{sa}, μ_{salt} for the complete microemulsion ($\mu_i = \partial G/\partial n_i$, where $G = \sum_1^4 F_i + p_1(V_1+V_2)$) and check under which conditions these μ_i combine to an equilibrium solution.

It is at this stage that difficulties are encountered which point to inadequacies in the model. According to equation (14) any value of a is allowed in the accessible range of γ. In particular if γ can be close to zero very large radii, i.e. very large water/sa ratios are allowed, but in the phase diagram such large w/sa ratios lead to demixing.

It appears necessary to add an explicit influence of the radius of curvature on the interfacial tension to the model and thus to give the interfacial layer a limited range of curvatures that are acceptable. Notions of this nature have been introduced by several authors. Bowcott and Schulman (1955) suggested that the interfacial layer may have two different interfacial tensions, one at the water side and one at the oil side, which would cause the film to bend (duplex film theory). Mitchell and Ninham (1981) have extended the idea, that the limitations to the shape of close packed surfactants determine the shape of micelles and vesicles, to microemulsions. Mukherjee *et al.* (1983) have made fairly elaborate calculations of the free energy of mixing of the hydrocarbon tails of surfactant and cosurfactant with the oil molecules for curved surfaces and the free energy of the electric double layer is also curvature dependent (Overbeek, 1978).

If we look at Fig. 2 and consider the actual numbers for the surface area per surfactant, we see that the ionic head-groups are not close packed (100 $Å^2$, whereas if present alone they pack to about 50 $Å^2$, (Van Voorst Vader, 1960)). With about three cosurfactant molecules per surfactant, the chains are nearly close-packed over the length of the co-chains (4 chains per 100 $Å^2$) and the outer fringe is again very open. The packing of the short chains and the packing of the polar heads will favour W/O curvature with a ratio of 3 or more for co/sa and favour O/W with a co/sa ratio of 2 or less with an uncertain region in between. A longer chain alcohol will promote W/O. The double layer free energy is always in favour of O/W, the more so the lower the electrolyte concentration. This double layer effect can be expressed quantitatively (Overbeek, 1978) and contributes a term: const/ac to the interfacial tension, where c is the total electrolyte concentration.

To give a quantitative expression for the crowding effect of the chains is more difficult.

The main conclusion at this stage must be that the thermodynamic theory is still developing. A few qualitative conclusions can nevertheless be drawn.

1) O/W microemulsions need less cosurfactant than W/O ones.
2) The formation of W/O microemulsions requires more cosurfactant, not only to decrease γ, but also to promote the curvature with its convex side towards the oil. This effect is expected to be more pronounced with longer chain cosurfactants.
3) Electrolyte promotes W/O.
4) Since γ depends explicitly on the radius a, there is a tendency towards monodispersity.
5) Low volume fraction (ϕ) microemulsions have a lower free energy of formation than high ϕ ones, since the entropy of mixing is relatively more important for low ϕ and small a.
6) At about 40 to 50% water volume in W/O emulsions ϕ (which includes the hydrocarbon tails at the interface) gets close to 74%. The advantage of forming droplets disappears since the free energy of mixing becomes very small. Then a lamellar (or hexagonal) phase is preferred, because in such a phase the hydrocarbon tails can be better packed. This occurs at a smaller water volume for small particles than for large ones, since the extra tail volume makes a relatively larger contribution with small droplets. This explains why the W/O microemulsion can take up most water at the lowest sa/w ratio, i.e. at the largest possible radius (see Figs 4 and 7).

NONIONIC SURFACTANTS

Nonionic surfactants e.g. polyoxyethylene (POE) alkylphenylethers also may form microemulsions. When these microemulsions are compared with ionic ones two striking differences are found:

1) No cosurfactant is needed even with pure specimens of nonionics (very often nonionics are mixtures with a range of chain lengths).
2) Since the solubility of nonionic surfactants is highly temperature dependent the temperature plays an essential role in microemulsion behaviour.

A POE-alkyl surfactant is soluble in water at low temperature and forms micelles of increasing size over a fairly wide

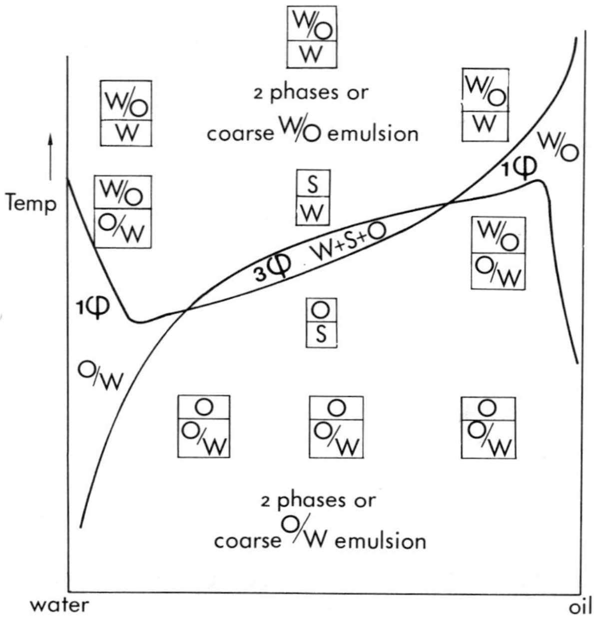

Fig. 8. Schematic phase diagram for a nonionic surfactant e.g. POE(n~9)-nonylphenylether in a fixed concentration e.g. 10% in mixtures of water and oil as a function of the temperature. The various one-, two- and three-phase regions are indicated. S is a "surfactant phase".

concentration region. At higher temperature the solubility of the POE moiety decreases until separation occurs in two liquid phases (cloud point). In hydrocarbon oil, on the contrary, the solubility is high at high temperature but decreases towards lower temperature.

The transition temperature depends on the hydrophilic-lipophilic balance (HLB) of the surfactant. A schematic phase diagram is given in Fig. 8 for an oil-water system containing a fixed percentage e.g. 10% of such a surfactant.

On the water side, at the lowest temperature, the surfactant solution solubilizes little oil; excess oil may be emulsified in a coarse O/W emulsion. At somewhat higher temperature the solubilization increases and fairly concentrated e.g. 10-20% O/W microemulsions may be formed. At still higher temperatures the solubility in water of the surfactant decreases rather suddenly and the system separates in a dilute O/W microemulsion and an oil phase, in which some water is solubilized in an oil-surfactant solution.

At the oil side of the diagram essentially the inverse processes occur.

But if oil and water are present in comparable amounts a narrow (in the temperature direction) three-phase region is found with a "surfactant phase" between surfactant-poor oil and water phases. In this middle region interfacial tensions are extremely low.

Microemulsions with a clear W/O or O/W character contain presumably spherical droplets (Robbins, 1977), but the surfactant phase may well be a bicontinuous structure, lamellar or otherwise.

Nonionic microemulsions are less sensitive to electrolyte than ionic ones, but the hydrophilic group is salted out by electrolyte and thus an addition of electrolyte works in the same direction as increasing the temperature or as switching to a more lipophilic surfactant (Robbins, 1977).

TECHNIQUES FOR CHARACTERIZATION OF MICROEMULSIONS

Microemulsions have been studied with a great variety of techniques, many of which are discussed in Prince's book (1977b). Here we only give a brief survey.

Phase diagrams are an indispensible tool in the study of microemulsions. The boundaries of the one-phase regions can be found from simple observation of mixtures of known composition or by a titration technique as depicted in Fig. 6. Determination of tie-lines in two-phase regions requires analysis of the phases. Given the small driving force for equilibrium, for example small differences in density, equilibration times are long and often run into many days, rather than hours or minutes.

The polarization microscope is needed to determine whether phases are anisotropic or not.

Interfacial tension measurements are required since low interfacial tension is a typical asset of microemulsions, but also to determine the area per surfactant or cosurfactant molecule and from there, using equation (6), $a = 3V/A$, the radius of the droplets. Low interfacial tensions require special techniques. The spinning-drop technique (shape of a drop in a cylindrical capillary rotating around its axis) is popular. Maintaining the temperature constant during the time needed for the freshly formed interface to come to equilibrium is one of the problems (Cayias *et al.*, 1975; Chatenay *et al.*, 1982).

Classical light scattering with a dilution technique which maintains the identity of the droplets gives the particle size and the interaction between particles. A hard sphere model is a good first approximation, but it needs refinements (Caljé *et al.*, 1977; Vrij *et al.*, 1978). Monodispersity or a narrow size distribution fits the data. The smallness of the particles usually allows the Rayleigh approximation ($a < \lambda/20$) to be used. In the transition region between O/W and W/O microemulsions critical phenomena accompanied by critical opalescence (Cazabat *et al.*, 1982) have been observed.

Dynamic light scattering or photon correlation spectroscopy has been used to determine diffusion coefficients (and thus the hydrodynamic radius of the particles) and their concentration dependence (Vrij *et al.*, 1978).

Small angle X-ray scattering (SAXS) has already been used by Schulman *et al.* (1948; 1951) to obtain information about particle size and shape. Small angle neutron scattering (SANS) allows separate information to be obtained about droplet core and surfactant layer by selective deuteration of water, surfactants or oil, since D scatters thermal neutrons much more than H (Fletcher *et al.*, 1982).

Viscosity with the use of Einstein's equation for the relative viscosity, η_r

$$\eta_r = 1 + 2.5\phi \tag{16}$$

or extensions thereof gives a check on the total hydrodynamic volume of the particles. More generally rheology and streaming birefringence are useful for determining hydrodynamic and other interactions and deviations from the sperical shape.

Sedimentation and ultracentrifugation rates and equilibria lead to hydrodynamic radii and particle mass (Hwan *et al.*, 1979; Dvolaitzky *et al.*, 1978).

Electrical conductivity is one of the classical methods to distinguish between O/W (high cond.) and W/O (low cond.) emulsions. In the microemulsion field the situation is more complex, since not too dilute W/O systems may show an increased conductivity as an indication of exchange of ions between droplets or of so-called "percolative behaviour", as a step towards a bicontinuous surfactant phase (Bennett *et al.*, 1982; Lagües *et al.*, 1978).

Measurements of dielectric dispersion give similar information (Boyle *et al.*, 1982; Boned *et al.*, 1979).

NMR measurements and self diffusion measurements, using radioactive labelling (Lindman *et al.*, 1982) give information on the mobility of various components. Slow diffusion indicates that a component is present in individual non-connected slowly moving particles. With NMR the complication arises that sharp signals, indicating high mobility, may be due to mobility within the entities just mentioned. A special NMR pulsed gradient spin-echotechnique allows, however, the diffusion coefficient to be obtained (Stibbs and Moseley, 1979; 1980).

Finally electronmicroscopy should be mentioned. Prince (1977a) shows small droplets of an O/W microemulsion after staining with OsO_4, but given the danger of artifacts (Biais *et al.*, 1981) electronmicroscopy is at best a tricky tool in this field.

APPLICATIONS

Applications of microemulsions are furthered by their spontaneous formation, their thermodynamic stability and thus their lack of aging. Applications may be based on the low interfacial tensions, on the possibility of preparing nearly homogeneous mixtures of oil- and water-soluble substances and also on the near uniform droplet size which occurs in several microemulsions.

The most spectacular use of microemulsions is found in the tertiary oil recovery (Bansal and Shah, 1977). In this application the low interfacial tension is essential. It allows residual oil in the reservoir to be pushed through the reservoir capillary structure without being hampered by the pressure drop at a curved oil-water interface with a normal interfacial tension, as schematically shown in Fig. 9. The method is still under development, but its huge potential in terms of money probably explains a good part of the very wide interest in microemulsions in the last ten years or so. The process is not simple on account of reservoir temperature, high salt content in the underground water, loss of expensive surfactant by adsorption in the reservoir, and by incomplete recovery.

Other applications (Prince, 1977d) are cutting oils (O/W), providing cooling and lubrication in one fluid, and dry-cleaning fluids (W/O) in which both oil- and water-soluble

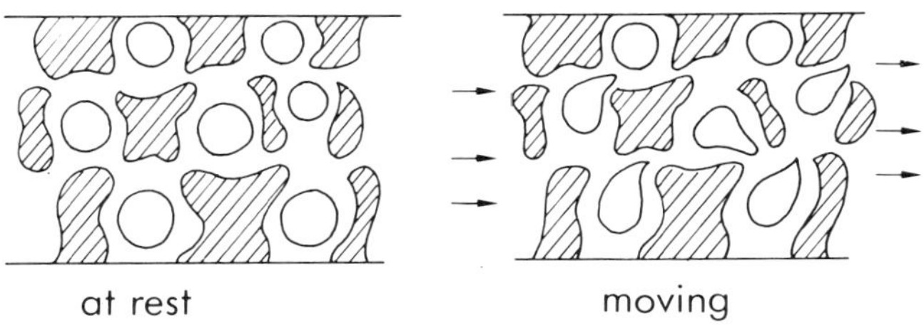

Fig. 9. In a porous oil bearing formation stagnant oil drops are spherical. In flow through narrow pores the drops must be deformed. This requires interfacial energy and an extra pressure drop, unless γ is very close to zero.

contaminants dissolve.

"Self emulsifiable oils" (Moilliet *et al.*, 1961) as used in agricultural and horticultural sprays, are possibly not strictly microemulsions at the final dilute O/W stage but they are often W/O microemulsions at the concentrated stage.

Probably the first commercial use of microemulsions - *avant la lettre* - is the use of Carnauba wax floor polishes in the 1930s (Prince, 1977e) prepared as a high temperature W/O microemulsion, which then is inverted into a translucent O/W type. The attraction of such emulsion-based floor and leather polishes is that they are easily applied (low viscosity) and dry to a high shine without the need for buffing.

Many cosmetic and pharmaceutical preparations are microemulsions.

The possibility of contacting oil- and water-soluble reactants at a large interface offers the possibility of greatly accelerating heterogeneous reactions, e.g. with lipids and water-soluble enzymes (Hilhorst *et al.*, 1982).

A completely different kind of application is found in pure science, where the (near) monodispersity is important (Agterof *et al.*, 1976; Caljé *et al.*, 1977; Vrij *et al.*, 1978). The droplets in such a monodisperse emulsion may be considered as huge molecules and the macroscopic behaviour of concentrated microemulsions can be used to test theories of hard or soft sphere fluids, where due to the size of the particles scattering and diffraction of visible light replaces X-ray techniques used for small molecules.

CONCLUSION

Microemulsions still offer worthwhile scientific challenges for many scientists and at the same time have a potential for applications which is by no means exhausted.

ACKNOWLEDGEMENTS

We thank O.M.I. (University Audiovisual Media Institute) for drawing the figures and Miss Renske Kuipers for typing the manuscript in a very short time.

REFERENCES

Adamson, A.W. (1967). "Physical Chemistry of Surfaces". Wiley-Interscience, New York. 2nd edn, pp.80-81.

Adamson, A.W. (1969). *J. Colloid Interface Sci.* 29, 261.

Agterof, W.G.M., van Zomeren, J.A.J. and Vrij, A. (1976). *Chem. Phys. Letters* 43, 363.

Bansal, V.K. and Shah, D.O. (1977). *In* "Microemulsions" (Ed. L.M. Prince), pp.149-173. Academic Press, New York, London.

Bennett, K.E., Hatfield, J.C., Davis, H.T., Macosko, C.W. and Scriven, L.E. (1982). *In* "Microemulsions" (Ed. I.D. Robb), pp.65-84. Plenum Press, New York, London.
Biais, J., Mercier, M. Bothorel, P., Clin, B., Lalanne, P. and Lemanceau, B. (1981). *J. Microscopy* 121, 169.
Boned, C., Peyrelasse, J., Clausse, M., Lagourette, B., Alliez, J. and Babin, L. (1979). *Colloid and Polymer Sci.* 257, 1073.
Bowcott, J.E. and Schulman, J.H. (1955). *Z. Elektrochemie* 59, 283.
Boyle, M.H., McDonald, M.P., Rossi, P. and Wood, R.M. (1982). *In* "Microemulsions" (Ed. I.D. Robb), pp.103-114. Plenum Press, New York, London.
Caljé, A.A., Agterof, W.G.M. and Vrij, A. (1977). *In* "Micellization, Solubilization and Microemulsions" (Ed. K.L. Mittal), Vol. 2, pp.779-790. Plenum Press, New York.
Carnahan, N.F. and Starling, K.E. (1969). *J. Chem. Phys.* 51, 635.
Cayias, J.L., Schechter, R.S. and Wade, W.H. (1975). *In* "Adsorption at Interfaces". *ACS Symposium Series* 8, 234.
Cazabat, A.M., Langevin, D., Meunier, J. and Pouchelon, A. (1982). *Adv. Colloid Interface Sci.* 16, 175.
Cebula, D.J., Mijers, D.Y. and Ottewill, R.H. (1982). *Colloid and Polymer Sci.* 260, 96.
Chatenay, D., Langevin, D. and Meunier, J. (1982). *J. Dispersion Sci. and Technology* 3, 245.
Danielsson, I., Friman, R. and Sjöblom, J. (1982). *J. Colloid Interface Sci.* 85, 42.
Dvolaitzky, M., Guyot, M., Lagües, M., Le Pesant, J.P., Ober, R., Sauterey, C. and Taupin, C. (1978). *J. Chem. Phys.* 69, 3279.
Fletcher, P.D.I., Robinson, B.H., Bermejo-Barrera, F., Oakenfull, D.G., Dove, J.C. and Steytler, D.C. (1982). *In* "Microemulsions" (Ed. I.D. Robb), pp.221-232. Plenum Press, New York, London.
Friberg, S. (1977). *In* "Microemulsions" (Ed. L.M. Prince), pp.133-148. Academic Press, New York, London.
Hermans, J.J. (1952). *In* "Colloid Science" (Ed. H.R. Kruyt), Vol. 1, pp.11-15. Elsevier, Amsterdam.
Hilhorst, R., Laane, C. and Veeger, C. (1982). *Proc. Natl Acad. Sci. USA* 79, 3927.
Hill, T.L. (1968). "Thermodynamics for Chemists and Biologists", pp.8-12. Addison-Wesley, Reading, Mass.
Hoar, T.P. and Schulman, J.H. (1943). *Nature* 152, 102.
Hwan, R.-N., Miller, C.A. and Fort, T. Jr (1979). *J. Colloid Interface Sci.* 68, 221.
Kerker, M. (1969). "The Scattering of Light", pp.30-39. Academic Press, New York, London.

Lagües, M., Ober, R. and Taupin, C. (1978). *J. de Physique* 39, L-487.
Lindman, B., Kamenka, N., Brun, B. and Nilsson, P.G. (1982). "Microemulsions" (Ed. I.D. Robb), pp.115-130. Plenum Press, New York, London.
Miller, C.A. and Neogi, P. (1980). *A.I.Ch.E.J.* 26, 212.
Mitchell, D.J. and Ninham, B.W. (1981). *J. Chem. Soc. Faraday 2* 77, 601.
Moilliet, J.L., Collie, B. and Black, W. (1961). "Surface Activity". Spon, London. 2nd edn, p.180, pp.251-256.
Mukherjee, S., Miller, C.A. and Fort, T. Jr (1983). *J. Colloid Interface Sci.* 91, 223.
Overbeek, J.Th.G. (1978). *Faraday Disc. Chem. Soc.* 65, 7.
Prince, L.M. (Ed.) (1977). "Microemulsions". (*a*) Frontispiece. (*b*) pp.6-17. (*c*) p.18. (*d*) pp.21-32. (*e*) p.23. Academic Press, New York, London.
Rayleigh, Lord (1871). *Phil. Mag.* 41, 107, 247, 447.
Reiss, H. (1975). *J. Colloid Interface Sci.* 53, 61.
Robbins, M.L. (1977). *In* "Micellization, Solubilization, and Microemulsions" (Ed. K.L. Mittal), Vol. 2, pp.713-754. Plenum Press, New York, London.
Ruckenstein, E. and Chi, J.C. (1975). *J. Chem. Soc. Faraday 2* 71, 1690.
Ruckenstein, E. and Krishnan, R. (1979). *J. Colloid Interface Sci.* 71, 321.
Ruckenstein, E. and Krishnan, R. (1980). *J. Colloid Interface Sci.* 75, 476; 76, 188, 201.
Schulman, J.H., Matalon, R. and Cohen, M. (1951). *Disc. Faraday Soc.* 11, 117.
Schulman, J.H. and Riley, D.P. (1948). *J. Colloid Sci.* 3, 383.
Schulman, J.H., Stoeckenius, W. and Prince, L.M. (1959). *J. Phys. Chem.* 63, 1677.
Stibbs, P. and Moseley, M.E. (1979). *Chem. Scripta* 13, 26.
Stibbs, P. and Moseley, M.E. (1980). *Chem. Scripta* 15, 176, 215.
Tolman, R.C. (1913). *J. Am. Chem. Soc.* 35, 307.
Vold, R.D. and Vold, M.J. (1983). "Colloid and Interface Chemistry". (*a*) pp.633-667. (*b*) pp.395-424. Addison-Wesley, Reading, Mass.
Van Voorst Vader, F. (1960). *Trans. Faraday Soc.* 56, 1067, 1078.
Vrij, A., Nieuwenhuis, E.A., Fijnaut, H.M. and Agterof, W.G.M. (1978). *Faraday Disc. Chem. Soc.* 65, 101.
Wagner, C. (1976). *Colloid and Polymer Sci.* 254, 400.

THE MICROSTRUCTURE AND THE RHEOLOGY OF SURFACTANT AND RELATED SYSTEMS

J.W. Goodwin

*Department of Physical Chemistry,
University of Bristol,
Cantock's Close, Bristol BS8 1TS, England.*

INTRODUCTION

When a stress, that is a force per unit area, is applied to any material, it will deform. The extent of the deformation relative to the original dimensions of the material is known as the strain. If the deformation is recovered on the removal of the stress then the material is elastic. However, if the components of the material have been able to diffuse a sufficient distance during the experiment to relieve at least part of the applied stress, then viscous flow will have occurred resulting in a permanent deformation. From this it is clear that the classification of a material as a solid or a fluid is governed by the decay of elastically stored energy. If the time scale for a significant amount of decay is short compared to the duration of the experiment, then we observe the material to respond as a viscous fluid, and conversely, if there is insignificant relaxation then we call the material an elastic solid.

The ratio of the characteristic or stress relaxation time of the material, t_r, to the experimental time, t_o, is known as the Deborah number, i.e. $D_e = t_r/t_o$, so that

$D_e \gg 1$ elastic response;
$D_e \ll 1$ viscous response; and
$D_e \sim 1$ viscoelasticity observed.

The common laboratory time scale is, of course, the one to which our bodies are sensitive and is $10^{-3} < t_o < 10^{+3}$ s. Dilute solutions of surfactant molecules and dilute dispersions of colloidal particles have stress relaxation times

SURFACTANTS
ISBN 0-12-682180-1

shorter than this range because of the rapid diffusive motion of the components. Both these systems are thus viscous liquids. However, when either of these systems is produced in a concentrated form the diffusivity can be markedly reduced due to strong interactions between the components, the systems then show viscoelastic behaviour. In the first part of this chapter dilute systems will be discussed but it should be remembered that "dilute" in this context means systems in which the interactions between the components are sufficiently weak so that free diffusion is not prevented. In terms of mass fraction many such systems would be classified as "concentrated". The latter part of the chapter will be devoted to concentrated systems. In addition, as the theme of this volume is surfactants and as these are normally used as stabilizers, only stable colloidal dispersions will be dealt with.

DILUTE SYSTEMS

The coefficient of viscosity is defined as the ratio of the applied stress to the resulting rate of deformation. The most common instruments produce a shearing action on the fluid and so we are referring to the shear viscosity, η /Ns m^{-2}; shear stress, τ/N m^{-2}; and shear rate $\dot{\gamma}$/s^{-1} where Newton's dot denotes the time differential operator $\frac{d}{dt}$. Then $\eta = \frac{\tau}{\dot{\gamma}}$.
Fig. 1 shows the commonly used viscometric flows.

The viscosity of dispersions of uncharged spherical particles was shown by Einstein (1906) to be described in the dilute limit by the equation

$$\eta = \eta_o[1 + 2.5\phi + O(\phi^2)] \quad (1)$$

where η_o represents the viscosity of the continuous phase or solvent and ϕ is the volume fraction of the dispersion. The coefficient, 2.5, is known as the intrinsic viscosity $[\eta]$. This was calculated from the dilation of the streamlines around the particles and the perturbation of the flow around one particle by the flow around other particles was ignored. As a result this equation can only be used at $\phi < 0.01$.

When surfactants are used to stabilize particles, sufficient material is added to the dispersion to give a high surface coverage. This ensures that there is no water penetration into the adsorbed layer. Cationic or anionic surfactant stabilized systems fall into this category. Alternatively, penetration may occur as with a nonionic or polymeric stabilizer system, but there is nearly always sufficient density in the layer to prevent free drainage. The viscosity of the dispersions with stabilizer present is higher than would be

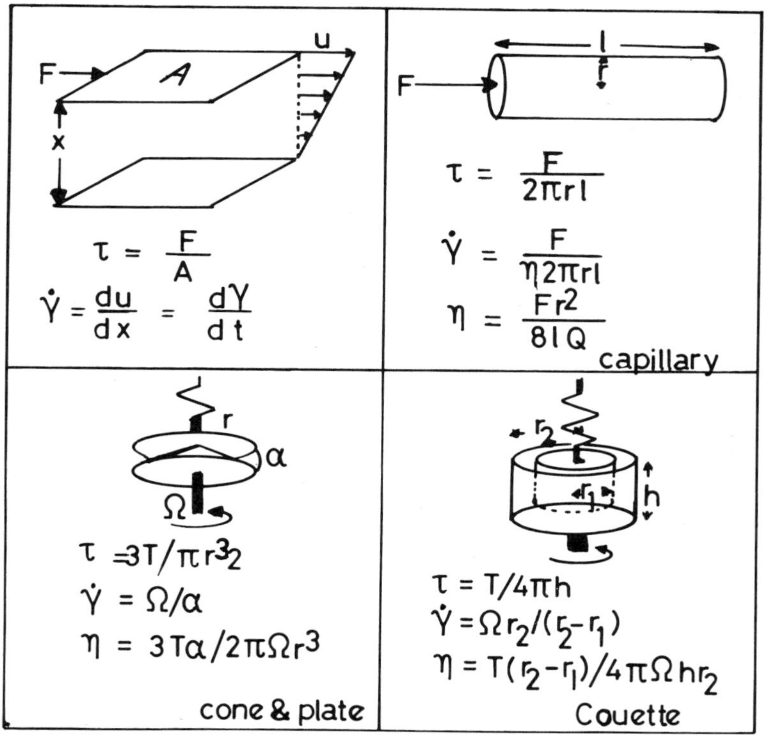

Fig. 1. Viscometric flows used for the determination of the shear viscosity coefficient. Q = volume flow rate/cm^{3}s^{-1}. T = measured torque/Nm.

anticipated from the particle concentration, due to the volume of layers. Hence the effective volume fraction, ϕ', becomes

$$\phi' = \phi(1 + \frac{\delta}{a})^3 \tag{2}$$

where a is the particle radius and δ is the hydrodynamic thickness of the layer. If equations (1) and (2) are being utilized to determine the dimensions of the stabilizer layer, only capillary viscometry can yield sufficient precision as $\frac{\delta}{a} << 1$.

Ionic surfactants produce a high potential at the surface of the disperse phase, whether we are considering micelles or adsorbed layers. The effect of shear on the diffuse part of the electrical double layer is to cause a distortion from its initial symmetry. More energy is continuously dissipated in such a system than if it were uncharged and so the result is an increase in viscosity. This is known as the "primary electroviscous effect".

Booth (1950) gave an analysis which is valid for values of the electrokinetic potential $\zeta < 25$ mV. In addition weak flows were assumed so that only small distortions of the diffuse layer from spherical symmetry were assumed. An analytical expression was derived for the intrinsic viscosity:

$$[\eta] = 2.5\ \{1 + (2\varepsilon\varepsilon_o\zeta)^2\ \frac{(\kappa a)(1 + \kappa a)^2 Z(\kappa a)}{\sigma\ \eta_o\ a^2}\} \qquad (3)$$

in which $(\varepsilon\varepsilon_o)$ is the permittivity, σ is specific conductivity of continuous phase, κ^{-1} is the characteristic decay length of the diffuse layer, i.e., the Debye length.

$$\kappa = \left[\frac{N_a\ e^2\ I}{0.05\ \varepsilon\varepsilon_o\ kT}\right]^{\frac{1}{2}}$$

Here N_a is the Avogadro number, k is the Boltzmann constant with T the absolute temperature, e is the electronic charge and I is the ionic strength. Booth's function, $Z(\kappa a)$, was given as a power series with the following limiting form for small κa values, i.e., for small particles and extended double layers as would be appropriate to micellar systems:

$$\text{for } \kappa a < 1; \quad Z(\kappa a) \cong \frac{1}{200\pi\ \kappa a} + \frac{11\ \kappa a}{3200\pi} \qquad (3a)$$

However, as most ionic surfactant systems have a higher electrokinetic potential than 25 mV, equation (3) is of limited use. Waterson and White (1981) have considered the general case and obtained a numerical solution. Figure 2 shows some of the values of intrinsic viscosity plotted as a function of ζ-potential.

The case of stronger flows has been considered by both Russel (1978a) and Lever (1979). Both of these results predict the occurrence of viscoelastic behaviour in that forces normal to the shear plane were predicted. This is the first indication of non-Newtonian behaviour.

At higher volume fractions, the disturbance of the fluid, as it flows around one particle interacts with that flowing around neighbouring particles and results in an increase in viscosity. An analysis of this effect in two viscometric flow fields has been given by Batchelor (1970; 1974; 1977). For shear flow

$$\frac{\eta(o)}{\eta_o} = 1 + 2.5\phi + 6.2\phi^2 + O(\phi^3) \qquad (4a)$$

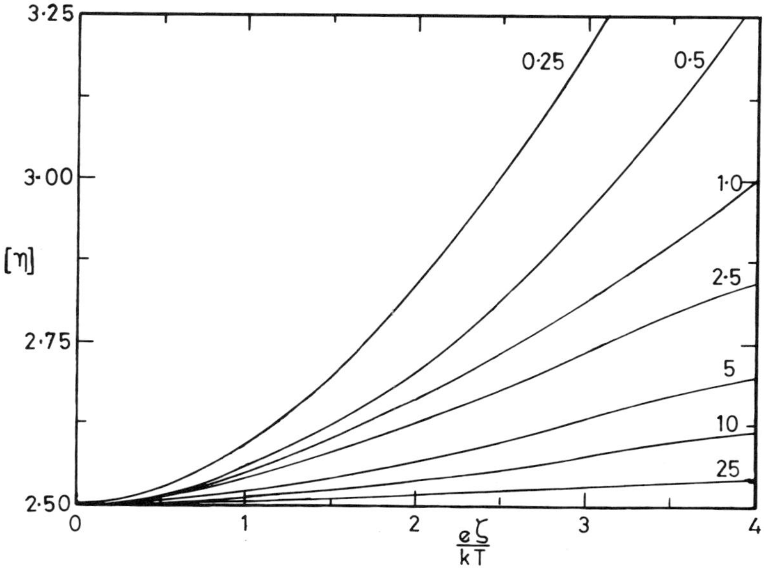

Fig. 2. The intrinsic viscosity as a function of reduced electrokinetic potential (eζ/kT) calculated for κa values from 0.25 to 25 using the analysis of Waterson and White.

and for extensional flow

$$\frac{\eta(o)}{\eta_o} = 1 + 2.5\phi + 7.6\phi^2 + O(\phi^3) \qquad (4b)$$

The symbol η(o) is used for the viscosity to indicate that this is the low shear limiting value, i.e., non-Newtonian or shear-thinning behaviour was predicted. This is an important result which has great implications for higher concentrations. The distribution of particles in the quiescent dispersion was uniform but when subjected to high shear flows a non-uniform distribution was produced. A tendency of the particles to form layers can result in a decrease in viscosity by reducing the collision frequency. The important concept here is the relaxation time of the structure which can be defined in terms of the time taken for a particle to diffuse a given distance. A particle radius is the simplest relevant distance to choose and this gives the relaxation time as

$$t_r = 6\pi\,\eta_o\,a^3/kT.$$

The characteristic time for the imposed convective motion is, of course,

$$t_o = \frac{1}{\dot{\gamma}}$$

The Deborah number in this case is better referred to as the reduced shear rate and is equal to

$$D_e = 6\pi\ \eta_o\ a^3\ \dot{\gamma}/kT.$$

When charged particles are considered an additional change in the microstructure occurs due to the high repulsive forces generated at small interparticle separations. Under quiescent conditions the pair distribution is altered so that there is a much reduced probability of finding adjacent particles with a surface-to-surface separation of $\frac{1}{\kappa}$. When the van der Waals attractive forces can be considered to be small, the closest distance of approach is governed by the Brownian motion and electrostatic forces. This condition has been analysed by Russel (1978b; 1980) who calculated the collision diameter of the particle, L, from the ratio of the electrostatic repulsion to the Brownian force, α, where

$$\alpha = \frac{16\pi^2\ \varepsilon\varepsilon_o\ \zeta^2\ a^4\ \kappa}{kT}\ \exp(2\kappa a).$$

For large repulsive forces, the analyses gave the collision diameter at the zero shear stress limit as

$$L_o = \frac{1}{\kappa}\ \ell n\left(\frac{\alpha}{\ell n(\alpha/Ln\ \alpha)}\right) \tag{5}$$

that is, the limits of L are:

$$2a \leqq L \leqq L_o$$

as $\infty \leqq \tau \leqq 0$

or $\infty \leqq \dot{\gamma} \leqq 0$.

This result indicates that shear thinning behaviour is to be expected from two-phase systems in which the disperse phase is charged and this latter effect is known as the "secondary electroviscous effect".

Many processes involve a very high shear regime followed by a low stress period. The simplest example would be filling a bottle from a pumped line (high τ and $\dot{\gamma}$), after which the creaming of bubbles or sedimentation of particles occurs (low τ and $\dot{\gamma}$). Printing, painting or surface coating processes provide equally good although more complicated examples. Therefore, the high and low stress limits to the viscosity are of great importance even though only a few of the laboratory instruments that are available cover a sufficiently wide range of stress levels to give a good estimate of both limits. From Russel's analysis the relative viscosity at low

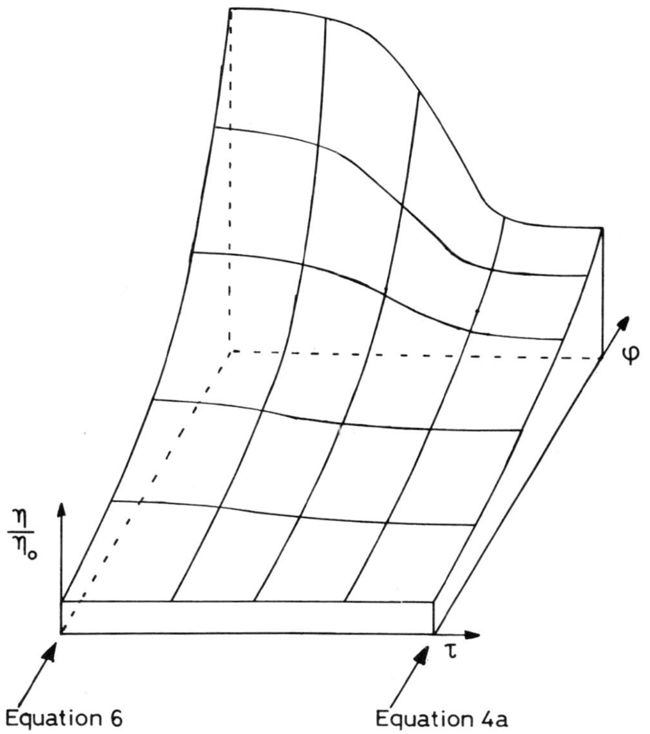

Fig. 3. The relative viscosity as a function of both applied stress and volume fraction for a two-phase system.

shear stress is given by

$$\frac{\eta(o)}{\eta_o} = 1 + [\eta]\phi + \left\{\frac{2}{5}[\eta]^2 + \frac{3}{40}\ln\left[\frac{\alpha}{\ln\alpha}\right]\left[\frac{L_o}{a}\right]^4(\kappa a)^{-1}\right\}\phi^2 \quad (6)$$

The value of $[\eta]$ would be that calculated from either the analysis by Booth (equation 3) or that by Waterson and White (see Fig. 2). Figure 3 shows the viscosity behaviour as a function of both ϕ and τ. However, the discussion so far has been limited to relatively dilute systems in that only two body interactions have been utilised. In concentrated systems multibody effects must be considered.

CONCENTRATED SYSTEMS

It is convenient to define a concentrated system as one in which spatial correlations exist between the components. In other words, the particles, whether they are solid, liquid or micellar units, are always interacting with their neighbours. This occurs when the time average surface-to-surface separation of the particles is of a similar magnitude to the range of the colloidal forces, whether electrostatic, steric, or van der Waals. When this occurs, a continuous three-dimensional structure is produced and this is analogous to the liquifaction of a gas. As the concentration is increased, highly ordered structures can be produced and the body-centred cubic, hexagonal, and lamellar mesophase structures of concentrated surfactant systems are a good example of this. Such structures resist deformation and hence can be highly elastic but, as the diffusive relaxation of the stress often has a time scale of a magnitude similar to many of our instruments, viscoelastic responses are frequently encountered. The problem, therefore, shifts from measuring the shear stress or shear rate dependence of viscosity to a rather broader description of the macroscopic properties, especially if we are to elucidate the subtle interplay between the microstructure and the material behaviour.

In simple terms the experimental objective is to define how much energy is stored elastically when a material is deformed and how much is lost by diffusive relaxation resulting in viscous flow. A continuous shear experiment only gives an estimate of the latter and so another mode of measurement must be sought. In order to keep the analysis of experimental data as simple as possible, only very small stresses and strains are used. This usually means that the viscous and elastic responses are linear, i.e., the viscosity is Newtonian and the elasticity is Hookean. Frequently this means strains of 1% or less with disperse systems.

MECHANICAL ANALOGUES

It is often useful to use a mechanical model as an aid to setting up equations to describe the material response. These models are only useful in that they enable a rheological equation of state (or constitutive equation) to be written down and subsequently tested. All of the models are made up of linear components, i.e., Hookean springs of modulus G or Newtonian dashpots containing a liquid with a coefficient of viscosity η. The main models are listed below with their constitutive equations.

Model	Constitutive equation

Newtonian liquid

η

$$\tau = \eta \dot{\gamma}$$

Hookean solid

G

$$\tau = G \dot{\gamma}$$

Maxwell model (liquid)

$G \leftarrow \dot{\gamma} = \frac{\dot{\tau}}{G}$

$\eta \leftarrow \dot{\gamma} = \frac{\tau}{\eta}$

$$\dot{\gamma} = \frac{\tau}{\eta} + \frac{\dot{\tau}}{G} \qquad (7)$$

i.e. $\tau = \eta \dot{\gamma} - \dot{\tau} \frac{\eta}{G}$

Stress same in both elements.

Strains additive and $\frac{\eta}{G} = t_M$ (Maxwell relaxation time)

The units of η are NS m^{-2} and those of G are N m^{-2}. The driving force for the relaxation is the spring and the dashpot viscosity controls the rate. Continuous constant stress ($\dot{\tau} \rightarrow 0$) gives flow at shear rate $\dot{\gamma}$.

Kelvin model (solid)

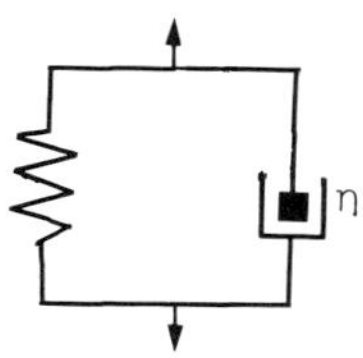

$$\tau = G \gamma + \eta \dot{\gamma} \qquad (8)$$

i.e. $\gamma = \frac{\tau}{G} - \dot{\gamma} \frac{\eta}{G}$

Strain same in both elements.

Stresses additive and $\frac{\eta}{G} = t_K$ (Kelvin retardation time)

The dashpot retards the spring motion, c.f. an automobile suspension unit.

Model

Constitutive equation

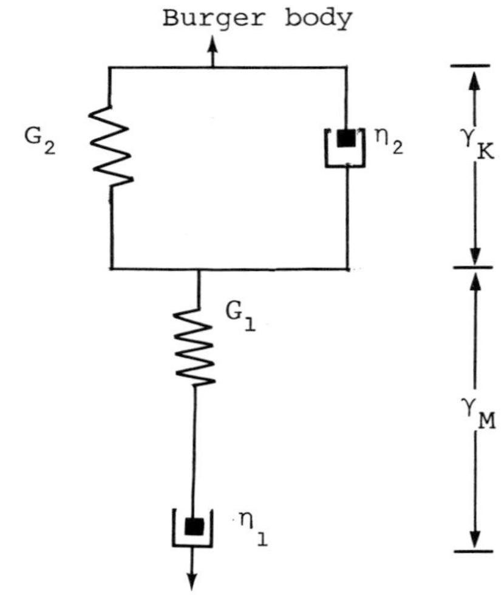

$$\ddot{\tau} + \left(\frac{G_1}{\eta_2} + \frac{1}{t_K} + \frac{1}{t_M}\right)\tau + \frac{1}{t_M t_K}\dot{\tau} = G_1\ddot{\gamma} + \frac{G_1}{t_K}\dot{\gamma} \qquad (9)$$

with $\dfrac{\eta_1}{G_1} = t_M$

and $\dfrac{\eta_2}{G_2} = t_K$

$\gamma = \gamma_K + \gamma_M$;

τ same in G_1, η_1 and Kelvin unit.

This model gives (i) an instantaneous elastic response from G_1; (ii) a retarded elastic response from G_2 and (iii) a continuous viscous response from η_1.

Other more complicated models can be devised by the addition of more elements. However, only one other will be considered here, that is, the generalized Maxwell model which has an infinite series of Maxwell units connected in parallel. This will be discussed later. In practical terms all that remains is to use the experimental stress loading with the constitutive equation chosen to approximate to the system in order to predict the material response. If a poor fit with experiment is obtained, then an alternative model is selected.

OSCILLATORY TESTS OR DYNAMIC MEASUREMENTS

By subjecting a material to a small amplitude and oscillating shear deformation, it is possible to measure the elastic response of a material at the same time as the viscous properties are measured. This is analogous to testing the strength of a scaffolding structure by shaking it. The

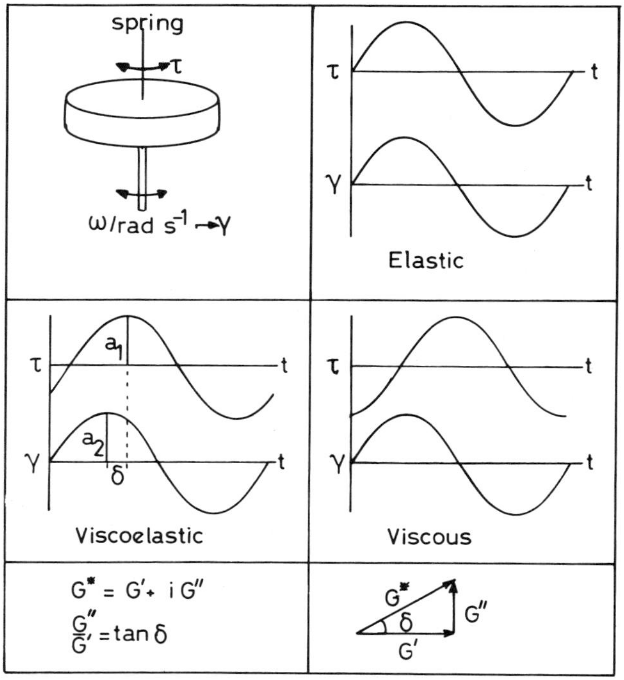

Fig. 4. Oscillatory or dynamic measurement of viscoelastic properties.

material is usually placed between two parallel discs, although a cone and plate or concentric cylinder geometry may also be used, and a rotational oscillation with a sine wave form is applied to one disc whilst the stress is measured by the deformation of the spring-mounted upper disc. This is shown schematically in Fig. 4. The parameters that are measured are the amplitude ratio, $\frac{a_1}{a_2}$, and the phase lag, δ. The time scale of the experiment is simply the inverse of the frequency, i.e. $\frac{1}{\omega}$, and this can easily be varied over a wide range to probe the change in Deborah number. The ratio of stress to strain gives the complex modulus, G*, which is the sum of the elastic or storage component (G') and the viscous or loss component (G"). Formally, in complex variable notation as a function of applied frequency this is

$$G^*(\omega) = G'(\omega) + i\,G''(\omega) \quad . \tag{10}$$

It should be noted that when

$$1 << D_e << 1 \quad ,$$

$$0 < \delta < \frac{\pi}{2} \quad .$$

Now, as a result of the strain changing sinusoidally, the stress will have a sinusoidal form and it is also convenient to write this in complex variable notation, i.e.

$$\tau(\omega) = \tau_o \exp(i\omega t) \tag{11}$$

where τ_o is the shear stress value at zero time.

For a material that behaves as a Maxwell model we can use equation (7) in conjunction with equation (11) simply by remembering the relationships $\gamma = \frac{\tau(\omega)}{G^*(\omega)}$ and $t_M = \frac{\eta}{G}$. Then, after differentiation of the stress with respect to time, the constitutive equation becomes

$$\frac{i\omega\ \tau_o\ e^{i\omega t}}{G^*(\omega)} = \frac{\tau_o\ e^{i\omega t}}{G\ t_M} + \frac{i\omega\ \tau_o\ e^{i\omega t}}{G} \quad .$$

This simplifies to

$$\frac{G\ i\omega t_M}{G^*(\omega)} = 1 + i\omega t_M$$

and after inversion and multiplication of both sides by the factor $(1 - i\omega t_M)$ to give us the complex modulus in terms of both real and imaginary parts we get

$$\frac{G^*(\omega)}{G} = \frac{i\omega t_M + \omega^2 t_M^2}{1 + \omega^2 t_M^2} \tag{12a}$$

and hence

$$\frac{G'(\omega)}{G} = \frac{\omega^2 t_M^2}{1 + \omega^2 t_M^2} \quad , \tag{12b}$$

$$\frac{G''(\omega)}{G} = \frac{\omega t_M}{1 + \omega^2 t_M^2} \tag{12c}$$

The response of this model is shown in Fig. 5 using equations (12b) and (12c) for the storage and loss components. The important features to note are that when $D_e = 1$ (i.e., $\log \omega t_M = 0$) the storage and loss moduli are equal, and that at high frequency the loss modulus becomes equal to the spring constant used in the model. However, the shape of the curves is usually much too narrow for a fit to real system behaviour. This is not surprising as the model assumes only one relaxation time and most systems will have a range of times corresponding to different diffusion processes, each

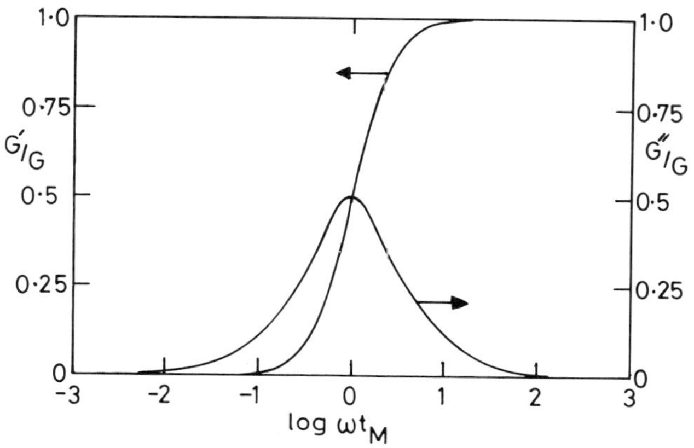

Fig. 5. Relative storage and loss moduli as a function of reduced frequency calculated for a Maxwell body.

with a Gaussian spread of times. The generalized Maxwell model shown in Fig. 6 is therefore more use experimentally. In this model each relaxation time corresponds to a particular spring-dashpot pair and the macroscopic response of the model, which is what we measure, is the sum of all the components given their fractional weighting. Replacing the sum by an integral we have

$$G'(\omega) = \int_{-\infty}^{+\infty} \frac{H(t)\ \omega^2 t^2}{(1 + \omega^2 t^2)}\ d\,\ell n\ t \qquad (13a)$$

$$G''(\omega) = \int_{-\infty}^{+\infty} \frac{H(t)\ \omega t}{(1 + \omega^2 t^2)}\ d\,\ell n\ t \qquad (13b)$$

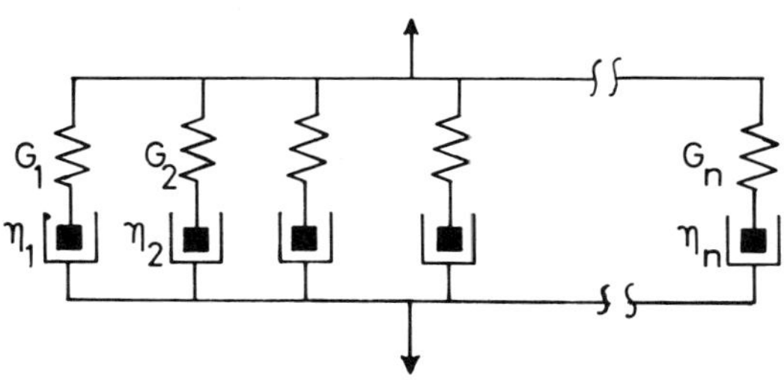

Fig. 6. The generalized Maxwell body.

The logarithmic form is used to enable as wide a range of frequencies as possible to be plotted and so the subscript M is dropped to indicate that the times have been reduced by division by 1 s. H(t) is known as the relaxation spectrum and is equal to the G value of each element giving a particular relaxation time multiplied by its fractional weighting. H(t) can be calculated from $G''(\omega)$, for example, by the Fourier transform of $G''(\omega)$ versus ω data. Integration under the curve of H(t) against t is the sum of all the spring constants, i.e., we have the high frequency limit to the modulus, $G(\infty)$

as $\omega \to \infty$

$$G(\infty) = \int_{-\infty}^{+\infty} H(t)\ d\ \ell n\ t \tag{14}$$

This can be measured directly if an experiment is carried out at a sufficiently high frequency that none of the components have an opportunity to relax, i.e., $D_e >> 1$. It is only for the most concentrated disperse systems that the relaxation times are sufficiently long for this to be attained with the available oscillatory instrumentation. However, other techniques, such as shear wave propagation, are available which operate at sufficiently high frequency.

The loss modulus divided by the frequency has units of viscosity and is known as the dynamic viscosity, η'. So that $\frac{G''}{\omega} = \eta'$. Rewriting equation (13b) in terms of the dynamic viscosity and taking the low frequency limit gives the zero shear viscosity, $\eta(o)$, thus

as $\omega \to 0$

$$\eta(o) = \int_{-\infty}^{+\infty} t.H(t)\ d\ \ell n\ t \tag{15}$$

This quantity can also be determined independently from creep measurements and its importance has already been discussed in relation to more dilute systems. It is appropriate at this stage to point out that stress relaxation measurements, which may be relatively simple to perform, can also be used to obtain a relaxation spectrum. The work of Ferry (1981) gives a useful review of the techniques by which the spectra can be obtained.

So, in summary, the characterization of a concentrated disperse system, whether of particles stabilized by surfactants or concentrated surfactant systems, requires some measurement of the relaxation spectrum in addition to the conventional shear stress - shear rate rheograms. After characterization, the question that immediately arises is "how do the macroscopic properties depend on the microstructure and hence the colloidal properties?" Only partial solutions are available so far.

RHEOLOGY AND THE MICROSTRUCTURE

As a convenient starting point for this discussion we can choose a continuation of ideas discussed with dilute systems. The dependence of the steady shear viscosity as a function of volume fraction can no longer be predicted from rigorous hydrodynamic analysis because of the difficulty in describing the multibody collision process. For spherical particles an alternative is to consider what happens as small increments in volume fraction are made to a dispersion and the clearest description of this approach has been given by Ball and Richmond (1980). This starts from the Einstein equation (equation 1) and differentiating to give the initial rate of change of viscosity

$$\frac{\partial \eta}{\partial \phi} = [\eta]\ \eta_o \quad .$$

Next we can use the mean field approximation that for any small addition of particles we can expect the same relative increase in viscosity as this, i.e., at any point we expect

$$\delta\eta = [\eta]\ \eta\ \delta\phi$$

where η is the viscosity of the system prior to the addition. Because the particles already in the suspension occupy a finite volume, the addition is to the <u>free volume</u> in the system. The excluded volume of a spherical particle is related to its radius by the expression $\frac{4}{3}\pi\, a^3 k_p$; where the value k_p is a function of the coordination number of the structure, i.e., for a coordination number of 12, $k_p = \frac{1}{0.74}$ from a face-centred cubic structure; and for a coordination number of 6, $k_p = \frac{1}{0.52}$ (simple cubic). A random microstructure results in a coordination number of 8 to 10 and $k_p \sim \frac{1}{0.62}$ and finally for a foam $k_p = 1$. The free volume is simply the difference between the total volume and the sum of the excluded volumes thus: $1 - N_p k_p \frac{4}{3}\pi\, a^3$. N_p is the number concentration of particles and so the free volume is $1 - k_p \phi$. The change in volume fraction is $\frac{\delta\phi}{1 - k_p \phi}$ and so expressing this in differential terms

$$\frac{\partial \eta}{\eta} = \frac{[\eta]\ \partial\phi}{1 - k_p \phi}$$

With the boundary conditions $\eta \rightarrow \eta_o$ when $\phi \rightarrow 0$ the solution is

$$\ln \eta = -\frac{[\eta]}{k_p} \ln (1 - k_p \phi) + \ln \eta_o$$

and so

$$\frac{\eta}{\eta_o} = (1 - k_p \phi)^{-[\eta]/k_p} \tag{16}$$

Equation (16) was first derived by Dougherty (in Krieger, 1972). Non-Newtonian behaviour can simply be regarded as a variation in k_p with shear stress. k_p can change if the distribution function for the microstructure changes and at high stresses there is only two-dimensional order instead of the three-dimensional symmetry that occurs at low stresses. When the disperse phase is charged, the excluded volume parameter is dependent on both the double layer dimensions and the applied stress. There are two approximate methods by which the problem can be approached which have been shown to be useful (Goodwin, 1982; 1983). Only the results for the high- and low-stress limit will be discussed here. The two methods are:

a) hard sphere approximation, and
b) activated diffusive relaxation.

The high-stress limit to the viscosity is the same for both models. At high stresses the microstructure has been shown to be formed into layers parallel to the shear plane with hexagonal order within the layers if the particles are spherical. This results in a maximum attainable volume fraction for rigid spheres of 0.605, i.e., $k_p = \frac{1}{0.605}$ and the high stress limit to the viscosity, $\eta(\infty)$, becomes with the intrinsic viscosity equal to 2.5

$$\frac{\eta(\infty)}{\eta_o} = (1 - 1.65\phi)^{-1.31} \tag{17}$$

The Hard Sphere Approximation

This is the simplest model in which the colloidal interaction force is approximated by a step function. For charged systems at low stress the analysis by Russel of the secondary electroviscous effect can be utilized. This gave the pair distribution function to be zero with a sphere of diameter L_o around each particle. By combination with the packing fraction a value of k_p can be obtained.

For rigid charged particles scattering studies indicate a face-centred-cubic symmetry (f.c.c.), but for viscous isotropic mesophases a body-centred symmetry (b.c.c.) is found and hence for f.c.c. $$k_p = \left(\frac{L_o}{2a}\right)^3 \frac{1}{0.74} \tag{18a}$$

for b.c.c. $$k_p = \left(\frac{L_o}{2a}\right)^3 \frac{1}{0.68} \tag{18b}$$

These values may be substituted into equation (16) to give the value of $\eta(o)$, i.e.,

$$\frac{\eta(o)}{\eta_o} = \left(1 - \left(\frac{L_o}{2a}\right)^3 \frac{1}{0.68}\right)^{-[\eta]\left(\frac{2a}{L_o}\right)^3 0.68} \tag{19}$$

for a viscous isotropic phase. However, it should be understood that $L_o = f(\phi)$ as at high disperse phase concentrations the counter ion concentration in the system is altering the value of κ (Benzing and Russel, 1981; Beresford-Smith and Chan, 1982) and hence L_o. However, at high volume fractions this model must become invalid as large double layer overlap occurs.

The Activated Diffusive Relaxation Model

This model has been described in detail previously (Goodwin, 1982; 1983) and only the salient features will be presented here. It is based on the approximate analysis by Eyring used to describe the viscosity of supercooled liquids. The model attempts to use a more realistic interaction potential than a step function and makes use of the normal double layer interaction energy. When an ordered liquid-like structure is stressed, the relaxation is by long-range diffusive motion of the disperse phase. This is restricted by the repulsive forces of neighbouring particles and there is therefore an activation energy barrier to this process. By assuming that there is no cooperative motion, the activation energy was easily calculated from the structure and the sum of the pair interaction energies. It was further assumed that there was a linear superposition of the stresses due to structural relaxation and the hydrodynamic interactions. This latter effect was calculated from equation (16) giving the final result

$$\frac{\eta(o)}{\eta_o} = \frac{\hbar}{16\,\pi\,\eta_o\,a^3}\exp(E^*/kT) + (1 - k_p\,\phi)^{-[\eta]/k_p} \tag{20}$$

E^* is the activation energy for the long-range diffusive motion, $\hbar$ is Planck's constant/2π, k is the Boltzmann constant and T is the absolute temperature.

At high concentrations cooperative motion must become of increasing importance and this will result in the viscosity increasing at a slower rate than predicted by equation (20). Bohlin (1980) has carried out an analysis in which there is a cooperative relaxation of the microstructure. This gives the low stress limit to the viscosity as

$$\eta(o) \cong G(\infty)\ t_r \frac{(\exp |\Xi|/kT - 1)^{(z-1)}}{z - 1} \tag{21}$$

where z is the coordination number of the microstructure, i.e., z = 2 for a lamellar phase, and $|\Xi|$ is the magnitude of the interaction energy between the different parts of the microstructure. This is, of course, different from the activation energy, E*, in equation (20). The value of Ξ must be determined from experimental data such as stress relaxation measurements. The relaxation time, t_r, is also determined from a stress relaxation experiment. This approach should be most useful at very high volume fractions. The problem remains with this model of the prediction of t_r and Ξ from the disperse phase properties.

Finally, the high frequency limit of the shear modulus, $G(\infty)|$, is of interest in that it can often be measured for concentrated disperse systems and it may also be predicted from the characteristics of the disperse phase for the simpler systems. Measurement of the shear wave propagation velocity is easily carried out using the "Shearometer" (Rank Brothers, Bottisham, Cambridge). This instrument works at a frequency in excess of 10^3 rad s^{-1} with an extremely small shear strain and so the wave rigidity modulus, $\tilde{G}$, which this instrument yields, is usually a good approximation to $G(\infty)$ for concentrated systems. For concentrated spherical colloidal systems it has been shown (Goodwin, 1982; 1983) that a simple model can be set up which describes the experimental behaviour. A static microstructure is assumed and, when this is distorted by a small strain, the restoring force can be calculated from the "spring constants" of the double layer interactions. The resulting expression was

$$G(\infty)| = \frac{3n}{32\ k_p\ d} \frac{\partial^2\ V(d)}{\partial\ d^2} \tag{22}$$

where d is the average centre-to-centre separation of the particles calculated from ϕ and the structure type. n is the coordination number. For structures with close-packed symmetry

$$n = 12; \quad k_p = \frac{1}{0.74} \text{ for f.c.c.,}$$

$$n = 10; \quad k_p = \frac{1}{0.68} \text{ for b.c.c.}$$

The foregoing indicates how some of the viscoelastic properties of concentrated systems may be related to the characterization of the disperse phase. More detailed statistical, mechanical analyses are becoming available but these usually mean that a computer simulation of the structures is required.

REFERENCES

Ball, R.C. and Richmond, P. (1980). *Phys. Chem. Liq.* 9, 99.
Batchelor, G.K. (1970). *J. Fluid Mech.* 41, 541.
Batchelor, G.K. (1974). *Ann. Rev. Fluid Mech.* 6, 227.
Batchelor, G.K. (1977). *J. Fluid Mech.* 97, 83.
Benzing, D.W. and Russel, W.B. (1981). *J. Colloid Interface Sci.* 83, 178.
Beresford-Smith, B. and Chan, D.Y.C. (1982). *Chem. Phys. Letters* 92, 474.
Bohlin, L. (1980). *J. Colloid Interface Sci.* 74, 423.
Booth, F. (1950). *Proc. Royal Soc.* A205, 533.
Einstein, A. (1956). "Investigations on the Theory of the Brownian Motion". Dover, New York.
Ferry, J.D. (1981). "The Viscoelastic Properties of Polymers". Wiley, New York. 3rd edn.
Goodwin, J.W. (Ed.) (1982). "Colloidal Dispersions". Royal Society of Chemistry, London.
Goodwin, J.W. (1983). *In* "Science and Technology of Polymer Colloids" Vol.II (Eds G. Poehlein, R.H. Ottewill and J.W. Goodwin). Martinus Nijhoff, Boston.
Krieger, I.M. (1972). *Adv. Colloid Interface Sci.* 3, 111.
Lever, D.A. (1979). *J. Fluid Mech.* 92, 421.
Russel, W.B. (1978a). *J. Fluid Mech.* 85, 673.
Russel, W.B. (1978b). *J. Fluid Mech.* 85, 209.
Russel, W.B. (1980). *J. Rheol.* 24, 287.
Waterson, I.G. and White, L.R. (1981). *J. Chem. Soc. Faraday 2* 77, 1115.

ADSORPTION AT THE AIR/LIQUID, LIQUID/LIQUID, AND SOLID/LIQUID INTERFACES

R. Aveyard

Department of Chemistry, University of Hull, Hull, England.

INTRODUCTION

The literature concerned with adsorption of surfactants at various types of interface is vast. The object in this chapter is not to attempt to give a catalogue of various results but to introduce some rudimentary ideas and broad principles. This is in some instances best done by reference to surface active materials, such as alkanols, which are not themselves surfactants, but the findings quoted can be expected to apply equally well to surfactants. Molecules of interest are amphiphilic, containing a water-soluble polar or ionic head-group and a water-insoluble tail which usually is (or contains) an alkyl chain. Surfactants are normally effective (in such processes as wetting, foaming, dispersion of solids in liquids, stabilization of emulsions, etc.) when present in only very low concentrations and, since all these processes involve interfacial effects, it is evident that surfactants are strongly adsorbed at liquid/vapour (L/V), liquid/liquid (L/L), and solid/liquid (S/L) interfaces. Frequently one of the major forces contributing to adsorption has its origins in the removal or partial removal of the nonpolar tail from contact with water. It is appropriate therefore to take a look first at the interactions between hydrocarbons and water, both in aqueous solution and at surfaces.

HYDROCARBON - WATER INTERACTIONS

Hydrocarbons are hydrophobic, i.e. they have a very low solubility in water but dissolve readily in many nonpolar

SURFACTANTS
ISBN 0-12-682180-1

liquids. Thus, although a surfactant below its c.m.c. dissolves as single molecules or ions, the solubility results from the presence of the polar (or ionic) group and there will be a tendency for the tail to 'escape' from the water. One means of escape is adsorption and another is micelle formation.

It is sometimes suggested that association of alkyl chains in micelle formation arises from the attractive London dispersion forces between the chains. The mutual dispersion forces between CH_2 groups however are expected to be of similar magnitude to those between a CH_2 group and a water molecule. The association of hydrocarbon tails in water, therefore, is seen to arise not from mutual attraction of nonpolar groups but rather from the unusually strong mutual attraction of water molecules. Indeed, hydrocarbon chains are 'squeezed' out of water even when, in the final state, they are not in mutual contact, as for example in adsorption into dilute monolayers at the L/V surface.

Liquid water at room temperature is an 'ordered' liquid there being a degree of tetrahedral arrangement of the molecules imposed by H-bonding. Introduction of a solute is expected to cause disruption of the local structure, and transfer of an alkane molecule from the vapour phase to water (i.e. 'hydration') would be expected to be accompanied by a more positive entropy change than for the corresponding transfer to a nonpolar solvent. In fact the hydration entropy of hydrocarbons is found to be large and <u>negative</u>. Further, the partial molar heat capacities of hydrophobic solutes in water are anomalously high. All this points to 'structure' being promoted rather than disrupted. Frank and Evans (1945) in a famous paper put it this way: "When a rare gas atom or a nonpolar molecule dissolves in water at room temperature, it modifies the water structure in the direction of greater 'crystallinity'; the water, so to speak, builds a microscopic iceberg around it ... As the temperature is raised these icebergs melt giving rise to the enormous partial molar heat capacity of these gases in water." It can be appreciated therefore that the impetus for the hydrophobic groups of surfactant molecules to be removed from water arises in the main from the positive entropy change (giving a negative contribution to the free energy change) which accompanies the transfer, be it to an adsorbed monolayer or a micelle interior.

The whole area of aqueous solutions of nonpolar molecules, and the interactions between these molecules in water ('hydrophobic interactions') is a subject of considerable interest and debate, not least because of its biological importance. It is fraught with difficulties, subtleties, and a stimulating measure of disagreement. The reader who wishes to get a flavour of the current state of the subject should browse through the recent books of Tanford (1980) and Ben-Naim (1980).

Although ill at ease in bulk aqueous solution, alkyl chains sit quite happily at the surface of water. Thus, alkane vapours adsorb at the water/vapour surface (Jones and Ottewill, 1955). The standard free energy of adsorption is linearly related to the alkyl chain length, N, implying that in dilute monolayers the chains assume configurations which are broadly parallel to the surface. Standard equilibrium enthalpies of adsorption, $\Delta_a H^o$, of alkanes are also known to vary linearly with N (Dorris and Gray, 1981), as shown in Fig. 1. It is illuminating to compare $\Delta_a H^o$ with molar enthalpies of cohesion, $\Delta_C^\alpha H$, of liquid alkanes, and with molar enthalpies of adhesion, $\Delta_A^{\alpha\beta} H$, between liquid alkanes and water. The work of adhesion, $W_A^{\alpha\beta}$, between phase α (alkane) and phase β (water) is defined as $\gamma^\alpha + \gamma^\beta - \gamma^{\alpha\beta}$ where $\gamma^{\alpha\beta}$ is the interfacial tension, and γ^α and γ^β are surface tensions (see Fig. 2). If α and β were to be the same material then of course $\gamma^{\alpha\beta} = 0$ and the process depicted in Fig. 2 (say for α) involves the work of cohesion, W_C^α, which is $2\gamma^\alpha$. From the

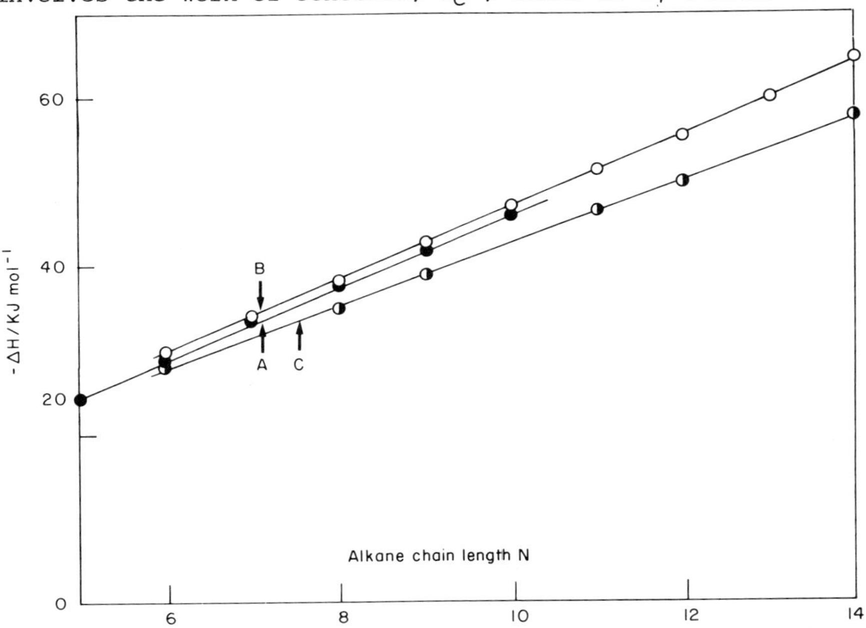

Fig. 1. Enthalpies of adsorption, adhesion, and cohesion as a function of alkane chain length N. Line A: enthalpy of adsorption, $\Delta_a H^o$. *Line B: enthalpy of cohesion,* $\Delta_C^\alpha H$. *Line C: enthalpy of adhesion,* $\Delta_A^{\alpha\beta} H$.

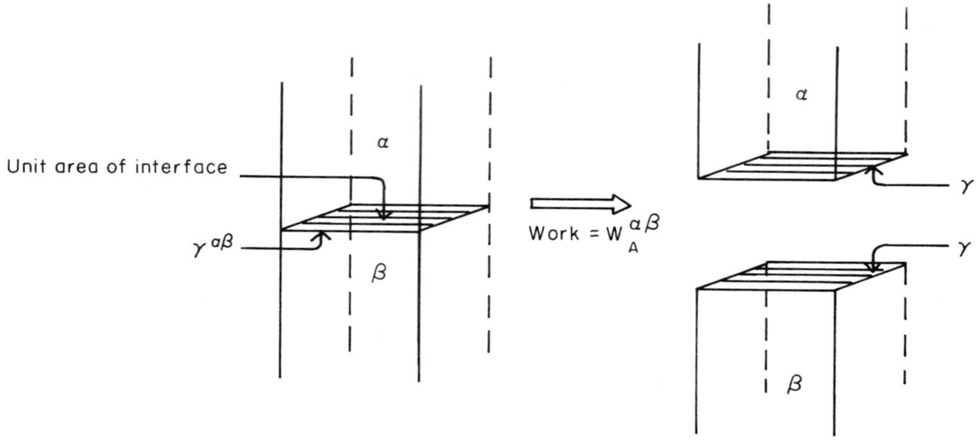

Fig. 2. Work of adhesion, $W_A^{\alpha\beta}$, *between immiscible phases* α *and* β.

temperature dependence of the free energies $W_A^{\alpha\beta}$ and W_c^{α} the corresponding enthalpies can be obtained. These unit area quantities can then be converted to (alkane) molar quantities if the molar surface areas of the alkanes are assumed. It has been argued elsewhere (Aveyard *et al.*, 1972) that alkane molecules are oriented parallel to alkane/water interfaces. Values of $\Delta_A^{\alpha\beta}H$, shown as a function of N in Fig. 1, have been calculated on this basis, as have the values of $\Delta_c^{\alpha}H$ for the alkanes. The striking observation is that both the absolute values and the chain length dependences of $\Delta_a H^o$, $\Delta_c^{\alpha}H$, and $\Delta_A^{\alpha\beta}H$ are very similar. Since cohesion within alkanes arises from dispersion forces the values of $\Delta_a H^o$ and $\Delta_A^{\alpha\beta}H$ are consistent with the proposition that both for dilute alkane monolayers on water, and for bulk liquid alkane in contact with water, the alkane - water interactions arise predominantly from dispersion forces. If the above arguments are accepted it would appear that there is no extensive structuring of interfacial water caused by contact with alkane molecules. This situation can be contrasted with that operating in bulk aqueous solutions, as already discussed.

We return to the theme of the transfer of hydrocarbon chains from water in the section on adsorption at L/V and L/L interfaces, where values of standard free energies of adsorption are used to support the intuitive picture of the disposition of surfactant molecules at liquid surfaces.

ADSORPTION AT L/V AND L/L INTERFACES (Aveyard and Haydon, 1973)

Adsorption at L/V and L/L interfaces occurs in many important processes such as emulsion stabilization, wetting, enhanced oil recovery, and microemulsion formation. Positive adsorption always leads to a lowering of surface or interfacial tension and, in the last two processes mentioned, adsorption leads ultimately to ultralow oil/water interfacial tensions (10^{-3} mN m^{-1} or less).

Determination of Surface and Interfacial Tensions and Adsorption (Adamson, 1982)

The extent of adsorption (i.e. the total surface concentration Γ) of a surfactant is normally determined indirectly from values of surface or interfacial tension γ. Changes in γ with concentration c are related to adsorption in an i-component system by the Gibbs equation (see Chapter 2)

$$-d\gamma = \sum_i \Gamma_i d\mu_i \qquad (1)$$

where μ_i is the chemical potential of the i-th species. The surface concentration is obtained from a gradient, $d\gamma/dc$ or $d\gamma/d\ell nc$ and if reasonably reliable values of Γ are to be calculated care must be exercised in the determination of γ. A variety of techniques is available and some appropriate methods, together with some cautionary notes, are mentioned below. The situation is a little confusing since experienced workers in this area have their own personal preferences and prejudices.

A very versatile method, applicable to L/V and L/L systems containing surfactants, is the drop-volume method in which the volume, V_f, of a drop of (the more dense) liquid which just becomes detached from a cylindrical tip, radius r, is accurately determined. When the maximum volume V_m is attained part of it will fall (volume V_f) and a fraction (volume V_r) will remain i.e. $V_f = V_m - V_r$ (Fig. 3, *a,b*). The surface (interfacial) tension is then obtained from the equation

$$\gamma = V_f \Delta\rho g / 2\pi r f(r/V_f^{\frac{1}{3}}) \qquad (2)$$

where $\Delta\rho$ is the difference in density between drop and surroundings, g is the acceleration due to gravity, and $f(r/V_f^{\frac{1}{3}})$ is the so-called Harkins-Brown correction factor (Adamson, 1982) which is a function of $r/V_f^{\frac{1}{3}}$. The method

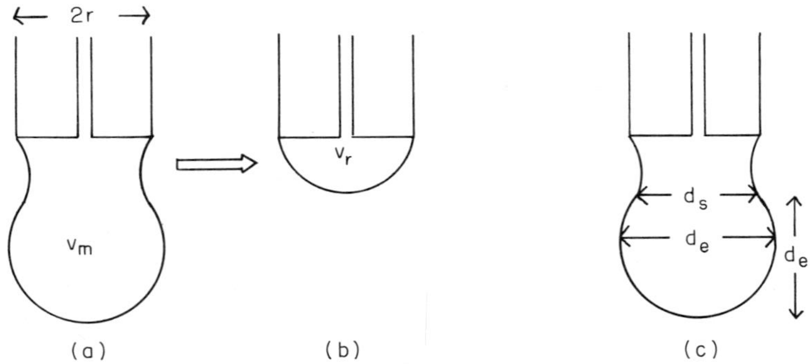

Fig. 3. Liquid drops hanging from cylindrical tips: (a) *and* (b) *represent the detachment process;* (c) *shows dimensions of a pendant drop (see text).*

becomes unsuitable if adsorption is slow since at the point of drop detachment the interface expands and the tension will rise. Ageing of the surface is likely to occur for very dilute solutions (such as those of some nonionic surfactants below their c.m.c.) or for those where surface active impurities are present. For such systems a static method must be used. One such method is that in which certain dimensions of an equilibrium drop hanging from a cylindrical tip are determined, usually photographically (pendant drop method). Convenient dimensions to measure are the maximum diameter, d_e, and diameter, d_s, at a height d_e above the apex of the drop (Fig. 3, *c*). The ratio $d_s/d_e = S$ is obtained; this shape factor is uniquely related to the quantity H (available in tabular form (Adamson, 1982), and the surface (or interfacial) tension is given by

$$\gamma = \Delta\rho g d_e^2/H \tag{3}$$

Another static approach which is often recommended is the Wilhelmy plate method. Here a thin plate of, say, glass or platinum (or even sometimes a filter paper) with a horizontal edge, and roughened surface (to give zero contact angle θ with the liquid) is lowered until it just touches the surface, when it is pulled down into the liquid. The liquid is then lowered until the lower edge of the plate is level with the unperturbed liquid surface, so that no buoyancy correction is subsequently needed. The tension is then given by

$$\gamma\cos\theta = \Delta W/p \tag{4}$$

where ΔW is the change in weight (i.e. force) observed on a microbalance resulting from the formation of the liquid meniscus at the contact of plate (perimeter p) and liquid. If

θ is not zero ($\cos\theta = 1$) then it is usually unknown and generally unreproducible, so that the method fails. The plate technique is effective for monitoring slow changes in γ, but nonetheless it is not easy to be certain that $\theta = 0$ and that the apparent change in γ is not in reality due (at least in part) to changes in θ.

Tensions of only a few $\mu N\ m^{-1}$ or less are sometimes produced by adsorption at L/L interfaces and these instances are of particular interest in connection with for example microemulsion formation and enhanced oil recovery. For such low tensions the above methods are inappropriate but the spinning-drop technique can be used. Here, a small drop of the less dense phase (usually the oil) is introduced into the more dense phase which fills a horizontal capillary tube rotating about its long axis. The oil drop becomes elongated and at sufficiently high frequencies takes on the form of a cylinder (radius r) with hemispherical ends. Under these conditions

$$\gamma = 0.25\ r^3 \Delta\rho\omega^2 \tag{5}$$

where ω is the angular velocity ($2\pi n$ where n is the frequency of rotation). Tensions of the order of 10^{-3} mN m^{-1} and less are readily determined by this method and commercial instruments are now available.

Having determined γ as a function of surfactant concentration in one (or both) of the adjacent phases, the surface concentration Γ_3 of surfactant can be determined using an appropriate form of the Gibbs equation. Thus, for example, for a system of a water-insoluble oil (component 1, which is the major component of phase α) in contact with water (component 2, being the major component of phase β) with an unionized surfactant (component 3) distributed between α and β, the Gibbs equation has the form

$$-d\gamma = [\Gamma_3 - x_3^\beta \Gamma_2 / x_2^\beta - x_3^\alpha \Gamma_1 / x_1^\alpha] d\mu_3 \tag{6}$$

where x are mole fractions. For strong adsorption from dilute solution x_3^α and $x_3^\beta << x_2^\beta \cong x_1^\alpha \cong 1$ and although Γ_1 and Γ_2 are not negligible compared to Γ_3, the terms in Γ_1 and Γ_2 in equation (6) may be neglected. Then, since $d\mu_3 = RTd\ell na_3$, where a_3 is the activity of the surfactant in solution,

$$\Gamma_3 \cong -(d\gamma/d\ell na_3)/RT \tag{7}$$

The forms of the Gibbs equation to be used in the calculation of surface concentrations of ionic surfactants adsorbed from aqueous solutions in the presence and absence of swamping inert electrolyte have been given in Chapter 2. It is common to set surfactant activity equal to concentration and this is usually an acceptable approximation.

Some Results

Broadly speaking there are two basic pieces of information which are sought, both at an academic and at a practical level. These are (a) the way in which the extent of adsorption depends on bulk concentration, and (b) the nature of the adsorbed film, e.g. the interaction between adsorbed surfactant molecules and the packing of the molecules within the film. Questions (a) and (b) can be discussed in terms of adsorption isotherms and surface equations of state, respectively, or alternatively the efficiency and effectiveness of adsorption, respectively.

Adsorption free energies Adsorption free energy normally refers to the difference in the standard chemical potential, $\mu^{\ominus}$, of the adsorbate at the interface and in bulk solution, i.e. $\Delta_a\mu^{\ominus} = \mu^{\ominus}$ (surface) $- \mu^{\ominus}$ (bulk) and may be obtained from

$$\Delta_a\mu^{\ominus} = -RT\ell n(\pi/x)_{x=0} \tag{8}$$

In equation (8) π, the surface pressure, is the lowering of γ caused by adsorption. The standard states for the surfactant are, for bulk solution, the hypothetical state in which $x = 1$, it being supposed the surfactant behaves as in infinitely dilute solution, and for surface, π = unity in an assumed ideal film. The ratio $(\pi/x)_o$ is a distribution ratio for surfactant between bulk and surface (at infinite dilution) and equation (8) is simply a form of the Van 't Hoff isotherm. Values of $\Delta_a\mu^{\ominus}$ (obtained using the same concentration scales and standard states) for different surfactants indicate the relative 'strengths' of adsorption into infinitely dilute films. They can also be used to substantiate the intuitive view as to how surfactants are disposed at L/V and L/L interfaces; this is now exemplified by reference to the adsorption of n-alkanols (Aveyard and Briscoe, 1970; Aveyard *et al.*, 1972; Aveyard and Chapman, 1975).

In Fig. 4, $\Delta_a\mu^{\ominus}$ (as a function of alkanol chain length N) is shown for adsorption (a) from solution in alkane to the alkane/water interface (curve A), (b) from solution in water to alkane/water interface (curve B), and (c) from solution in water to the air/water (A/W) surface. The approximate constancy of $\Delta_a\mu^{\ominus}$ for adsorption from the oil phase (curve A) indicates the alkanol chains remain in an essentially hydrocarbon environment at the interface. Adsorption from water to the oil/water interface gives a linear dependence of $\Delta_a\mu^{\ominus}$ on N, reflecting the transfer of each successive CH_2 from water to interfacial oil. The chain length dependence of the standard free energy of transfer of alkanols from water to bulk alkane (curve D) is almost exactly equal to that for

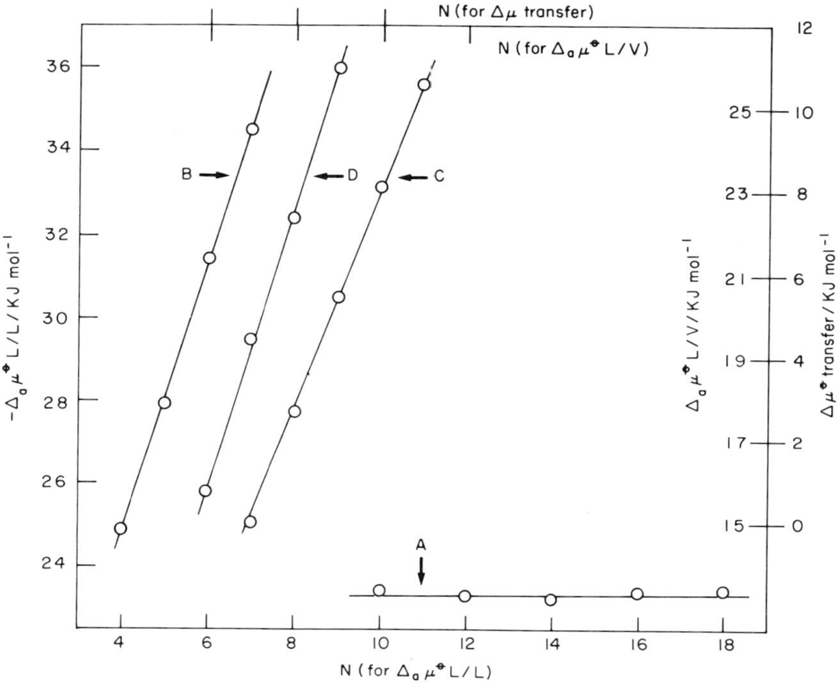

Fig. 4. Standard free energies of adsorption and transfer of alkanols, chain length N. A: from dodecane to dodecane/water interface. B: from water to dodecane/water interface. C: from water to A/W surface (from J. Clint et al., *1968). D: transfer from water to dodecane. All at 20°C.*

the adsorption process. Curves A, B and D taken together therefore indicate that the state of the hydrocarbon chain at the alkane/water interface is very similar to that in bulk alkane. The free energies of adsorption to the A/W surface are also linearly related to N but $\Delta_a\mu^\ominus$ per CH_2 group is less than for adsorption to the alkane/water interface. Presumably at the A/W surface, surfactant chains are oriented parallel to the surface and, as discussed earlier, have dispersion forces with the underlying water molecules. The chains have no nearest neighbour molecules on the air side of the surface, however, giving rise to the smaller adsorption free energy.

Surface equations of state The above considerations relate to infinitely dilute surfactant monolayers. What happens as the film concentration increases can be understood by considering the shape of the surface pressure (π)-surface area ($A = 1/\Gamma_3$) isotherms. An equation which describes (π,A) behaviour is called a surface equation of state (Aveyard and

Briscoe, 1970; Aveyard *et al.*, 1972; Aveyard and Chapman, 1975). Infinitely dilute films of nonionic surfactants such as those discussed above can be expected to obey the ideal 'two-dimensional gas' equation, $\pi A = RT$ per mole, as can very dilute films of ionic surfactants on a swamping inert electrolyte subphase. On a subphase of water, however, a 1,1-electrolyte surfactant monolayer will be expected to obey $\pi A = 2RT$ (Robb and Alexander, 1968).

The ideal equation of state can be derived for a model in which the adsorbed molecules are vanishingly small and have no mutual attractive forces. However, as the film concentration increases, the finite size of the molecules becomes significant compared to A/N_o. If this is the only significant additional effect, the ideal equation can be modified to give the Volmer equation

$$\pi(A - A_o) = RT \text{ per mole} \tag{9}$$

where A_o is the molar co-area of the adsorbate which can be taken as representative of the geometrical cross-sectional area of adsorbed molecules (Aveyard and Haydon, 1973). This excluded area gives rise to an increase in π for a given A. Lateral attractive forces within a film will give a negative contribution to the pressure. Such lowering of π is observed in films at A/W surfaces where the hydrophobic moieties interact mutually through dispersion forces. Similar attraction is not in evidence for films at L/L interfaces since the tails are surrounded by, say, alkane molecules and chain - chain interactions are indistinguishable from chain - alkane interactions. Thus the chains are always involved in lateral interactions but these do not <u>change</u> significantly with A. These points are illustrated in Fig. 5, *a*. Alkanols of different chain length adsorbed at the alkane/water interface all give the same π,A curve, suggesting an essentially vertical orientation of the chains. The data are well described by equation (9) with $A_o/N_o = 0.25\ nm^2$, consistent with the notion of vertical chains. Although data are not yet available, it is likely that nonionic surfactants of the type $C_nH_{2n+}(OCH_2CH_2)_yOH$ for example will obey equation (9) at the oil/water interface but with a larger A_o reflecting the larger head-group size.

Results for heptanol at the A/W interface show the importance of lateral chain - chain interactions in reducing π. There is no way of simply representing π,A data here as there is for the L/L interface, although a two-dimensional analogue of the van der Waals equation has sometimes been employed. One of the problems is that as Γ_3 changes so do the configurations of the chain, and hence the value of A_o; in a close-packed monolayer, the cohering chains will be more nearly

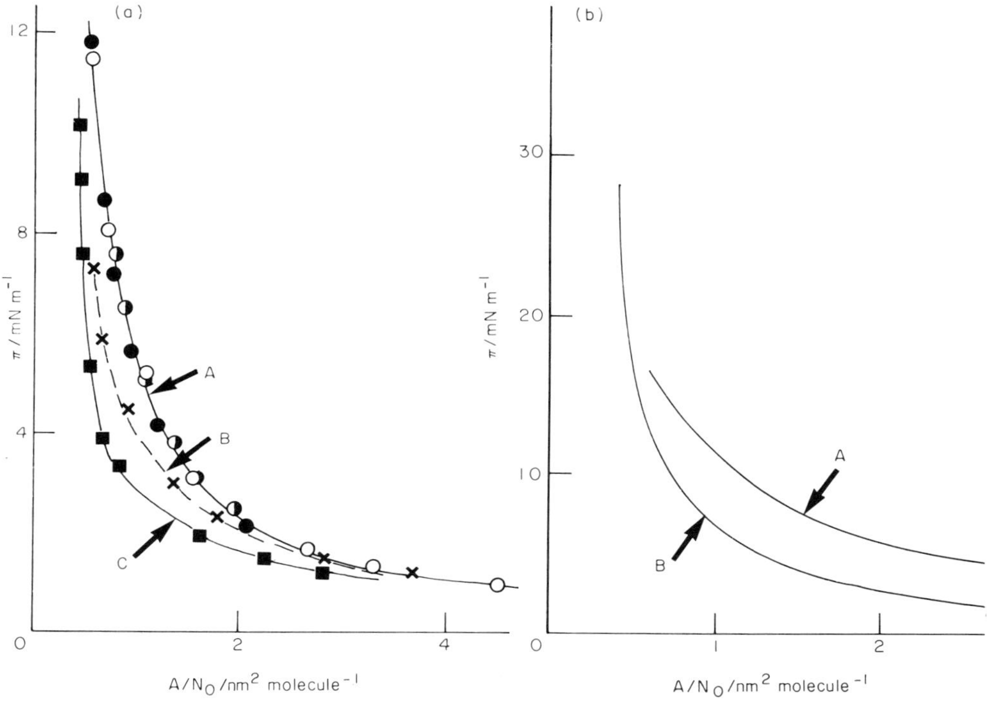

Fig. 5. Surface pressure π *as a function of area A. (a) Results for alkanols at 20°C. Curve A:* ◑ *butanol,* ● *pentanol,* ○ *octanol, all at dodecane/water interface. Curve B: X butanol at air/water surface,* --- $\pi A = RT$. *Curve C: heptanol at air/water surface. (b) Curve A: sodium sebacate. Curve B: neutral sebacic acid. Both at heptane/water interface at 35°C; results from Pal* et al. *(1975).*

vertical than horizontal. In this context it is also worth remarking that, for obvious reasons, one cannot use a value of A_o appropriate for a surfactant adsorbed at an oil/water interface for the same material present at the A/W surface. The data for butanol at the A/W surface, which obey the ideal equation very closely, possibly indicate how attractive and repulsive (A_o) terms can effectively cancel each other over a range of area.

Finally, it is clear that if a head-group is charged a further, repulsive, term is to be included in the surface pressure. This effect is shown in Fig. 5, *b* where the π,A curves for neutral sebacic acid and the sodium salt of this acid, both at the heptane/water interface, are shown.

Effectiveness and efficiency A more practically oriented method of looking at the strength of adsorption and the nature of concentrated surfactant monolayers is now briefly mentioned. Rosen (1978) points out that useful quantities are (a) the concentration of surfactant in solution needed to give a fixed lowering of tension and (b) the maximum value of adsorption attainable, Γ_m. The former is a measure of the <u>efficiency</u> of the surfactant adsorption, and the latter is an indication of the <u>effectiveness</u> of adsorption. The maximum adsorption Γ_m is attained at about the c.m.c. and is calculated from γ using the Gibbs equation. Rosen lists Γ_m for a very wide range of surfactants; values understandably depend, *inter alia*, on surfactant molecular geometry and the orientation in saturated films. Straight chain compounds with a terminal head-group tend to have large Γ_m whereas Γ_m is much smaller for large molecules with polar or ionic groups at each end. For ionic surfactants Γ_m is increased by the presence of electrolyte in the subphase. For polyoxyethylenated surfactants, RE_y, the size of the polar group (i.e. the value of y) determines Γ_m, the larger y, the smaller Γ_m.

For the efficiency of adsorption Rosen (1978) argues that - log (concentration, C_{20}) of surfactant required to produce a surface pressure of 20 mN m^{-1}, i.e. pC_{20}, is a good measure, since for many surfactants when π = 20 mN m^{-1} the surface is between 84 and 100% saturated. The value of pC_{20} is related to $\Delta_a\mu^\ominus$ and, like plots of the latter against alkyl chain length, plots of pC_{20} versus N are linear. A large number of pC_{20} values are listed by Rosen.

Adsorption leading to ultralow tensions The surface pressures at surface saturation (i.e. for c $\geqq$ c.m.c.) for most surfactants at the A/W surface lie between about 20 and 40 mN m^{-1}, so that the lowest γ is around, say, 30 mN m^{-1}. At the oil/water interface γ_i can be substantially lower. For example for solutions of Aerosol OT (sodium salt of dioctylsulphosuccinate), just above the c.m.c. in 0.13M-NaCl, in contact with heptane, the tension is as low as 0.2 mN m^{-1}. In some instances, however, alkane/water tensions can be reduced even more to, say, 10^{-3} mN m^{-1} or less. As mentioned earlier, attainment of such low values of γ_i is of crucial importance in microemulsion formation and enhanced oil recovery.

Microemulsions are often 4-component systems containing a cosurfactant in addition to surfactant. But some (apparently) pure anionic surfactants at very low concentrations in dilute NaCl solution can give ultralow alkane/water tensions for certain well-defined values of surfactant concentration, alkane chain length, temperature and salt concentration (Doe *et al.*, 1977). Equally, and more reproducibly, we have found that ultralow γ_i can be produced in alkane/water systems containing pure nonionic surfactants of the type $C_{12}H_{25}(OCH_2CH_2)_yOH$ (Aveyard and Lawless, unpublished). Some sample results of γ_i as a function of temperature, obtained using the spinning-drop technique (see earlier), are shown in Fig. 6. We note that, in the region of the tension minimum,

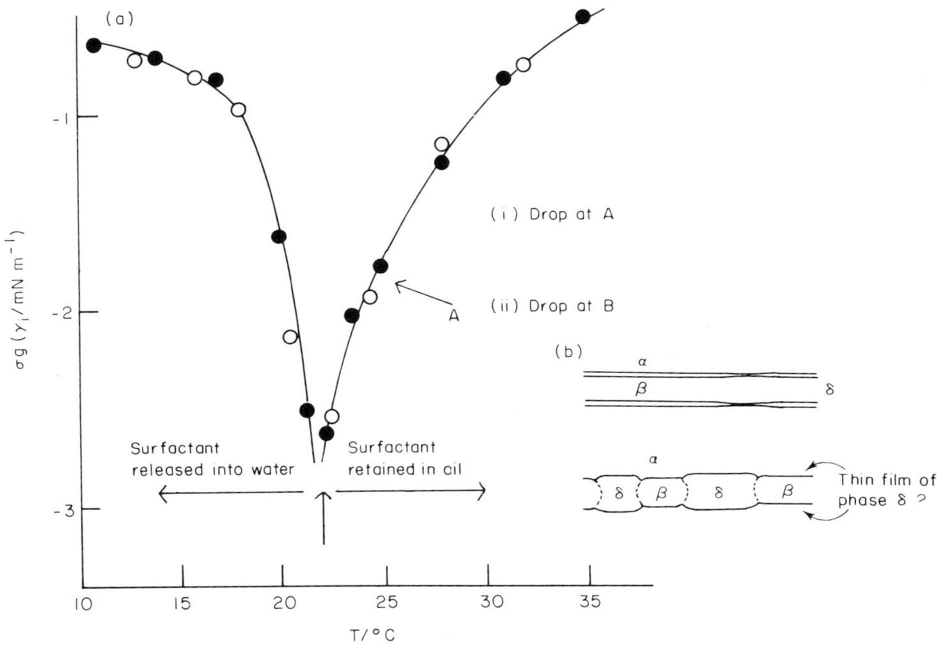

Fig. 6. Effect of temperature on γ_i between 5×10^{-4}M $C_{12}E_4$ in 5 g ℓ^{-1} aqueous NaCl and 5% W/W $C_{12}E_4$ in dodecane. (a) ● *for new oil samples at each T,* ○ *for 35°C sample cooled down.* (b) *Appearance of oil drop in spinning-drop apparatus at two temperatures.*

a third phase (δ) is formed of density intermediate between the densities of the two major (α and β) phases. This is quite evident from observation of the oil (α) drop in water (β) in the spinning-drop apparatus. As the temperature is decreased the drop becomes coated with a fluid phase δ, as shown for point A in Fig. 6. In the region of the minimum (B) the drop forms discrete segments of δ and α, the latter possibly being coated with a thin (not visible) film of the δ-phase. At temperatures below that of the minimum the surfactant resides mainly in the aqueous phase whereas at higher temperatures the surfactant is largely in the oil phase. The temperature at the minimum is evidently the phase inversion temperature (PIT) and in a small range around the PIT the surfactant is largely in the δ-phase. There is a remarkably close correspondence between these results and the equilibrium phase diagrams.

An interesting and important point, as yet unanswered, concerns the nature of the various interfaces which exhibit such low tensions. In situations such as that depicted in Fig. 6, *b*(i) the measured tension is the sum of $\gamma^{\alpha\delta}$ and $\gamma^{\beta\delta}$ both of which must be very low. In Fig. 6, *b*(ii) the α-phase may be covered by a thin film of δ and if so the same comment applies. But at some stage before the δ-phase spontaneously forms, a very low γ_i is presumably exhibited by the αβ interface containing a close-packed monolayer. We are currently investigating these and other questions.

ADSORPTION AT S/L INTERFACES

It is perhaps not surprising, in view of the practical importance in, for example, the control of colloid stability and wetting behaviour, that the literature on the adsorption of surfactants from solution onto solids is very extensive. There is a complexity of possible interactions involved in these systems since the surfactant head-group can be anionic, cationic, zwitterionic or nonionic. The solid surface proper can carry a net positive or negative charge and may be, broadly speaking, either a hydrophilic 'high energy' surface or hydrophobic 'low energy' surface. The state of the surface, and in some cases the surfactant, can be critically dependent on the solution conditions (e.g. pH, temperature). Clearly under such circumstances no unified general approach to understanding the adsorption process is available. In what follows a simple framework is presented in terms of which adsorption in any particular case might be usefully considered.

Electrical Nature of S/L Interfaces

Possibly the most commonly encountered S/L interfaces involve solids in contact with aqueous solutions and these are the only systems considered here. Often, the most important factor which governs the adsorption of ionic surfactants is the electrical interaction between ion and surface. There is generally an uneven distribution of charge normal to a S/L interface. The solid surface itself may contain ionogenic groups such as -COOH (as for example on some carbons and polymers) which at high pH will be ionized. The charge on inorganic oxide surfaces (e.g. those of Al_2O_3, TiO_2, SiO_2) is also influenced by pH, a workable model being

$$MOH_2^+ \underset{H^+}{\rightleftharpoons} MOH \underset{H^+}{\rightleftharpoons} MO^-$$

For some solids such as silver halides and $BaSo_4$ ions of one or other charge may preferentially dissolve leaving the surface with the opposite charge. Alternatively, a solid may preferentially adsorb one of the ion types from solution.

It is clear that any net charge on the solid surface (charge density σ_o) must be effectively neutralized by opposite charge in the solution close to the interface. The arrays of positive and negative charges so formed are referred to as the electrical double layer. A simple representation of the interface is shown in Fig. 7. Surfactant or other ions may be 'specifically' adsorbed (i.e. by non-electrical interactions) in which case they may be unsolvated or only partially solvated. The centres of the ions (or ionic groups) lie in the plane B, termed the <u>inner Helmholtz plane</u> (IHP), where the charge density is σ_β. If not specifically adsorbed, the centres of the (solvated) ions or ionic groups can approach the surface no closer than C, the <u>outer Helmholtz plane</u> (OHP) or <u>Stern plane</u>. Outwards, into the solution from the OHP, exists the <u>diffuse</u> part of the double layer which consists of a mixture of positive and negative ions; σ_d is the net charge in a prism, say, of unit cross-section extending out from the OHP, strictly to infinity. The distribution of ions here results from the opposition of thermal effects, tending to distribute ions evenly, and coulombic forces tending to create an uneven distribution. As a result of the requirement for electrical neutrality $\sigma_o + \sigma_\beta + \sigma_d = 0$. The condition where $\sigma_o = 0$ is called the <u>point of zero charge</u> (pzc) and when the solid surface plus adsorbed ions has zero

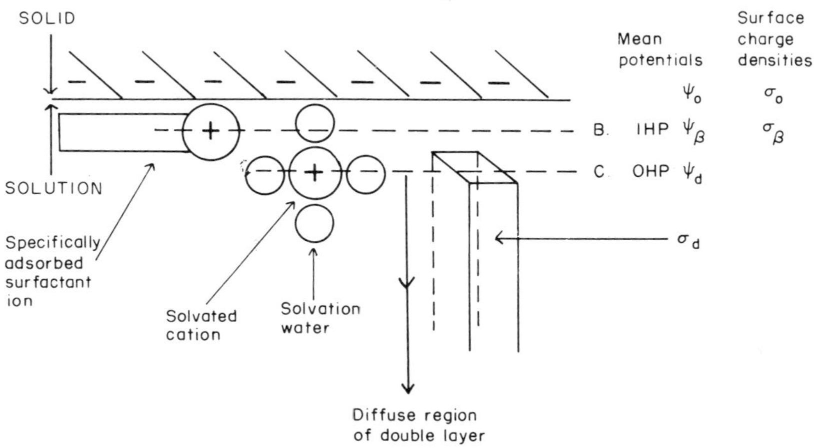

Fig. 7. Model for the electrical double layer at the S/L interface.

net charge (so that $|\sigma_o| = |\sigma_\beta|$) the isoelectric point (iep) is achieved. The potential ψ_d (Fig. 7) is equal (or very close) to the zeta potential ζ. At the iep, ζ (and ψ_d) = 0 and in the absence of specific adsorption (i.e. when $\sigma_\beta = 0$), the pzc and iep coincide (Hunter, 1981).

Measurement of Adsorption

Various methods have been used to measure the extent of adsorption but by far the most common involve the determination of the reduction in solution concentration of the surfactant upon addition of the solid. Clearly, for such a technique to be viable, the solid sample must present a sufficient area of surface to the solution for a readily determinable decrease in solution concentration to occur at equilibrium. The problem then reduces to selecting a sufficiently sensitive analytical technique for the determination of surfactant concentration (Rosen and Goldsmith, 1972).

In two-component solutions both components will be present at the interface and therefore the change in solution composition reflects the (positive or negative) adsorption of both. The isotherm representing the adsorption, termed a composite isotherm is (Aveyard and Haydon, 1973)

$$n_o \Delta x_2/m = n_2^s x_1 - n_1^s x_2 \tag{10}$$

where n_o is the total number of moles of solution before adsorption, m is the mass of the solid, Δx_2 is the change in surfactant mole fraction resulting from adsorption, n_1^s and n_2^s are the numbers of moles of solvent (1) and surfactant (2) on the surface of unit mass of solid at equilibrium, and x_1 and x_2 are solution mole fractions at equilibrium. For the dilute solutions of interest here $x_2 << x_1 \cong 1$ so to a good approximation

$$n_o \Delta x_2/m = n_2^s \tag{11}$$

In the case of adsorption of surface active ions it should be noted that n_2^s is made up from contributions from specific adsorption in the IHP (giving rise to the charge density σ_β), and adsorption in the diffuse region of the double layer (contributing therefore towards the value of σ_d). Adsorption in the diffuse region is non-specific and the surfactant ion simply competes with any other ions of the same charge. Thus, in the presence of a swamping concentration of indifferent electrolyte, σ_d has a negligible contribution from the surfactant ions. In the absence of indifferent electrolyte, however, the amount may well become significant when the charge on the substrate surface and surfactant ion are opposite. Obviously these problems do not arise for nonionic surfactants, although other difficulties can be encountered since very often multicomponent surfactants are used in commercial systems.

Interactions Involved and some Results

It will be assumed that the various contributions to the total interactions responsible for adsorption, and hence to the adsorption free energy ΔG_{ads}, are additive. Broadly, we may divide the contributions into those due to electrical interactions, giving rise to ΔG_{elec}, and all others referred to collectively as 'specific', giving rise to ΔG_{spec}. Specific interactions are sometimes referred to as 'chemical' interactions but this is not meant to imply that chemical bonds have been formed, although sometimes they may be (Parfitt and Rochester, 1983).

Electrical contributions The electrical part of the free energy may be, somewhat arbitrarily, divided into a coulombic contribution ΔG_{coul}, and one, ΔG_{dip}, arising from the reorganization of dipoles (on the water and ionic or nonionic surfactant) consequent upon adsorption. The coulombic term promotes adsorption when the surface active ions have the opposite charge to the net surface charge $(\sigma_o + \sigma_\beta)$. The IHP (Fig. 7) contains the centres of the charged head-groups at a density σ_β and mean electrical potential ψ_β. Thus, at low levels of adsorption such that $|\sigma_o| > |\sigma_\beta|$, and when σ_o and σ_β have opposite signs, ΔG_{coul} $(= ze\psi_\beta)$ is negative since z, the valence (including sign) of the surface active ion has the opposite sign to ψ_β. Similarly, if $\sigma_o + \sigma_\beta$ has the same sign as z, ΔG_{coul} is positive and (further) adsorption is opposed. Such would be the case (in the absence of other specifically adsorbed ions) for sufficiently low surface coverage of a negatively charged surface by a surface active cation. Further, adsorption is opposed electrically for oppositely charged solid surface and surface active ion when the iep is crossed. At the iep, $\Delta G_{coul} = 0$ and, neglecting contributions from ΔG_{dip}, $\Delta G_{ads} = \Delta G_{spec}$.

To exemplify in part the above considerations results for the adsorption of sodium dodecylsulphonate onto alumina as a function of pH (Fuerstenau, 1971) are shown in Fig. 8, *a*. The pzc (where $\sigma_o = 0$) is at pH = 9.1 for alumina (see above). The positive surface charge decreases as the pzc is approached and the adsorption of the surfactant anion falls to virtually zero at the pzc. Obviously in this example ΔG_{spec} is approximately zero in the range of coverage considered.

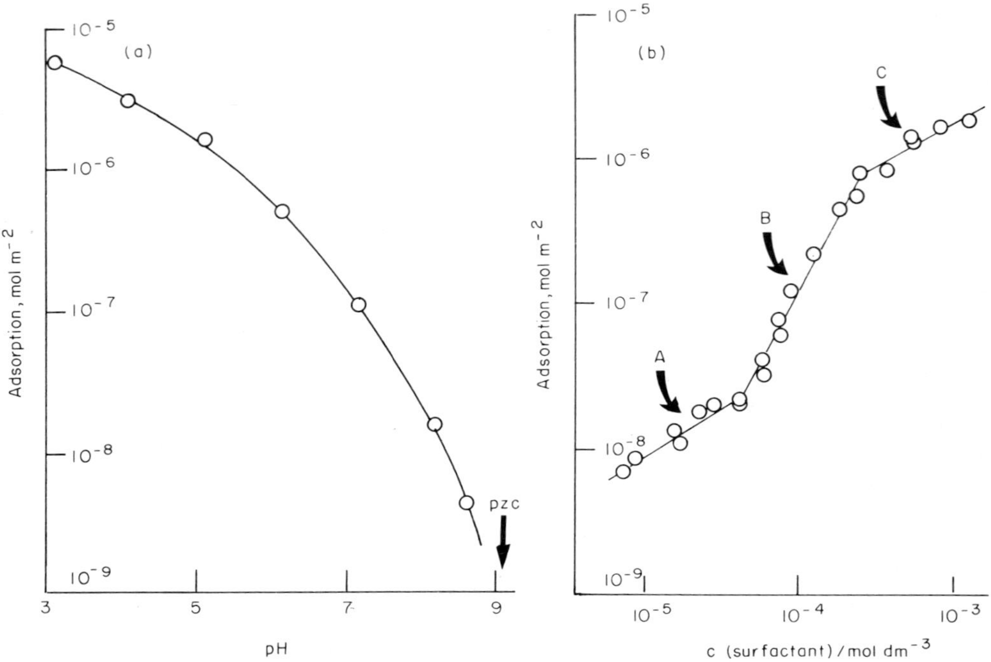

Fig. 8. Adsorption of sodium dodecylsulphonate from aqueous NaCl on alumina as a function of (a) *pH,* (b) *surfactant concentration.* (a) *Surfactant concentration = 1 × 10^{-4}M, ionic strength = 2 × 10^{-3}M.* (b) *Ionic strength = 2 × 10^{-3}M, pH = 7.2. (Results redrawn from Fuerstenau, 1971.)*

Specific contributions Contributions to ΔG_{spec} are possible from various combinations of interactions involving surfactant chain, surfactant head-group, and solid surface. At relatively high surface coverages mutual hydrophobic interactions between chains of adsorbed surfactant molecules are likely to occur. Such interactions which, as discussed earlier, are mainly entropic, are attractive and can lead to enhanced adsorption. This is illustrated in Fig. 8, *b*, again for the adsorption of sodium dodecylsulphonate on alumina (Fuerstenau, 1971). The initial (slow) rise in adsorption as surfactant concentration increases (region A) has been ascribed to simple exchange of surfactant ions with the indifferent electrolyte ions in the double layer. At higher concentrations the adsorption is seen to increase more rapidly (region B) and it is suggested that this is due to the chain - chain interactions in the adsorbed layer. At higher concentrations still (region C) the rate of increase in adsorption is reduced as a result of the onset of unfavourable electrical effects. Fuerstenau and his co-workers (1971) have suggested that chain - chain interactions lead to the formation of so-called <u>hemimicelles</u> at the surface, although the presence of such structures is not universally accepted, nor is it necessary to postulate their presence in order to understand the enhanced adsorption (Parfitt and Rochester, 1983).

Chain - surface attractive interactions are expected to be slight in the case of polar surfaces; this appears to be the case for oxide surfaces (as illustrated above where ΔG_{spec} was seen to be negligible at low coverages on alumina). For more hydrophobic surfaces, however, such as those of various polymers (e.g. polystyrene) and some carbons, hydrophobic chain - surface interactions exist and this can give sharp increases in adsorption in the initial parts of isotherms, as for example in the adsorption of alkyltrimethylammonium ions on (negatively charged) polystyrene latex particles (Connor and Ottewill, 1971). Finally there is the possibility that there are interactions, other than coulombic, between head-groups and surface. These include solvation effects, hydrogen bonding, or even covalent bonding.

It will be appreciated from all the above considerations that the adsorption of an ionic surfactant at a charged interface can be a complicated affair and observed isotherms can take on a variety of shapes depending on the exact nature of the system. With nonionic surfactants coulombic forces are absent but as mentioned many such surfactants are ill-defined mixtures, which complicates matters. For careful studies, however, some pure single-component materials are available, such as compounds of the type $C_nH_{2n+1}(OCH_2CH_2)_yOH$. The head-group can thus be much larger than is generally encountered in ionic surfactants and can exceed the size of the hydrophobic tail. In such materials the size of both chain and head-group can be varied in a regular manner. Clunie and Ingram (Parfitt and Rochester, 1983) have summarized the situation for the adsorption of well-defined nonionic surfactants. Strong adsorption at very low solution concentrations is observed on a wide variety of adsorbents. Below the c.m.c. surfactant displaces water from the surface and close to the c.m.c. a filled monolayer is achieved in which surfactant molecules are disposed with their long axes parallel to the surface. At higher solution concentrations, but in the region of the c.m.c. still, adsorption again increases and the adsorbate molecules become more closely packed. On nonpolar substrates surfactant molecules now tend to be vertically oriented with head-groups directed towards the solution so that the film covered surface becomes hydrophilic. For surfaces which originally contain hydrophilic sites, however, surfactant molecules adsorb (vertically) with their head-groups interacting with the polar sites (possibly through hydrogen bonding) and the hydrocarbon chains are directed to the water. A second layer can then be adsorbed, the two layers mutually interacting hydrophobically through their chains. The polar groups of the second layer, which are directed towards the solution, again render the final film-covered surface hydrophilic.

The writer is by no means an expert in the area of adsorption onto solid surfaces and before attempting to write the above summary he felt that the situation was hopeless but not serious; he now inclines to the view that things are serious, but not hopeless.

REFERENCES

Adamson, A.W. (1982). "Physical Chemistry of Surfaces". Wiley-Interscience, New York. 4th edn, Chapter 2.

Aveyard, R. and Briscoe, B.J. (1970). *Trans. Faraday Soc.* 66, 2911. (1972) *J. Chem. Soc. Faraday 1* 68, 478.

Aveyard, R. and Chapman, J. (1975). *Can. J. Chem.* 53, 916.

Aveyard, R. and Haydon, D.A. (1973). "An Introduction to the Principles of Surface Chemistry". Cambridge University Press. Chapters 1 and 3.

Aveyard, R. and Lawless, T.A. Unpublished work.

Aveyard, R., Briscoe, B.J. and Chapman, J. (1972). *J. Chem. Soc. Faraday 1* 68, 10.

Ben-Naim, A. (1980). "Hydrophobic Interactions". Plenum, New York.

Clint, J. *et al.* (1968). *J. Colloid Interface Sci.* 28, 522.

Connor, P. and Ottewill, R.H. (1971). *J. Colloid Interface Sci.* 37, 642.

Doe, P.H., Wade, W.H. and Schechter, R.S. (1977). *J. Colloid Interface Sci.* 59, 525.

Dorris, G.M. and Gray, D.G. (1981). *J. Phys. Chem.* 85, 3628.

Frank, H.S. and Evans, M.W. (1945). *J. Chem. Phys.* 13, 507.

Fuerstenau, D.W. (1971). "The Chemistry of Biosurfaces" (Ed. M.L. Hair), Vol. 1, Chapter 4. Marcel Dekker, New York.

Hunter, R.J. (1981). "Zeta Potential in Colloid Science". Academic Press, London.

Jones, D.C. and Ottewill, R.H. (1955). *J. Chem. Soc.* 4076.

Pal, R.P. *et al.*(1975). *J. Colloid Interface Sci.* 52, 46.

Parfitt, G.D. and Rochester, C.H. (1983). "Adsorption from Solution at the Solid-Liquid Interface". Academic Press, London.

Robb, I.D. and Alexander, A.C. (1968). *J. Colloid Interface Sci.* 28, 1.

Rosen, M.J. (1978). "Surfactants and Interfacial Phenomena".

Wiley-Interscience, New York.

Rosen, M.J. and Goldsmith, H.A. (1972). "Systematic Analysis of Surface-Active Agents". Wiley-Interscience, New York. 2nd edn.

Tanford, C. (1980). "The Hydrophobic Effect: Formation of Micelles and Biological Membranes". Wiley-Interscience, New York. 2nd edn.

EMULSIONS AND FOAMS

B. Vincent

School of Chemistry, University of Bristol,
Cantock's Close, Bristol BS8 1TS, England.

INTRODUCTION

This chapter deals with two closely-related classes of colloidal systems: dispersions of liquids in liquids (emulsions) (L/L) and dispersions of gases in liquids (foams) (G/L). Such systems occur widely in nature and in a large range of technologies (e.g. foodstuffs, pharamaceuticals, cosmetics, agricultural products, detergents). Surfactants play an important role in both their formation and their subsequent stability. It is the latter aspect which will be given most consideration here, partly because the physico-chemical processes involved are better understood and partly because it is this aspect which is perhaps more important technologically: one is, in the main, either interested in producing emulsions or foams of (controlled) stability, or breaking down unwanted ones. Examples of the latter would be the dispersed water present in crude oil, or the foams/emulsions formed in effluent streams.

In the case of emulsions most attention has been paid to water plus organic liquid systems, although there have been some studies of two immiscible organic liquids (e.g. Sharma, 1975). Similarly for foams the majority of studies have been on aqueous foams although non-aqueous foams have received growing attention recently.

For a comprehensive review on emulsions the reader is referred to the book edited by Becher (1983): "Encyclopedia of Emulsion Technology". With regard to foams, the book by Bikerman (1973) is still perhaps the most up-to-date complete text, as such, but an excellent, more recent review article

SURFACTANTS
ISBN 0-12-682180-1

on the "Dynamic Properties of Free Liquid Films and Foams" has been given by Lucassen (1981).

FORMATION

Both emulsions and foams can be formed by comminution and by condensation methods. The former involves the intimate mixing of two fluid phases, by stirring, extrusion, ultrasonics, etc. In all cases the hydrodynamic processes involved are complex and highly method-dependent; the flow conditions may be either laminar or turbulent. A steady-state situation may be reached between droplet or bubble formation, and recoalescence. Walstra (1983) has considered the various comminution processes involved in emulsion formation in the book by Becher, referred to above, and Bikerman (1973) has considered the same in the case of foams. Surfactants play a key role in these processes. In fact pure liquids will not form stable foams. Although adsorption of the surfactant at the liquid/gas or liquid/liquid interface reduces the interfacial tension (γ), and hence the amount of work ($\gamma\Delta A$) required to increase the surface area by ΔA, the primary role played by the adsorbed surfactant is its ability to sustain interfacial tension gradients at the interface which oppose film thinning and recoalescence. This subject will be a recurring theme in this chapter.

The amount of energy required to form droplets by comminution processes is not simply the thermodynamic energy. Droplets are formed by the local distortion of interfaces. Any such distortion involves a local increase in curvature and hence a local increase in the Laplace pressure difference, Δp,

$$\Delta p = \gamma[1/r_1 + 1/r_2] \tag{1}$$

where r_1 and r_2 are the principal radii of curvature of the local interfacial deformation. In laminar flow this deformation is produced by viscous forces exerted by the surrounding bulk liquid, i.e. $\eta.dv/dz$ (where η is the viscosity of the continuous liquid phase and dv/dz is the velocity gradient). To cause deformations the local viscous forces (per unit area) must be at least of the same order of magnitude as the local Laplace pressure differences. Walstra (1983) has estimated that the energy input to create the necessary velocity gradients in an emulsion could be at least 10^3 times the thermodynamic energy ($\gamma\Delta A$) referred to above. The excess energy is, of course, dissipated locally as thermal energy. Again the presence of surfactant lowers the energy requirement, through a lowering in γ.

Condensation methods of producing emulsion droplets or gas bubbles involve the creation of supersaturation conditions in a homogeneous solution (e.g. by a change of concentration, temperature or pressure). An example here would be taking the lid off a can of beer and the subsequent bubble release. Again the main roles played by any surfactant present are the lowering of the interfacial tension during nucleation and the establishment of interfacial tension gradients during growth. A reduction in γ leads to a reduction in the critical nucleus size for spontaneous growth to occur, but the expansion of the bubble or droplet is opposed by the interfacial dilational elasticity (modulus, $\varepsilon = d\gamma/d\ln A$). This effect may be small at very low surfactant concentrations where the adsorbed amount, and therefore ε, are both small and also at high surfactant concentrations ($\sim$ c.m.c.) where there is a sufficient flux of surfactant molecules towards the interface to maintain ε virtually constant during expansion of the interface and hence, ε is again small. At intermediate concentrations ε may be appreciable, however.

The relationship between droplet (or gas bubble) size and surfactant type and concentration is complex, although, in general, the higher the concentration the smaller the average size. However, droplet size is only poorly correlated with γ, as measured under static conditions.

With respect to emulsions when two liquids are mixed by comminution methods a key question concerns which phase will be the dispersed phase and which will be the continuous one. A number of factors have been identified: 1) the volume ratio of the two bulk phases; 2) the viscosities of the two phases; 3) the nature and concentration of the surfactant; 4) the temperature.

For two pure liquids factors (1) and (2) are clearly the relevant ones. When two such phases, e.g. a hydrocarbon and water, are mixed, say by stirring, a continuum structure, illustrated schematically in Fig. 1 is formed in which "blobs" of hydrocarbon in water and "blobs" of water in hydrocarbon are both present. The important criteria as to which type of emulsion (albeit transient) is observed are: (a) what type of "blobs" are more common - clearly this will be largely controlled by the volume ratio, and (b) what types of "blobs" are more persistent - this will be determined, in the main, by the rate of thinning of the thin liquid films between the blobs. This in turn depends on the viscosity of the liquid comprising the film. Abramazon and Kryzhanovskaya (1975) have studied the preferred emulsion type in a number of pure organic liquids plus water. For example, with bromobenzene (viscosity 1.89 cP) and water (viscosity 1 cP), only for water volume fractions greater than $\sim$0.95 would it appear that water-continuum droplets form preferentially.

Both effects (1) and (2) may be overridden when surfactants

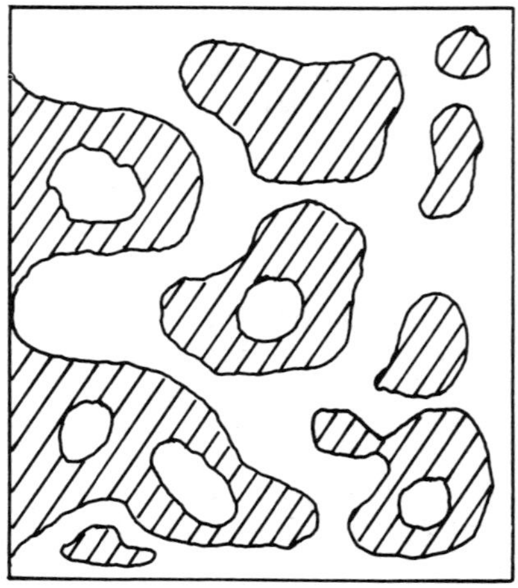

Fig. 1. Schematic representation of the bicontinuous structure produced during stirring of two immiscible liquids.

are present. Several "rules of thumb" have been proposed to predict the expected behaviour, ranging from the qualitative idea that the continuous phase will be the one in which the surfactant is more soluble, to more elaborate, semi-quantitative proposals based on the H.L.B. concept. These are useful, but not all-embracing. A more comprehensive guide is the form of the three- (or more, as appropriate) component/temperature phase diagram for the system under study. By way of illustration the phase diagram for the system water plus cyclohexane plus the nonionic surfactant $CH_3(CH_2)_8$-⟨◯⟩-$E_{8.6}$ ($E = CH_2CH_2O$), established by Shinoda and Kunieda (1970), is reproduced in Fig. 2. Here the surfactant concentration is fixed at 5 wt %. On stirring, cyclohexane-in-water emulsions tend to form in the lower two-phase region, and water-in-cyclohexane ones in the upper two-phase region. This reflects the fact that in the lower two-phase region most of the surfactant is present in the aqueous phase (in the form of micelles), whilst in the upper two-phase region most of the surfactant is present in the cyclohexane phase (in the form of inverted micelles). Clearly phase inversion may occur either by changing the temperature at a fixed cyclohexane:water ratio, or by changing the cyclohexane:water ratio at a fixed temperature (within a given range). This gives rise to the concept of a phase inversion temperature (P.I.T.) and of a phase inversion composition (P.I.C.). The location/extent of the phase inversion "region" in the phase diagram will vary somewhat with

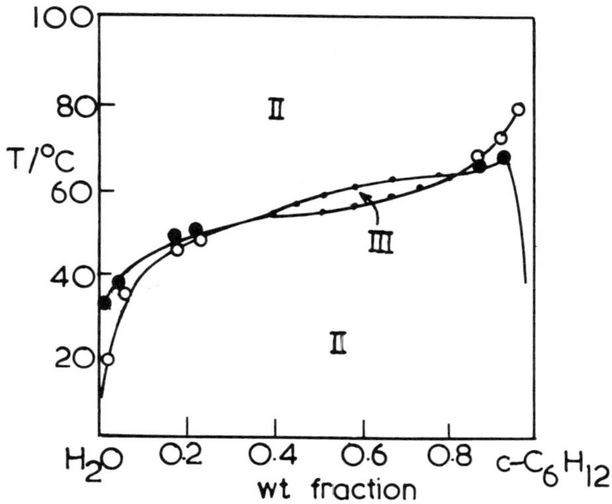

Fig. 2. Temperature-water/hydrocarbon composition phase diagram for the system: water + cyclohexane + 5 wt % $CH_3(CH_2)_8$-C_6H_4-$E_{8.6}$ *(E =* CH_2CH_2O*). Roman numerals indicate the number of coexisting phases (Shinoda and Kuneida, 1973).*

surfactant concentration, but more significantly with surfactant type. Thus, for example, an increase in ethylene oxide content of the nonylphenyl ethoxylate molecules (i.e. a higher H.L.B. value) tends to displace the phase inversion region to higher temperatures.

These phenomena can be rationalized in terms of the preferred curvature imposed on the interface by the adsorbed surfactant. The interface will tend to take up the curvature which minimizes the tangental pressure gradients in the interface; these are associated with the mutual interactions (excluded volume, van der Waals, H-bonding and electrostatic for charged surfactant head-groups) between the aqueous-phase molecules, the hydrocarbon-phase molecules, the surfactant hydrophobic moeities and hydrophilic moieties. Thus, in the case of the system depicted in Fig. 2, at low temperatures (say ∿20°C) the strong hydration of the ethylene oxide moieties leads to a preferred curvature which is concave with respect to the oil phase. With increasing temperature the hydration weakens, the extent of the curvature becomes smaller and, at a given temperature, changes sign (i.e. becomes concave with respect now to the <u>water</u> phase). Clearly in the neighbourhood of the phase inversion region, where the preferred curvature is weak, both hydrocarbon-in water and water-in-hydrocarbon droplets may be sustained, and under these conditions multiple emulsions may sometimes be formed.

The most usual method of preparing double emulsions is a two-step procedure, as described for example by Matsumoto *et al.* (1976). The first step involves the formation of a water-in-hydrocarbon emulsion using a low H.L.B. surfactant, e.g. "Span 80" (sorbitan monooleate); the second step is to mix this emulsion with an aqueous solution of a high H.L.B. surfactant, e.g. "Tween 20" (an ethoxylated derivative of sorbitan monooleate). Multiple emulsions of this type have found numerous applications, for example, in controlled drug release.

It is of interest to consider the three-phase region that separates the two two-phase regions in Fig. 2, at weight ratios cyclohexane:water $\sim$1:1. Shinoda and Kunieda (1973) and Saito and Shinoda (1970) suggest that the three phases present here are a small volume fraction of the cyclohexane phase, plus a small volume fraction of the aqueous phase, plus a very large volume fraction of what they term the "surfactant phase". This is a liquid crystalline (probably neat) phase. For the lamellar structures occurring in neat phases the interfaces are essentially flat although they may contain relatively long wavelength "ripples". Shinoda and Friberg (1975) have discussed many examples of hydrocarbon plus water plus single surfactant (or mixed surfactant) systems whose phase diagrams exhibit, in some cases extensive, liquid crystalline regions. In many cases where the hydrocarbon, aqueous and liquid crystalline phases coexist it is possible to form emulsion droplets which are surrounded by a "coating" of the liquid crystalline phase. For example, when systems such as p-xylene plus water plus a nonylphenylethoxylate surfactant were studied using a polarizing microscope, birefringent halos were observed surrounding the droplets (Friberg and Solyon, 1970).

The role of liquid crystalline structures in foam formation has received relatively little attention, but Friberg and Saito (1976), have shown, for example, that in systems such as water plus cetyltrimethylammonium bromide plus hexanol very stable foams may be formed in those regions of the three-component phase diagram where the aqueous and liquid crystalline (neat) phases coexist.

Finally, in this section on formation, we return to a discussion of the thermodynamics of emulsion formation, and the possibilities for "spontaneous" emulsification. This has already received attention in the chapter by Overbeek *et al.* on "Microemulsions" where it was shown that very low (near zero) interfacial tensions are required. It is important to stress that even if the equilibrium interfacial tension

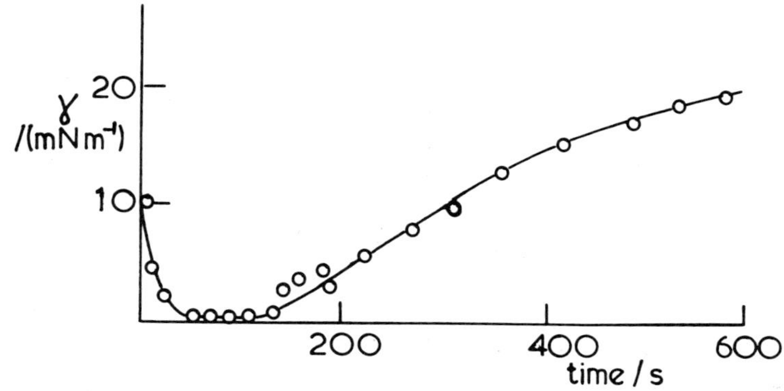

Fig. 3. Variation of the n-hexadecane/aqueous sodium dodecyl sulphate solution interfacial tension (γ) as a function of time after injection of n-pentanol into the hydrocarbon phase (Gerbacia and Rosano, 1973).

exceeds the necessary critical value, spontaneous emulsification may occur if the interfacial tension falls transiently below the critical value. Such a situation may arise, for example, when a surfactant mixture is used in which the second or cosurfactant species partitions between the two liquid phases. An example of this would be in the use of so-called "emulsifiable concentrates" in agricultural dispersions. Here the water-insoluble active species is dissolved in an oil phase, together with relatively high concentrations of a surfactant and cosurfactant (e.g. hexanol). On addition to water a portion of the cosurfactant molecules transverses the oil/water interface, transientally lowering the interfacial tension such that spontaneous emulsification can occur. This effect has been demonstrated by Gerbacia and Rosano (1973) who monitored the interfacial tension, using an automated Wilhelmy plate apparatus, between n-hexadecane and an aqueous solution of sodium dodecyl sulphate on adding pentanol to the hydrocarbon phase. The results are shown in Fig. 3. It may be seen that for a period of $\sim$1 min γ falls transiently to a very low value whilst the alcohol transverses the interface, but then rises to the equilibrium value of $\sim$20 mN m^{-1}. Spontaneous emulsification was also observed with this system.

STABILITY AND BREAKDOWN

Since, as we have seen, the formation of an emulsion or dispersion of gas bubbles in the majority of cases involves

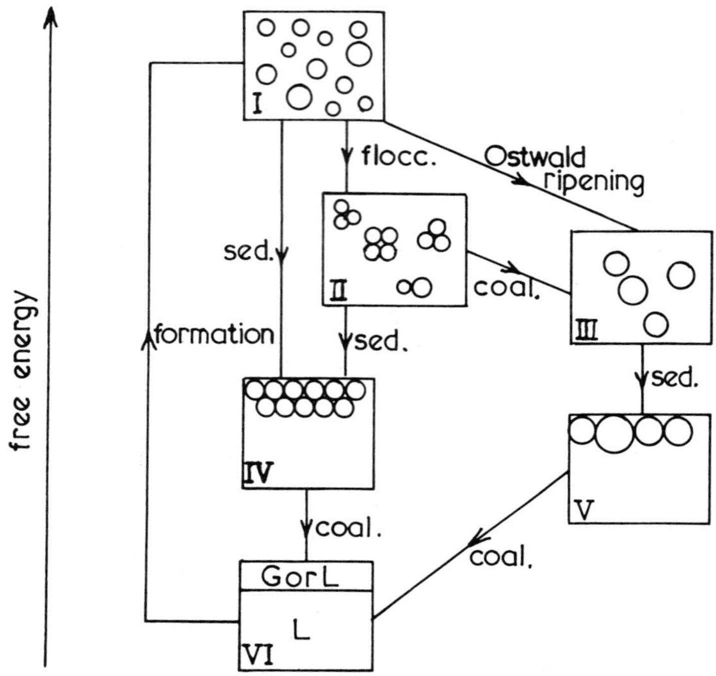

Fig. 4. Schematic representation of the breakdown processes involved in G/L or L/L dispersions.

an increase in the free energy of the system, the breakdown process, i.e. the reversion to two separate fluid phases, must necessarily be thermodynamically spontaneous. The situation is summarized schematically in Fig. 4 where some of the various possible breakdown mechanisms are depicted, i.e. molecular diffusion (Ostwald ripening), (reverse) sedimentation, flocculation and coalescence. In practice, particularly with emulsions, these procedures may occur simultaneously, as well as sequentially. This makes their study in real systems difficult. However, for discussion purposes we shall consider them separately.

Although each process, in general, leads to a decrease in free energy, as we shall see, significant energy barriers may have to be overcome so that any one of the states depicted in Fig. 4 can be a metastable state in which the system resides for a considerable period of time.

It should also be noted that Fig. 4 is not universal, in the sense that it encompasses all emulsion and foam systems.

Indeed we have already seen that in some cases, state I (i.e. the dispersed state) may be in fact the lowest free energy state (i.e. the thermodynamically stable state); this is normally the case with microemulsions, for example, at least when they are true equilibrium swollen micellar systems.

Molecular Diffusion: Ostwald Ripening

No two liquids are completely immiscible, similarly all gases have a finite solubility in liquids. Hence, in all G/L or L/L dispersions (state I, Fig. 4) transfer of molecules between bubbles or droplets through the continuous phase must be considered. Consider Fig. 5 which depicts two, initially equal size, gas bubbles or liquid droplets. The internal pressure in the droplets is the same, and given by

$$p = p_o + 2\gamma/a \tag{2}$$

where p_o is the external pressure and a is the droplet radius. Suppose that some fluctuation occurs in the system such that the left-hand droplet/bubble grows, whilst the right-hand one diminishes, by some small amount. A thermodynamically unstable situation now exists: the internal pressure in the left-hand droplet is now less than that in the right-hand one, and the left-hand one will grow spontaneously at the expense of the right-hand one by diffusion through the continuous phase. For an initially polydisperse system of droplets/bubbles this unstable situation is present at the start. This process is called "Ostwald ripening". Its rate will depend, *inter alia*, on the solubility of the gas/internal phase liquid in the external phase. An equivalent interpretation is in terms of the relative solubilities, S_1 and S_2, of the two droplets, radii a_1 and a_2, i.e. the Kelvin equation

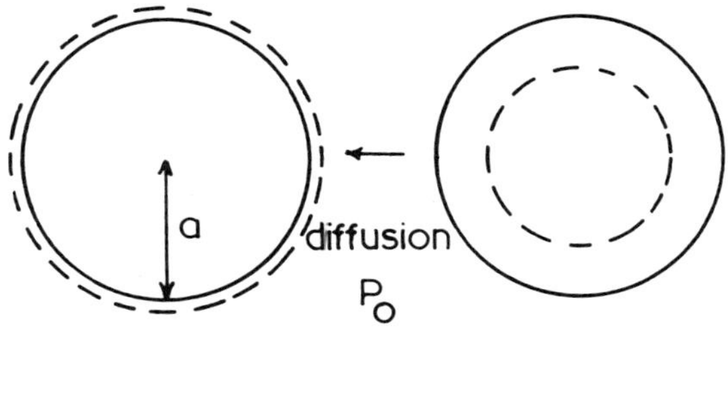

Fig. 5. Ostwald ripening through molecular diffusion.

$$\frac{RT}{M} \ln \left(\frac{S_2}{S_1}\right) = \frac{2\gamma}{\rho} \left(\frac{1}{a_1} - \frac{1}{a_2}\right) \tag{3}$$

where R is the gas constant, T the absolute temperature, and M and ρ are the molar mass and density of the solute species involved.

The presence of adsorbed surfactants not only reduces γ and therefore the pressure (solubility) difference between different size droplets/bubbles; diffusion of the solute across the interface may be sterically impeded by the adsorbed surfactant (although the flux through the continuous phase may be enhanced at surfactant concentration in excess of the c.m.c., if appreciable solubilization of the solute occurs). Again, however, it is the fact that the adsorption of surfactant can lead to interfacial tension gradients that is the main retarding force in Ostwald ripening processes. A droplet or bubble is in mechanical equilibrium if $dp/da > 0$. It follows from this condition and equation (2) that for mechanical equilibrium,

$$\frac{d\gamma}{d\ln a} > \gamma$$

or, since $A = 4\pi a^2$, that

$$\frac{2d\gamma}{d\ln A} > \gamma$$

i.e.

$$2\varepsilon > \gamma \tag{4}$$

where ε is the interfacial dilational modulus (see earlier). In the absence of surfactant condition (4) cannot be sustained ($\varepsilon = 0$), but it may be in the presence of adsorbed surfactant. We saw earlier that ε goes through a maximum with increasing surfactant concentration; at high surfactant concentrations ($\sim$c.m.c.) the flux of surfactant to the interface may be sufficient that the interface remains fully saturated with adsorbent during expansion (effectively $\varepsilon \to 0$, again).

An example of Ostwald ripening in emulsion systems has been given by Davis and Smith (1976). They studied the change in the average droplet size ($\bar{a}$) with time (t) for various hydrocarbon-in-water emulsions, at 0.2 volume fraction of the hydrocarbon, in the presence of 10^{-2} mol dm^{-3} sodium dodecyl sulphate, at 25°C. Some of the results are given in Table 1.

It may be seen that $\bar{a}_t/\bar{a}_o$ increases with increasing solubility (S_∞) in water. It is argued that flocculation and coalescence (see below) are not the primary breakdown mechanisms, since the electrophoretic mobilities (u) (and

TABLE 1

Ostwald Ripening in Hydrocarbon/Water Emulsions

	$\bar{a}_t/\bar{a}_o$*		u	a		S_∞
Hydrocarbon	t=2 days	t=10 days	$/(\mu m\ s^{-1}\ V^{-1} cm)$	$/nm^2$	$t_{\frac{1}{2}}/s$	$/(mol\ dm^{-3})$
$n-C_{16}$	1.0	1.0	8.9	0.54	3.0	2.6×10^{-8}
$n-C_{12}$	1.0	1.1	9.1	0.52	5.9	4.7×10^{-8}
$n-C_8$	1.1	1.5	9.7	0.50	14.6	5.8×10^{-6}
$n-C_6$	1.7	4.0	10.0	0.49	24.3	1.1×10^{-4}
Cyclohexane	1.7	3.5	9.7	0.51	3.0	6.5×10^{-4}
Toluene	2.1	4.8	9.4	-	8.1	5.6×10^{-3}
Benzene	2.6	5.1	8.8	0.73	2.0	2.3×10^{-2}

* $\bar{a}_o$ was $\lesssim 0.5\ \mu m$

hence the energy barrier to flocculation) are similar, as are the areas per adsorbed molecule (a), except for benzene, suggesting that the barriers to coalescence are similar in each case.

In fact, in a parallel series of experiments, in which the half-lives ($t_{\frac{1}{2}}$) of droplets at the bulk hydrocarbon/water interface were measured, the results indicated that, for the homologous series of the nC_6-C_{16} hydrocarbons, the lower members of the series were the most stable to coalescence (see Table 1).

Flocculation

The flocculation of emulsion droplets or gas bubbles (i.e. state I → state II, Fig. 4) has many features in common with that occurring in particulate dispersions and the reader is referred to the chapter by Tadros on "Suspensions" in this volume.

Let us consider what happens when two droplets of bubbles approach each other (Fig. 6). We may distinguish two cases: a) in which no deformation from spherical geometry occurs; b) where flattening (or dimpling*) of the two interfaces in the interaction zone occurs. The observed behaviour depends on the balance of forces between the droplets/bubbles, as a function of separation, and on their mutual rate of approach. The long-range forces operating are the London dispersion

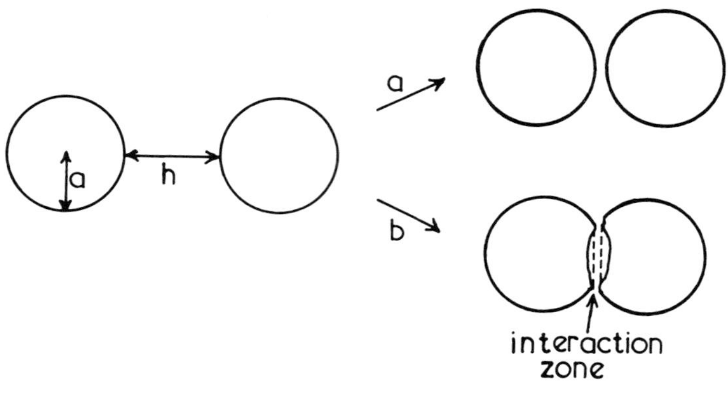

Fig. 6. The approach of two droplets or bubbles, (a) without deformation, (b) with deformation.

---

* Dimpling occurs because the rate of flow of liquid out of the film is greater at the centre than closer to the interface.

forces (electromagnetic), together with the electrical double-layer interactions (electrostatic) in the case of adsorbed <u>ionic</u> surfactants. It is the attractive forces which tend to cause distortion, since the attractive force increases the greater the contact area between the droplets. Any distortion will be opposed by bulk compressional forces (if volume changes occur, as might be the case for gas bubbles) and/or by interfacial dilational forces associated with the expansion of the interfacial area in the contact zone.

If the bubbles/droplets approach slowly enough then equilibrium conditions will be maintained (with respect to surfactant adsorption, etc.), but this may not be the case in a typical Brownian encounter. The rate of approach is governed not only by the diffusion constants of the bubbles/droplets (and hence their size, and the bulk viscosity of the liquid medium), but also by extra hydrodynamic interactions associated with the drainage of the liquid out of the interaction zone. Interfacial viscosities will be involved here. The situation is complex but clearly depends on the nature of the interfacial film involved and therefore on the type of surfactant or surfactant mixture used. Again this aspect will be considered further later.

For droplets which do not undergo significant distortion on approach, the pairwise interactive energies can be discussed in terms of the D.L.V.O. theory of colloid stability derived for particulate dispersions. Hence, the London dispersion interaction energy $V_A(h)$ is given (to a first approximation) by,

$$V_A = -\frac{Aa}{12h} \tag{5}$$

where A is the net Hamaker constant and a is the droplet/bubble radius. One form of the equation for electrical double-layer interaction energy $V_E(h)$ is given by,

$$V_E = 2\pi\varepsilon a\psi_d^2 \exp(-\kappa h) \tag{6}$$

where ε is the bulk medium permittivity, κ is the Debye-Hückel decay length and ψ_d is the potential in the Stern layer. To a first approximation ψ_d may be replaced by the electrophoretic (zeta) potential of the bubbles or droplets.

At short distances other forces may come into play. For adsorbed surfactants there may be: 1) steric forces associated with any oligomeric chains protruding into the film (e.g., ethylene oxide condensates for aqueous films having adsorbed nonionic surfactant); 2) solvation forces associated with the water molecules around the head-groups in the case of an adsorbed ionic surfactant.

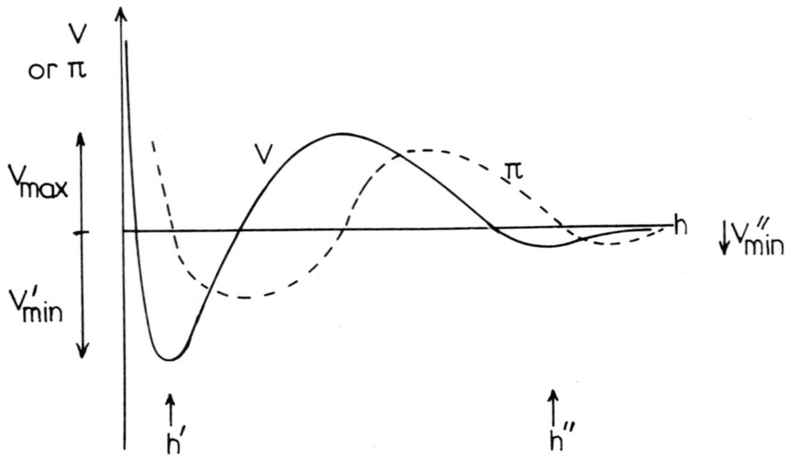

Fig. 7. Net interdroplet or interbubble interaction energy (——) or disjoining pressure (---) as a function of separation as given by D.L.V.O. theory.

The general form of the net interaction energy (strictly free energy) V(h) - droplet/bubble separation (h) curve is depicted in Fig. 7 (full line). In the absence of surface charge groups and, therefore, electrical double-layer interactions, or at high bulk electrolyte concentrations where the latter are suppressed, droplets or bubbles coagulate into the relatively deep primary minimum, V'_{min}, (e.g. for cyclohexane-in-water droplets of radius 1 μm, V'_{min} is ∿50 kT). For charged droplets, at low electrolyte concentrations, the droplets have to surmount the energy barrier (V_{max}) before coagulation into V'_{min} can occur. The height of V_{max} is controlled essentially by the surface charge density (potential) and the bulk electrolyte concentration. If V_{max} is sufficiently high (say > ∿15 kT) then coagulation into the primary minimum is effectively prevented, i.e. a metastable state is reached. However, flocculation into the secondary minimum (V''_{min}) is possible. Since V''_{min} is usually relatively shallow (say < ∿ 5 kT) flocculation of this type is generally reversible, and therefore a pseudo-equilibrium state may be attained. Srivastava and Haydon (1964; 1972) have demonstrated the existence of such flocculation equilibria in emulsion systems. The equilibrium was shown to be between singlet droplets and doublets in this case, although by analogy with the weak flocculation observed with sterically-stabilized particulate dispersions (Long *et al.*, 1973) equilibrium between singlet particles and large flocs should also be observable in some cases. There have, however, been very few studies to date of reversible flocculation with emulsion systems. The weak flocculation theory of Long *et al.* (1973) would also predict a critical droplet number concentration (depending on V''_{min} and droplet size), below which the droplets were thermodynamically stable with respect to aggregation into the secondary minimum (assuming no primary minimum aggregation occurred).

The kinetics of weak reversible flocculation have not been fully worked out, but approximate equations for the effective rate constant (k) have been given by Richmond and Smith (1975) and by Cowell and Vincent (1983). These are of the form,

$$k = k_o\ [1 - \exp(V_{min}/kT)] \qquad (7)$$

where k_o is the rate constant for irreversible flocculation in the absence of an energy barrier. k_o is typically $2\text{-}3 \times 10^{-12} cm^3\ s^{-1}$, with water as the contiuum phase (viscosity, $\eta = 1$ cP). Increasing the viscosity of the continuous phase reduces k_o and hence k.

The presence of a high concentration of surfactant in the continuum phase, particularly in liquid crystalline form, effectively increases η. For example Barry *et al.* (1968; 1970; 1971; 1972) have shown that hydrocarbon-in-water emulsions, prepared at ∿70 °C with mixtures of an ionic surfactant (anionic or cationic) and a long-chain alcohol (e.g. C_{16}, C_{18} or a mixture of the two), are very stable when cooled to around room temperature. At ∿70 °C the alcohol is largely present in the hydrocarbon phase, but on cooling it partially transfers into the aqueous phase leading to the build-up of gel-like (liquid crystalline) structures, which greatly reduce aggregation. On long-term standing, some weakening of the gel structure was observed due to partial crystallization of the alcohol molecules out of solution. Correlations between emulsion stability and the room-temperature ternary phase diagram for the ionic surfactant plus alcohol plus water system were observed.

(Reverse) Sedimentation

The accumulation of gas bubbles or liquid droplets at the top of the continuum phase leads to foam or cream formation, respectively (state IV, Fig. 4). With emulsion droplets creaming may be preceded by, or accompanied by, flocculation, coalescence or even Ostwald ripening, but this is unlikely to be the case in general for gas bubbles. A situation analogous to state IV may, of course, be achieved initially when a sufficiently large volume fraction (say >0.5) of the gas or dispersed liquid is mixed into the continuum liquid phase.

In the foam or cream the bubbles or droplets lose their spherical geometry and form closely-packed polyhedra, separated by liquid films of the continuum phase. The stability of the system then depends on the drainage/coalescence properties of these films; this will be treated in the following section. Here we consider the factors affecting the rate of cream/foam formation from dilute dispersions.

The sedimentation velocity (v) of particles in dilute dispersions is normally given by Stokes' Law, i.e.

$$v = \frac{2\ \Delta\rho\ g\ a^2}{9\eta} \tag{8}$$

where $\Delta\rho$ is the density difference between the internal and external phases and g is the acceleration due to gravity. Even for dilute L/L or G/L dispersions equation (8) has to be modified to take into account the hydrodynamic deformation, internal liquid motion, etc. In concentrated dispersions interparticle interactions also have to be considered. Moreover, in concentrated dispersions, droplets or bubbles tend to move upwards *en masse*, rather than as discrete entities, i.e. so-called "hindered settling".

Just as "structuring" of the continuum phase leads to a reduction in the flocculation rate, through an effective increase in η, so similar effects will also lead to a reduction in the (upward) sedimentation rate of bubbles or droplets (equation (8)).

Coalescence

Coalescence may occur either between aggregated droplets (bubbles) in the continuous phase (i.e. prior to sedimentation: state II to III, Fig. 4) or in the cream/foam state (state IV to VI, Fig. 4). We shall deal here specifically with the latter case, although similar principles apply to the former. Figure 8 depicts the situation at the boundary between the foam or cream, and the bulk continuous phase. Again consideration of both the static and dynamic aspects involved is required.

Static aspects The forces acting in a foam or a cream are of exactly the same type as those considered above in the pairwise flocculation of two dispersed bubbles or droplets. The disjoining pressure $\Pi(h)$ is the net force (per unit area) resulting from those forces (dispersion, electric double layer, steric, etc.) acting across the film (i.e. normal to the interfaces), and is simply defined as $\Pi(h) = -dV/dh$. The

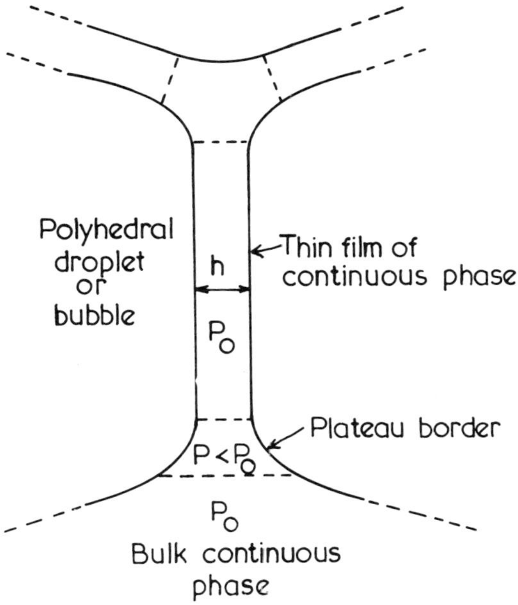

Fig. 8. Schematic representation of the boundary region between a foam or cream and the bulk continuous phase liquid.

form of this curve is shown in Fig. 7 by the dotted line. The positions corresponding to the two minima in V(h) lead to metastable states where $\Pi = 0$. The thin film may reside (at least temporarily) in either one of these states at the corresponding separation h" or h'. Π also equals 0 in the bulk region (Fig. 8) since the separation here is beyond the range of the interdroplet forces. However, in the transition region (Plateau border) between the film and the bulk solution Π is not zero.

The capillary pressure in the film itself is also zero (for plane parallel interfaces, i.e. no curvature) but will be non-zero in the Plateau border region, where it is given by equation (1). For mechanical equilibrium in the Plateau border region,

$$\Pi + \Delta p = 0 \tag{9}$$

Equation (9) neglects any contribution from gravitational forces. Although, as we shall see, these are important in the initial drainage of thick films, they become negligible once the film has thinned sufficiently such that the disjoining pressure and capillary forces become significant.

Dynamic aspects The liquid films of continuous phase formed in a cream or foam experience two stages in their lifetimes when catastrophic breakdown can occur leading to coalescence. In the early stages, whilst the films are thick, gravitational forces will be primarily responsible for their drainage. Steady drainage in a film requires a high dilational elastic modulus (ε) for the two interfaces concerned, which has to be provided by adsorbed surfactant molecules as discussed earlier. In that case solvent flow in the boundary regions of the film (i.e. close to the two interfaces) is impeded by the tendency for interfacial tension gradients to be created. Hence a velocity profile across the film results (i.e. similar to capillary flow). If ε is too low, on the other hand, drainage is too rapid resulting in film collapse and rupture.

For thick films undergoing steady drainage some idea of the drainage rate can be obtained from the arguments presented by Mysels *et al.* (1959) for a single, supported soap film. The flow rate/unit length (Q) of liquids of viscosity η and density ρ, flowing downward under gravity between two vertical, parallel, rigid plates, separation h, is given by,

$$Q = \rho g h^3 / 12\eta \tag{10}$$

and hence its mean velocity,

$$Q/h = \rho g h^2 / 12\eta \tag{11}$$

Thus, for water at 25°C, the mean velocity is 5 mm min^{-1} for h = 10 μm and 0.05 mm min^{-1} for h = 1 μm. Even the most slowly draining films (high ε) are reduced to ~10 μm very quickly (< ~1 min), but thereafter drainage is much slower.

Mysels *et al.* (1959) showed that for a supported film, with no influx of liquid from the top, the conservation equation,

$$\partial h/\partial t = -\partial Q/\partial z$$

(where z = vertical distance down the film), together with equation (10) leads to,

$$h = (4\eta z/\rho g t)^{\frac{1}{2}} \tag{12}$$

i.e. implying a parabolic shape to the film and $h \propto t^{-\frac{1}{2}}$.

Clearly in a foam or an emulsion cream consisting of a series of interconnected liquid films the situation is rather more complex; drainage will occur from the top downward. Moreover, once a given film has reached the "thin" state (characterized by interference colours and eventually a "black" appearance), further drainage is governed by the capillary forces and takes place through the interconnected Plateau border channels. It is then that the film enters its second period of danger with respect to its stability.

Instabilities in thin films, residing in either the

secondary or primary minima (i.e. at h" or h', Fig. 7), occur through random thickness fluctuations. The existence of such fluctuations in soap films has been demonstrated by light scattering (Rijnbout *et al.*, 1974). The condition for spontaneous growth of such a random fluctuation, leading to rupture, is

$$\frac{\partial \Pi_A}{\partial h} > \frac{\partial \Pi_R}{\partial h} \tag{13}$$

where Π_A is the contribution to Π arising from attractive forces and Π_R that from repulsive forces. Scheludko (1962) and later Vrij (1966) and Vrij and Overbeek (1968) showed that this leads to the concept of a critical rupture thickness (h_{crit}) for thin films. Sheludko considered the thickness fluctuations as cylindrical micro-corrugations, whilst Vrij assumed that the fluctuations were the result of independent standing sine waves whose amplitudes were time-dependent exponential functions.

Later Lucassen *et al.* (1970) extended the fluctuation theory to include the interfacial dilational elasticity (ε) arising from adsorbed surfactant molecules, and to consider the case of thin films bounded by liquids. Qualitatively, the dilational elasticity comes into play through the so-called "Gibbs-Marangoni" effect which tends to stabilize films against fluctuations and rupture. If a film is subjected to a local stretching as a result of some thickness fluctuation, the consequent local increase in surface area is accompanied by a decrease in the surface excess of adsorbed surface active agent in that region, and hence a local increase in interfacial tension, opposing the stretching (Gibbs effect). Since a finite time is usually required for surfactant molecules to diffuse to this region of the interface to restore the original surface tension (Marangoni effect), the fluctuation will tend to dampen rather than grow. Hence, rupture is prevented. As we have seen earlier, since ε often goes through a maximum value with increasing surfactant concentration, the Gibbs-Marangoni effect may be most effective at stabilizing films against rupture at intermediate surfactant concentrations.

Weakening of the Gibbs-Marangoni effect opposing coalescence is one means of effectively breaking down unwanted emulsions or foams. This may be achieved, for example, by displacing the surface active material stabilizing the films in the emulsion or foam by a more strongly surface active material but one which leads to a weaker interfacial dilational elastic modulus, ε. Thus, for example, poly(alkylsiloxanes) and polyamides are often used as antifoamers in water-boilers. Other additives, e.g. tributylphosphate, may act by adsorbing so rapidly at the interface that the local interfacial tension differences cannot be sustained in a fluctuation (Marangoni effect). Spraying of a foam with, for example, ether is often an effective way of breaking it. This presumably works because the penetration of the ether through the interface displaces the surfactant from that region, which also carries some of the underlying liquid molecules with it. Small regions of the films are thus thinned and left virtually devoid of adsorbed surfactant.

With emulsions additives such as ethylene oxide/propylene oxide multiblock copolymers are often used in the de-emulsification of water droplets in crude oils. These presumably also act through a weakening of the Gibbs-Marangoni effect.

A good example of the importance of the role of interfacial elastic properties in governing emulsion stability is afforded by the work of Saito and Shinoda (1970) who studied the effect of temperature on the stability of emulsions formed by water plus cyclohexane plus nonylphenylethoxylate nonionic surfactants of the type described earlier in the section on formation. For the system containing equal weight fractions of the two liquids plus 3 wt % of the surfactant having an average of 6.2 moles of ethylene oxide per chain (as opposed to 5% of the one containing an average of 8.6 moles of ethylene oxide considered previously, i.e. a lower H.L.B. value), the phase inversion temperature (P.I.T.) was found to be $\sim$25°C (as opposed to $\sim$55°C in the earlier case). Below the P.I.T. cyclohexane-in-water emulsions were formed, whilst above the P.I.T. water-in-cyclohexane emulsions were formed. The maximum stability to coalescence of the latter system was found to occur some 30 or so degrees <u>above</u> the P.I.T. Here presumably ε is a maximum. At very high temperatures in excess of the P.I.T. the adsorption of the surfactant is relatively sparse (rather high γ values) leading to low ε values. In the vicinity of the P.I.T., on the other hand, the adsorption is high (γ approaches very low values) but the interfacial "structure" is weak. There is no preferred curvature and hence fluctuations in curvature are easily accommodated, leading to rupture. Phase inversion, which occurs on heating an emulsion through the P.I.T., must of course, proceed via a coalescence mechanism.

Finally, another commonly used method of breaking emulsions stabilized by adsorbed surfactants involves contacting the droplets with an immersed solid surface (e.g. fibres or filters). When a droplet contacts such a surface the subsequent stability of the droplet depends on thinning of the (now asymmetric) liquid film between the droplet and the solid surface. The rupture step is now equivalent to the spreading of the internal liquid phase over the solid surface rather than coalescence. The subject of wetting is dealt with in detail in the chapter by Blake. Suffice to say here that the static aspects of the film stability are described in terms of the adsorption of the surfactants at the three interfaces involved (Young's equation), but the dynamic aspects are again controlled, at least in part, through the dilational elasticity of the liquid/liquid interface.

REFERENCES

Abramazon, A.A. and Kryzhanovskaya (1975). *Colloid J. USSR* 37, 1035.

Barry, B.W. (1968). *J. Colloid Interface Sci.* 28, 82.

Barry, B.W. (1970). *J. Colloid Interface Sci.* 32, 551.

Barry, B.W. and Saunders, G.M. (1970). *J. Colloid Interface Sci.* 34, 300.

Barry, B.W. and Saunders, G.M. (1971). *J. Colloid Interface Sci.* 35, 689; 36, 130.

Barry, B.W. and Saunders, G.M. (1972). *J. Colloid Interface Sci.* 38, 626; 41, 331.

Becher, P. (Ed.) (1983). "Encyclopedia of Emulsion Technology". Marcel Dekker, New York.

Bikerman, J.J. (1973). "Foams". Springer-Verlag, Berlin.

Cowell, C. and Vincent, B. (1983). *J. Colloid Interface Sci.* 95, 573.

Davis, S.S. and Smith, A. (1976). *In* "Theory and Practice of Emulsion Technology" (Ed. A.L. Smith), p.285, Academic Press, London.

Friberg, S. and Solyon, S. (1970). *Kolloid Z.Z. Polymere* 236, 173.

Friberg, S. and Saito, H. (1976). "Foams". Academic Press, London.

Gerbacia, W. and Rosano, H.L. (1973). *J. Colloid Interface Sci.* 44, 242.

Long, J.A., Osmond, D.W.J. and Vincent, B. (1973). *J. Colloid Interface Sci.* 42, 545.

Lucassen, J. (1981). *In* "Anionic Surfactants: Physical Chemistry of Surfactant Action" (Ed. Lucassen-Reynders) *Surfactant Science Series* 11, 217.

Lucassen, J., van den Tempel, M., Vrij, A. and Hesselink, T.Th. (1970). *Proc. Kon. Ned. Akad. Wetensch. (B)* 73, 109.

Matsumoto, Y., Kita, A. and Youezawa, D. (1976). *J. Colloid Interface Sci.* 57, 353.

Mysels, K.J., Shinoda, K. and Frankel, S. (1959). "Soap Films". Pergamon Press, Oxford.

Richmond, P. and Smith, A.L. (1975). *J. Chem. Soc. Faraday 2* 71, 468.

Rijnbout, J.B., Donners, W.A.B. and Vrij, A. (1974). *Nature* 249, 242.

Saito, H. and Shinoda, K. (1970). *J. Colloid Interface Sci.* 32, 647.
Sharma, M.K. (1975). *J. Colloid Interface Sci.* 53, 340.
Sheludko, A. (1962). *Proc. Kon. Ned. Akad. Wetensch.* 65, 87.
Shinoda, K. and Friberg, S. (1975). *Adv. Colloid Interface Sci.* 4, 281.
Shinoda, K. and Kunieda, H. (1973). *J. Colloid Interface Sci.* 42, 381.
Srivastava, S.N. and Haydon, D.A. (1964). *Trans. Faraday Soc.* 60, 971.
Srivastava, S.N. and Haydon, D.A. (1972). *Z. Physik. Chem.* 249, 101.
Vrij, A. (1966). *Disc. Faraday Soc.* 42, 23.
Vrij, A. and Overbeek, J.Th.G. (1968). *J. Amer. Chem. Soc.* 90, 3074.
Walstra, P. (1983). *In* "Encyclopedia of Emulsion Technology" (Ed. P. Becher), Chapter 2, p.57. Marcel Dekker, New York.

SUSPENSIONS

Th.F. Tadros

I.C.I. Plant Protection Division,
Jealott's Hill Research Station,
Bracknell, Berkshire RG12 6EY, England.

INTRODUCTION

Surfactants, both ionic and nonionic, play a major role in the preparation of solid in liquid dispersions (suspensions), the control of their properties and the maintenance of their long-term physical stability (Tadros, 1980). Suspensions, particularly concentrated ones, are the end product of many industrial preparations: for example, paints, dyestuffs, printing inks, cosmetics, agricultural and food products, pesticides, pharmaceuticals, etc. These systems cover a wide range of particle dimensions from the colloid range (1nm - 1μm), i.e., colloidal suspensions, to coarse suspensions (>1μm) which can be tens or even hundreds of μm in size. Both types of suspensions may undergo a number of physical changes on standing, which lead to the formation of secondary units from primary particles. Various types of aggregates may form which may be classified into tight clusters (particles strongly held together by van der Waals forces), loose metastable structures usually referred to as flocs or floccules, and agglomerates. On the other hand, suspensions which are kept stable in the colloid sense (as a result of strong repulsive forces between the particles) may undergo settling, forming compact layers (referred to as "cakes" or "clays") when the particles are more dense than the medium and have a size that is outside the colloid range.

The maintenance of particles in a suspension as individual primary units or their combination into aggregates, agglomerates or flocs depends on the interaction forces that operate between them. Moreover, the process of settling under gravity in a concentrated suspension also depends on interparticle

SURFACTANTS
ISBN 0-12-682180-1

forces, as does the final state of the sediment (loose or compact). The interparticle forces are governed to a large extent by the nature of the particles and their surfaces. It is clear that surfactants which are either incorporated into the particles, or simply adsorbed on their surface, will play a major role in the control of the interaction forces between the particles.

In this chapter, the role of surfactants in the preparation of suspensions will first be described. This is then followed by a section on their application in the control of physical stability. Three aspects will be considered, namely, stability in the colloid sense, caking or claying, and the role played by surfactants in Ostwald ripening. Finally, the role of surfactants in controlling the flow characteristics (rheology) of suspensions will be briefly described.

ROLE OF SURFACTANTS IN SUSPENSION PREPARATION

Generally speaking there are two main procedures for the preparation of suspensions. The first depends on the "build-up" of particles from molecular units, i.e., the so-called "condensation" methods, which involve two main processes: nucleation and growth. In this case it is necessary first to prepare a molecular (ionic or atomic) distribution of the insoluble substances; then by changing the conditions precipitation is caused leading to the formation of nuclei that grow to the particles in question. In the second procedure, usually referred to as a dispersion process, larger "lumps" of the insoluble substance are subdivided by mechanical or other means, into smaller units. The role of surfactants in the preparation of suspensions by the above mentioned procedures is briefly described below.

Role of Surfactants in Condensation Methods

To understand the role of surfactants in the condensation method, it is essential to consider the major processes involved, namely, nucleation and growth. Nucleation is the spontaneous process of the appearance of a new phase from a metastable (supersaturated) solution of the material in question (Gibbs, 1906). The initial stages of nucleation result in the formation of very small nuclei where the surface to volume ratio is very large and hence the role of specific surface energy is very important. With the progressive increase of the size of the nuclei, this ratio becomes small and eventually larger crystals appear, with a corresponding reduction in the role played by the specific surface energy. It is possible to distinguish between homogeneous nucleation where the nuclei form solely as a result of supersaturation and heterogeneous nucleation induced by the presence of foreign materials. It is clear that the addition of surfactants, which can either adsorb on the surface of a nucleus or act as centres for inducing nucleation, can be used to control the process of nucleation and the stability of the resulting nuclei. This is due to their effect on the specific surface energy on the one hand and their ability to incorporate the material in the micelles on the other.

According to Gibbs (1906) and Volmer (1939), the free energy of formation, ΔG, of a spherical nucleus, is given by the expression

$$\Delta G = 4\pi r^2 \gamma - \frac{4\pi r^3 \rho}{3M} RT \ln (S/S_o) \qquad (1)$$

where r is the radius of the nucleus, γ is the specific surface energy, M is the molecular weight, ρ is the density and (S/S_o) is the relative supersaturation, R the gas constant and T is the absolute temperature. Equation (1) clearly shows that the free energy of formation of a nucleus depends on two main parameters, γ and (S/S_o), both of which are influenced by the presence of surfactants; γ is influenced in a direct way by the adsorption of surfactant on the surface of a nucleus, whereas (S/S_o) could be affected by the presence of surfactant micelles in the supersaturated solution.

Surfactants play a major role in the preparation of suspensions of polymer particles by heterogeneous processes. This may be illustrated by considering the various procedures used, namely, emulsion, dispersion and suspension polymerization. In emulsion polymerization, the monomer is emulsified in a non-solvent (usually water) using a surfactant, whereas the initiator is dissolved in the continuous phase. The role of surfactants in this process is obvious since nucleation occurs in the monomer swollen soap micelles. Indeed the number of particles formed and their size depend to a large extent on the nature of the surfactant used and its concentration (which determines the number of micelles formed) (Blakely, 1975). Dispersion polymerization differs from emulsion polymerization in that the reaction mixture, consisting of monomer, initiator and a solvent (aqueous or nonaqueous) for both is usually homogeneous, but as polymerization proceeds polymer separates out and the reaction continues in a heterogeneous manner (Barrett, 1975). Usually a "protective agent" such as a polymeric surfactant or block or graft copolymer is added to stabilize the particles once formed. Again the number of polymer particles formed in a given volume of polymer dispersion depends directly on the concentration of surfactant or dispersant used. Clearly, the properties of the final dispersion depend on the nature and concentration of the surfactant used to stabilize the particles formed. Usually dispersants are specifically designed to achieve the desired properties. The most effective materials are those of the block and graft type consisting essentially of two polymeric compounds, one soluble and one insoluble in the continuous medium. The soluble part provides the stabilizing moiety (steric stabilization) whereas the insoluble group (Barrett, 1975) forms the "anchoring" component. The size of the macromolecular surfactant to be designed depends on a number of requirements which need to be satisfied for effective dispersion. It should be mentioned that graft dispersants can form aggregates or micelles in the dispersion medium as a result of self-association of the anchoring component. As long as such a micellar solution is stable and no precipitation occurs, the dispersant remains effective, although its activity may be influenced by the extent of association. With an anchor chain with a molecular weight of $\sim$1000 the minimum molecular weight of the soluble chain should be between 500 and 1000 (Barrett, 1975).

Suspension (pearl, bead) polymerization is similar to emulsion polymerization, except the initiator is dissolved in the emulsion monomer droplets. Thus the monomer droplets are gradually converted into insoluble polymer particles but no new particles are formed in the continuous phase. This process usually leads to the formation of coarse (>1μm) suspensions. As with emulsion polymerization, the role of surfactants in this process is crucial.

Role of Surfactants in Dispersion Methods

Dispersion methods are used for the preparation of suspensions of preformed particles. The term dispersion is used to refer to the complete process of incorporating the solid into a liquid such that the final product consists of fine particles distributed throughout the dispersion medium. Dispersion methods usually lead to suspensions with a wide distribution of particle sizes. In most cases the average size is outside the colloid range (>1μm), although of course in the comminution process a significant proportion of fine particles (<1μm) is also formed.

The role of surfactants in the dispersion process can be seen from a consideration of the stages involved (Patton, 1964). Three steps have been considered by Parfitt (1967; 1981): wetting of solid powder by the liquid; breaking of the aggregates and agglomerates; and comminution of the resulting particles. It is obvious that the particles formed should be maintained in the dispersed state during the above-mentioned processes.

Wetting of Powders Powders of organic or inorganic materials are produced by various manufacturing processes in which various drying methods are used, e.g., oven drying, spray drying, fluid-bed drying, etc. In some cases the powder is also subjected to a dry grinding process producing particles of various sizes and shapes. Generally, the particles in a dry powder either form aggregates in which the particles are joined together at their crystal faces or agglomerates in which the particles touch only at edges or corners, forming looser, more open structures.

The wetting of a powder by a liquid is the process in which the solid - air interface is replaced by a solid - liquid interface. For complete wetting, it is essential to replace all the adsorbed air and other contaminants. Moreover, it is also essential to wet the external and internal surfaces and displace any air entrapped between the particles or present in the pores of the agglomerated particles. The surface of a solid is usually rough, containing many irregularities, which have a major influence on its wetting. Moreover, many powders are porous in nature, containing pores of various sizes.

The wetting of a solid by a liquid is described in terms of the equilibrium contact angle of the liquid on the surface and the appropriate surface tensions (see Chapter 10). For the complete wetting of a solid particle by a liquid one may distinguish between three types of wetting process: adhesional wetting, W_a, when 1 cm^2 of the surface is brought into contact with 1 cm^2 of the liquid; immersional wetting, W_i, when the solid particles are immersed in the liquid; spreading wetting, W_s, when the liquid spreads over the surface of the partially immersed particles. The work of dispersion, W_d, is given by the sum of W_a, W_i and W_s, i.e.,

$$W_d = W_a + W_i + W_s \tag{2}$$

and from a consideration of the interfacial tensions and application of the Young equation (see Chapter 10),

$$W_d = -6 \gamma_{LV} \cos \theta \tag{3}$$

where γ_{LV} is the surface tension of the liquid (liquid/vapour interface) and θ is the contact angle of the liquid on the surface of the solid.

Thus, the wetting of a solid by a liquid depends on two measurable parameters, γ_{LV} and θ. In practice, since surface active agents are added in sufficient quantities, θ is often very close to zero and spontaneous dispersion of a powder in a surfactant solution is the rule rather than the exception. The addition of the surface active agent also causes a reduction in γ_{LV} and if adsorbed on the surface of the solid also reduces γ_{SL}. Both effects result in enhanced wetting.

The next stage to consider is the wetting of the internal surface, which implies penetration of the liquid into channels between and inside the agglomerates. This process is more difficult to define, although one may make use of the equations derived for capillary phenomena to predict whether penetration is spontaneous or non-spontaneous. To force a liquid into a capillary tube of radius r, a pressure P is required such that,

$$P = (-2\gamma_{LV}\cos\theta)/r \quad = \quad -2(\gamma_{SV} - \gamma_{SL})/r \tag{4}$$

where θ is the contact angle of the liquid with the capillary wall. Hence, penetration is spontaneous as long as $\theta < 90°$. Equation (4) shows that for spontaneous penetration, γ_{SL} has to be made as small as possible by sufficient adsorption of surface active agents. However, when $\theta = 0$, P is proportional to γ_{LV}, i.e., a large surface tension is desirable. Changes in θ accompany the increase in γ_{LV} and assuming the only variable tension to be γ_{LV}, P will tend to remain constant and the limiting value increases proportionally to $(\gamma_{SV}-\gamma_{SL})$. Hence it is desirable to maximize $(\gamma_{SV}-\gamma_{SL})$ and keep γ_{LV} as small as possible. Since addition of surface active agents affects both γ_{LV} and γ_{SL}, the best agent for providing such opposing effects may be difficult to find and will vary from one system to another.

Another important factor in the wetting process is the kinetics of penetration of the liquid into the channels between and inside the agglomerates. The rate of penetration has to be as fast as possible to aid the dispersion process. The rate of penetration of liquid into a powder can be qualitatively treated by the Rideal-Washburn equation (1921; 1922) assuming the powder to be represented by a bundle of capillaries, i.e.,

$$\frac{d\ell}{dt} = \frac{r\ \gamma_{LV}\ \cos\ \theta}{4\eta\ell} \tag{5}$$

where ℓ is the distance the liquid has travelled along the pore at time t, r is the radius of the capillary tube and η is the viscosity of the liquid. Thus, the rate of penetration is directly proportional to the radius of the capillary tube, $\gamma_{LV}\cos\theta$ and inversely proportional to η. Thus, to aid penetration, γ_{LV} should be made as high as possible and θ as low as possible. However, high γ_{LV} and low θ are normally incompatible and a low contact angle is certainly very important. Moreover, to increase penetration the powder has to be made as loose as possible.

Breaking of Aggregates and Agglomerates As mentioned above, most powders consist of clusters of particles which may be joined at their crystal faces forming compact aggregates or at their corners or edges forming agglomerates. In the latter case, the particles are held together by weak forces and therefore require little energy to disperse. In some cases, the penetration of the liquid into the channels between the particles in agglomerates may provide sufficient pressure to bring about the disintegration of the agglomerates. If this is not sufficient, the breaking of agglomerates may be achieved by the use of high shear mixers. However, with aggregates, the bonding forces between the original primary particles is so strong that the aggregates can only be disrupted by considerable forces. Moreover, many materials that are produced by precipitation may be "cemented" together and in all cases aggregated particles require mechanical methods for their disruption. This will be discussed below under the heading comminution.

Comminution of Particles As mentioned above, mechanical energy is required to destroy the aggregates and break down the single crystals into smaller units. Comminution or milling (the generic term for size reduction) is a complex process and there is little fundamental information on the mechanism of size reduction. Any comminution process involves rupturing of chemical bonds to create new surfaces and, therefore, any process that enhances rupturing of bonds will facilitate size reduction. It has also been argued that reduction of surface energy may facilitate comminution, although this cannot be easily proved in practice since in any wet milling process the energy input is far greater than that required to create a new surface. However, it has been argued that reduction in surface energy retards the rejoining of the ruptured surfaces and prevents particle agglomeration, both of which retard the process of grinding. The role of surface active agents in the process of comminution has been considered by Rehbinder and co-workers (1957; 1958) who claimed that adsorption of surface active agents at structure defects in the surface (which results in a reduction of the interfacial energy at the boundary between solid and medium) should aid the process of deformation or destruction of that surface. Unfortunately, there are not sufficient experimental data to prove or disprove this effect (which is sometimes referred to as the Rehbinder effect).

ROLE OF SURFACTANTS IN CONTROLLING THE PHYSICAL STABILITY OF SUSPENSIONS

As mentioned in the Introduction, three main processes need to be controlled in a suspension. The first and most important is the control of particle aggregation which determines the final properties of the suspension such as its flow characteristics, its behaviour on application, and its state on standing. Particle aggregation is determined by the interaction forces that exist between them which may be classified into electrostatic, van der Waals and steric forces. Depending on the magnitude of these forces, various states may be reached varying from totally deflocculated (where the repulsive forces exceed the attractive forces by orders of magnitude) to weakly flocculated and strongly flocculated (coagulated) systems. The second process that needs to be controlled is that of "caking" or "claying" which occurs with coarse suspensions when these are stable in the colloid sense. Finally, with suspensions with a wide distribution of particle sizes, a process in which there is a shift in distribution towards larger sizes (as a result of the preferential dissolution of small particles and the subsequent growth of the larger ones) i.e., the so-called Ostwald ripening process, needs to be controlled. A brief account of these three major processes is given below.

Control of Particle Aggregation

The maintenance of particles in a suspension as individual primary units and their combination to form secondary units such as aggregates, agglomerates, or flocs depend on the interaction forces between them. The various states that need to be reached depend on the stages encountered and the final property that is needed for application and long-term physical stability. For example, during the preparation of a suspension by condensation or dispersion methods, it is necessary to keep the particles as individual units, since their aggregation may lead either to handling problems (as a result of an increase of viscosity) or prevention of further processes, e.g., size reduction. On the other hand, having prepared a stable suspension it may be necessary in some cases to reduce the stability somewhat depending on the subsequent processing stages. For example, a state of weak flocculation is needed to prevent claying or caking with coarse suspensions. Also, if the particulate solid needs to be separated from the medium, it is necessary to induce aggregation either to increase the sedimentation rate or for ease of filtration. All these processes may be controlled by the use of surface active agents of various types. This is best understood by discussing the various interactions described below.

Electrostatic Interactions Electrostatic interactions result when using ionic surfactants of the anionic, cationic and zwitterionic type. As a result of their adsorption on the particle surfaces, an electrical double layer develops at the solid/solution interface (see Chapter 7). The double layer consists of the surface charge resulting from the adsorbed surfactant ion (e.g., SO_4^- charges when using sodium dodecylsulphate or $-N^+$ charges when using tetra-alkyl ammonium salts) and a compensating diffuse layer of ions protruding into the bulk solution.

When two particles, each carrying a diffuse layer of counterions, approach each other, e.g., in a Brownian encounter, repulsion results from overlap of the diffuse layers. As a result of this interaction the free energy of the system increases, i.e., the interaction of the double layers will give rise to a repulsion between the surfaces of the particles. For spherical particles, the repulsive energy is given by the following approximate expression (due to Derjaguin, 1939), when the range of interaction is small compared with the radius of the particles,

$$G_E = 2\pi a\ \varepsilon_r\ \varepsilon_o\ \psi_o^2\ \ln\ [1 + \exp\ (-\kappa h)] \tag{6}$$

where ε_r is the dielectric constant of the medium, ε_o is the permittivity of free space, ψ_o is the surface potential, h is the surface - surface separation, and κ is the reciprocal of the "thickness" of the double layer, given by

$$\kappa = \left(\frac{2Z^2e^2C}{\varepsilon_r \varepsilon_o kT}\right)^{\frac{1}{2}} \tag{7}$$

where Z is the valency of the counterions, C the electrolyte concentration in bulk solution, e the electronic charge, k the Boltzmann constant and T the absolute temperature.

Equation (6) is valid only for low surface potentials (i.e., $Z\psi_o < 25$ mV), for spheres whose radius is large in comparison to κ^{-1} (i.e., $\kappa a \gg 1$), and for weak interactions between the spheres (i.e., $\kappa h > 1$). It is clear from equation (6) that G_E depends on ψ_o, a and κ. The surface potential is determined by the number of surfactant ions adsorbed per unit area. The higher ψ_o, the higher the repulsion. On the other hand, G_E is reduced in magnitude by an increase in κ (i.e., increase in electrolyte concentration).

It should be mentioned that G_E is positive as long as the diffuse layers are of the same sign. If these layers are of opposite sign, then G_E becomes attractive. This is the case, for example, when a dispersion stabilized by an anionic surfactant is mixed with one stabilized by cationic surfactant. As a result of double layer attraction, flocculation occurs (usually referred to as heteroflocculation) and the phenomenon may be used, for example, for solid/liquid separation.

Van der Waals Interaction The van der Waals interactions between molecules are of three main types: dipole - dipole (Keesom) interactions, dipole - induced dipole (Debye) interactions, and dispersion (London) interactions. The dispersion forces arise from charge fluctuations within a molecule associated with the motion of its electrons. Although the London force between two molecules is short range (being inversely proportional to the seventh power of their separation), the net force between them is of much longer range.

For spherical particles of radius a *in vacuo* the van der Waals interaction energy G_A is given by the approximate expression (Hamaker, 1937),

$$G_A = -\frac{A_{11}a}{12\ h} \tag{8}$$

which is applicable to the case h << a. A is the Hamaker constant and is given by the expression,

$$A_{11} = \pi^2 q_1{}^2 \beta_{11} \tag{9}$$

where β_{11} is the London interaction constant for two molecules of type 1, q_1 is their number density; β_{11} is related to the polarizability of the atoms or molecules concerned (it has a value in the range 10^{-78} to $10^{-76}\mathrm{Jm}^6$).

Equation (8) refers to particles *in vacuo*. If they are suspended in a continuous phase medium of material 2, then the net interaction is reduced. In this case, the Hamaker constant must be replaced by the net Hamaker constant A_{12}, i.e.,

$$A_{12} = A_{11} + A_{22} - 2A_{12} \cong (A_{11}{}^{\frac{1}{2}} - A_{22}{}^{\frac{1}{2}})^2 \tag{10}$$

In the presence of an adsorbed surfactant layer, the van der Waals interaction has to be further modified since the layer may have a Hamaker constant, A_{33}, which is different from that of the medium and/or the particle. In this case (Vold, 1961),

$$G_A = -\frac{1}{12}\left[(A_{22}{}^{\frac{1}{2}} - A_{33}{}^{\frac{1}{2}})^2\left(\frac{a+\delta}{a}\right) + (A_{33}{}^{\frac{1}{2}} - A_{11}{}^{\frac{1}{2}})^2\left(\frac{a}{(\Delta+2\delta)}\right) + \frac{4a\ (A_{22}{}^{\frac{1}{2}} - A_{33}{}^{\frac{1}{2}})\ (A_{33}{}^{\frac{1}{2}} - A_{11}{}^{\frac{1}{2}})\ (a+\delta)}{(\Delta+\delta)\ (2a+\delta)}\right] \tag{11}$$

(where δ is the thickness of the adsorbed surfactant layer and Δ is the surface - surface separation of the layers). It is clear from equation (11) that the effect of the surfactant layer on G_A depends on the relative magnitude of A_{11}, A_{22} and A_{33} (Vincent, 1973). For the case $A_{11} > A_{22}$ (which is the case in most situations) the presence of a surfactant layer leads to a reduction in G_A when $A_{33} > A_{22}$. However, when $A_{33} < A_{22}$ then the adsorbed layer results in a net increase in G_A.

Steric Interactions Steric interactions result when using long chain surfactants which form "thick" adsorbed layers. These arise when using nonionic surfactants of the ethoxylate type, i.e. R-(CH_2CH_2O)-OH, containing several ethylene oxide units or macromolecular surfactants of the A-B or A-B-A block type, e.g., Pluronics HO-$(CH_2CH_2O)_X$-$(CH_2$-$CH(CH_3)$-$O)_Y$-$(CH_2CH_2O)_X$-OH.

In this case steric interactions arise as soon as the adsorbed layers interfere with each other. This is schematically shown in Fig. 1.

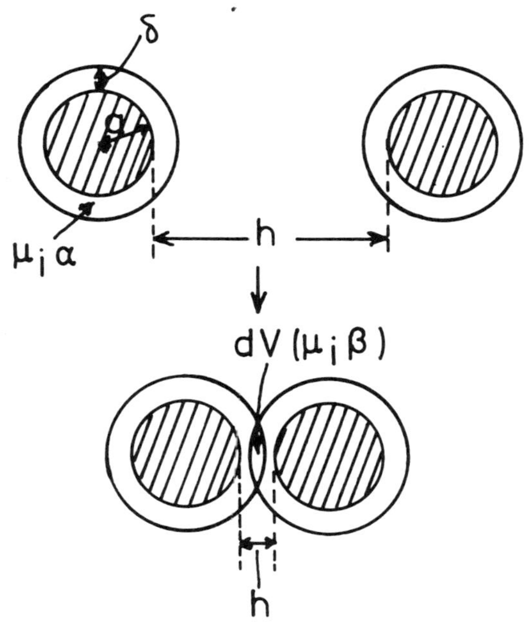

Fig. 1. Schematic representation of overlap of adsorbed layers.

There are two contributions to steric interactions. The first contribution arises from the build-up of segment concentration in the interaction zone which leads to an increase in local osmotic pressure and free energy. In the overlap region dV (Fig. 1) the chemical potential of the solvent μ_i^β is now lower than that in the rest of the adsorbed layer μ_i^α, since its activity or concentration is less. Solvent molecules will, therefore, tend to diffuse from the bulk solution, tending to force the particles apart. This contribution to steric interaction is usually referred to as the osmotic or mixing term, G_M. Clearly G_M operates as soon as the adsorbed layers overlap, i.e., when $h < 2\delta$. With further overlap compression of the chains may occur and the configurational entropy of the chain is reduced since the total volume available to the chain is reduced. This is especially the case when $h < \delta$. This interaction (which constitutes the second contribution to G_S) is known as the volume restriction G_{VR} or elastic term G_{EL}. Thus,

$$G_S = G_M + G_{VR} \tag{12}$$

Various theories exist for the evaluation of G_M and G_{VR} and these have been recently reviewed (Vincent and Whittington, 1982; Tadros, 1982; 1983). For the case of adsorbed surfactant layers of uniform segment concentration in the adsorbed layer, simple expressions for G_M and G_{VR} can be derived. For example, G_M can be adequately calculated using the simple Fischer theory (1958)

$$\frac{G_M}{kT} = \frac{4\pi\phi_2^2}{3\,v_1}\,(\tfrac{1}{2} - \chi)\,(\delta - \frac{h}{2})^2\,(3a + 2\delta + \frac{h}{2}) \tag{13}$$

where ϕ_2 is the volume fraction of surfactant in the adsorbed layer, v_1 is the volume of a solvent molecule and χ is the chain-solvent interaction parameter. It is clear from equation (13) that G_M is positive as long as $\chi < 0.5$, i.e., the chains are in good solvent conditions. When $\chi = 0.5$, i.e., under the θ conditions for the chain, $G_M = 0$, whereas when $\chi > 0.5$, i.e., under worse than θ conditions, G_M is negative and the mixing interaction is attractive.

G_{VR} can be calculated, for simple surfactant chains, using the Mackor theory (1951) which has been further modified by Mackor and van der Waals (1952) to take into account the lateral interaction between the chains. At low coverage,

$$G_{VR} = N_s\,kT\,\theta_\infty\,[1 - (h/\delta)] \tag{14}$$

where N_s is the number of adsorbed chains per unit area and θ_∞ is the fraction of surface covered by an adsorbed layer with the surfaces at infinite separation.

Total Interaction and Description of the State of the Suspension The total interaction of particles containing adsorbed surfactant layers is given by

$$G_T = G_E + G_A + G_S \qquad (15)$$

Two general cases may be considered, the first where there is little contribution from G_S, i.e., the case when using ionic surfactants. In this situation G_T is given by the sum of G_E and G_A in the well-known theory of Derjaguin, Landau, Verwey and Overbeek (DLVO theory) (1948). The form of the energy-distance curve is illustrated in Fig. 2, which shows the various contributions.

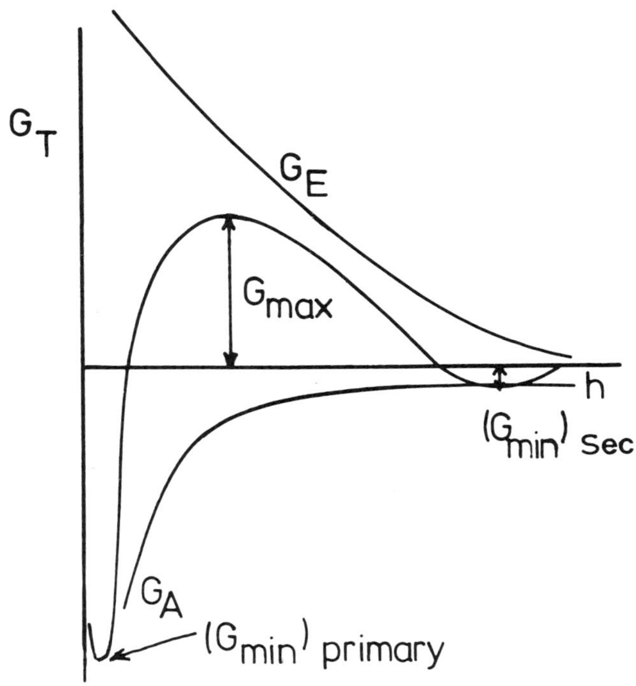

Fig. 2. Total interaction energy-distance curve according to the DLVO theory.

The energy-distance curve is characterized by three main parameters: G_{max}, $(G_{min})_{primary}$ and $(G_{min})_{sec}$ the magnitudes of which are determined by the surface potential (or the measured ζ potential), electrolyte concentration, particle size and Hamaker constant. Depending on the relative magnitude of the parameters, various states of the suspension are reached. For example, if G_{max} is high (>> kT) and $(G_{min})_{sec}$ is shallow, a deflocculated suspension is produced. This is the case when the surface charge is high (as a result of adsorption of ionic surfactant molecules oriented perpendicular to the surface) and the electrolyte concentration is low. On the other hand, if G_{max} is small (< 5-10 kT) or absent, the suspension becomes coagulated (since the particles are able to surmount the energy barrier and enter the deep primary minimum). This is the case when there are only few surfactant molecules (which are probably parallel to the interface) on the surface and/or in the presence of a high concentration of electrolyte. The third case is that in which G_{max} is reasonably large (> 10 kT) whereas $(G_{min})_{sec}$ is fairly deep (1-5 kT). This results in a state of weak flocculation, whereby the particles are now held by a relatively weak attraction at separation distances around the location of $(G_{min})_{sec}$. This may arise with coarse anisometric suspensions at intermediate electrolyte concentrations. Under these circumstances, the suspension may form a three-dimensional gel network structure, which can be fluidized by shaking. This state may be used for the prevention of caking.

The second case to consider is that in which there is a contribution from G_S as, for example, is the situation when using nonionic or macromolecular surfactants. Here G_S dominates the interaction for $h < 2\delta$. At separation distances larger than 2δ, there is no contribution from G_S and the interaction is determined by G_E and G_A. The electrostatic interaction arises from the residual double layer interaction (as a result of the residual charges on the particle surface) which is considerably modified by the presence of adsorbed layers (Vincent and Whittington, 1982). This electrostatic interaction becomes relatively unimportant when thick layers are present since it only operates at relatively large separation distances. Moreover, with thick adsorbed layers the contribution from G_A is also relatively small. Thus, with systems containing adsorbed nonionic or macromolecular surfactant, G_T may be described in terms of G_S and G_A and the form of the energy - distance curve is illustrated in Fig. 3 which also shows the various contributions.

Unlike the DLVO type systems, the energy - distance curve shows only one minimum, G_{min}, whose depth depends on the adsorbed layer thickness (and of course on the Hamaker constant and particle size). Provided that G_{min} is shallow (when thick adsorbed layers are present) a highly deflocculated suspension results since G_T shows a very steep rise with reduction of h below 2δ. Thus, provided that the surfactant molecules are strongly adsorbed, steric stabilization offers a powerful means of producing deflocculated systems. Since

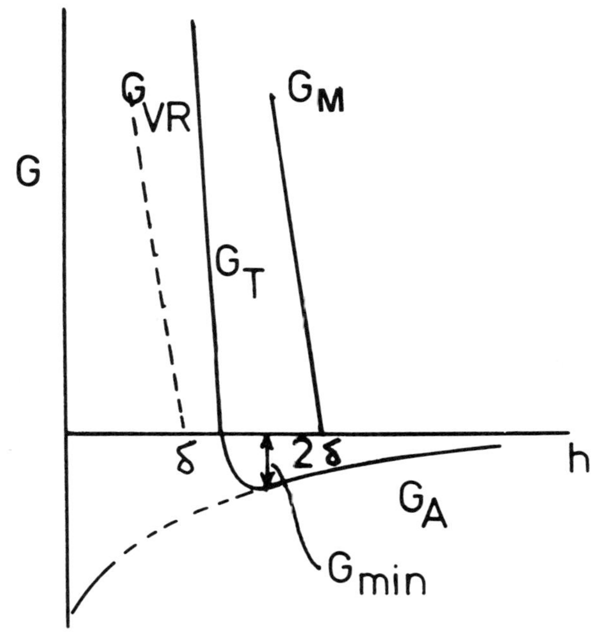

Fig. 3. Total interaction energy - distance curve for sterically stabilized systems.

anchoring of the chains is important, the merit of using A-B or A-B-A block copolymers is now obvious. By choosing the B chains with very low solubility in the medium and having high affinity to the surface, it is possible to "anchor" the chain to the surface leaving the A chains (which are chosen to be highly solvated by the medium) dangling in solution to provide the steric barrier.

As mentioned before (see equation 13) steric interactions can be made attractive by reducing the solvency of the chain to the point where $\chi \geqq 0.5$. This is illustrated in Fig. 4 which shows the effect of changing the χ parameter from a value < 0.5 to a value > 0.5.

When $\chi > 0.5$, G_M becomes negative and a deep minimum is produced in the G-h curve. This condition results in catastrophic flocculation of the suspension (usually referred to as incipient flocculation). Changes in the χ parameter with ethoxylated surfactants are produced by temperature changes and/or addition of a non-solvent for the chain (e.g., high concentrations of electrolyte). Thus, a suspension stabilized by ethoxylated surfactant may undergo flocculation at

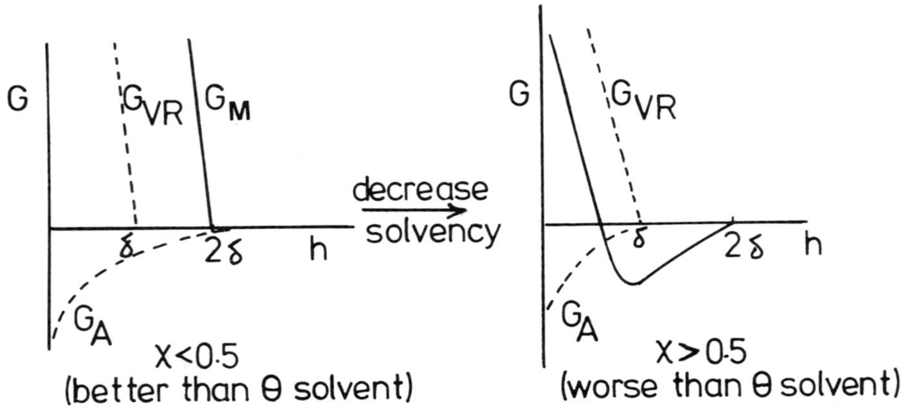

Fig. 4. Influence of change of solvency on the energy - distance curve.

temperatures above the θ-temperature of the polyethylene oxide chain and/or when sufficient electrolyte is added to reach the θ-condition of the chain.

Control of Claying or Caking of Suspensions

Claying is the result of settling of the particles (which have sizes outside the colloid range and are more dense than the medium) under gravity to the bottom of the container. The repulsive forces necessary to ensure stability in the colloid sense enable the particles to move past each other to form a compact layer. As a consequence of the dense packing and small spaces between the particles, such "clays" are difficult to redisperse. As mentioned above, it is possible, at least in principle, to prevent the formation of compact layers by controlling the interparticle interaction. For example, with suspensions containing ionic surfactants, compact sediments may be prevented by secondary minimum flocculation. The latter may be produced with coarse suspensions by controlling the depth of the secondary minimum (Fig. 2). This may be achieved, for example, by addition of electrolytes.

With systems containing nonionic or macromolecular surfactants, compact sediments may also be prevented by controlling the adsorbed layer thickness. As shown in Fig. 3, for a given system, G_{min} depends on the magnitude of δ. Thus, by reducing δ, G_{min} becomes deeper reaching values that are sufficient to induce weak flocculation. This has been illustrated before with polystyrene latex suspensions, using poly (vinyl alcohol) fractions with various molecular weights. When the molecular weight of the polymer was reduced from 43,000 to 8000 (adsorbed layer reduced from 38 nm to 7 nm), flocculation was obtained, producing a "loose" structure and this could be used to prevent claying.

Although the above methods are sound in principle, their practical application is not easy since a number of parameters need to be controlled simultaneously. For that reason, several other, more practical methods are used for prevention of claying. These methods, which are beyond the scope of this chapter, have been described before (Tadros, 1980). Basically, one adds to the continuous medium another disperse system, which is capable of forming a three-dimensional network preventing extensive settling of the coarse particles. The most commonly used systems are those based on sodium montmorillonite, oxides, polymeric thickeners or a combination of these. Of relevance in the present context is the use of liquid crystalline phases of surfactants to prevent claying of suspensions. As mentioned in Chapter 3 surfactants form mesophases or liquid crystalline phases, which have liquid-like disorder in one or more dimensions and crystal-like order in others. Because of this hybrid structure liquid crystals are non-rigid fluids, with rheological parameters falling between those of crystals and of liquids. The type of liquid crystal that is of particular use for structuring suspensions, is the hexagonal phase which is a highly viscous structure based on a close-packed array of cylindrical elements of infinite length. Such systems are capable of controlling the sedimentation of particles, in view of their ability to show viscoelastic behaviour. Due to their anisotropic nature (rod-like micelles), they show pronounced viscoelasticity (Hoffmann, 1982). In a shear field, the viscosity of the solution increases strongly when the rods begin to overlap.

Control of Crystal Growth (Ostwald Ripening)

Ostwald ripening occurs as a result of the differences in solubility between small and large crystals (Ostwald, 1900),

$$\frac{RT}{M} \ln \frac{S_2}{S_1} = \frac{2\sigma}{\rho} \left(\frac{1}{r_1} - \frac{1}{r_2}\right) \tag{16}$$

where S_1 and S_2 are the solubilities of crystals of radii r_1 and r_2 respectively, σ is the specific surface energy, ρ is the density and M is the molecular weight of the soluble molecules, R is the gas constant and T the absolute temperature.

Thus, with time the smaller crystals dissolve preferentially and the solute molecules so created become deposited on the large crystals. The growth of crystals may create undesirable changes in suspensions. As a result of drastic change in particle size distribution, the settling of the particles may be accelerated, leading to caking and cementing together of particles in a sediment. Moreover, increase in particle size is undesirable for many applications, e.g., in pharmaceuticals and pesticides, whereby the biological efficacy of the drug or pesticide may be affected.

Surfactants play a major role in the above process of Ostwald ripening which can be understood if one considers the main processes involved. The latter consist of two main steps (Tape and Kahlweit, 1961; Nancollas and Purdie, 1964), namely the diffusion of solute to the surface of the growing crystal which is followed by incorporation of molecules in the structure of the crystal lattice. A surfactant may affect the rate of crystal growth by modifying one or both of these steps. In the first place, the presence of surfactant influences the diffusion process in several ways by affecting the transport away from the boundary layer (Merill and McBain, 1966), although its addition is not likely to change the rate of dissolution proper (passage from the solid to the dissolved state in the immediately adjacent layer). The dissolution rate can thus be greatly accelerated (Bates *et al.*, 1966). On the other hand, since surfactants form micelles that are capable of solubilizing the solute, the diffusion coefficient D of the solute in a micellar solution will be reduced. However, as a result of solubilization the concentration gradient, $\partial c/\partial x$, will greatly increase (in direct proportion to the extent of solubilization). Since the flux, J, of the solute in mol cm^{-2} s^{-1} is governed by the product of D and $\partial c/\partial x$, i.e.,

$$J = -D\ (\partial c/\partial x) \tag{17}$$

the net effect depends on the relative effect on D and $\partial c/\partial x$. Since D is roughly inversely proportional to the radius of the diffusing particle, it is only reduced by less than a factor of 10 (corresponding to a micelle having a volume 1000 times larger than that of the solubilizate). On the other hand, $\partial c/\partial x$ can be increased by orders of magnitude as a result of solubilization. Thus, the net effect will be an increase in the flux, when surfactant micelles are present. Thus, solubilization by surfactant micelles may, under certain conditions, enhance transport by diffusion, thus enhancing crystal growth (Tadros, 1973).

Surfactants can also influence the second step, namely surface deposition and incorporation in the crystal lattice. Adsorption of a surfactant on the surface of a crystal can drastically change the specific surface energy and make it, effectively, inaccessible to the solute. Moreover, if the surfactant is preferentially and strongly adsorbed at one or more faces of the crystal (for example, by electrostatic attraction between, say, a highly negative face of the crystal and a cationic surfactant), surface nucleation is no longer possible at this particular face(s). Growth will then take place at the remaining faces which are either bare or incompletely covered with the surfactant. This will result in a change in crystal habit.

From the above discussion it can be seen that surface active agents, if properly chosen, may be used for crystal growth inhibition in suspensions and control of habit formation.

INFLUENCE OF SURFACTANTS ON THE FLOW BEHAVIOUR (RHEOLOGY) OF SUSPENSIONS

Whether suspended in Newtonian or non-Newtonian fluids, suspensions usually exhibit non-Newtonian flow, except perhaps at low concentrations of the disperse phase (Mewis and Spaull, 1976). For example, rigid spheres suspended in a Newtonian fluid produce significant non-Newtonian stresses as a result of their interaction (Batchelor, 1974). On the other hand, in a non-Newtonian medium, the presence of particles will enhance the non-Newtonian behaviour of the suspending medium. Whether the suspension is stable or unstable in the colloid sense, aqueous or non-aqueous, the rheology of the system must be governed by the volume fraction of the disperse phase (which controls the range of interaction), the size of the flow units and the interparticle interaction. Most suspensions encountered in practice (particularly those where a "structure build-up" is necessary, for example, for prevention of claying) show complex rheological behaviour which exhibits various responses ranging from simple liquid-like (viscous), plastic, elastic to mixed viscous and elastic (i.e., viscoelastic). Moreover, above a certain critical volume fraction, ϕ (typically 0.4 for near-spherical particles) a shear-thickening or dilatant region develops at high shear rates, which becomes steeper with increasing ϕ.

Having now briefly described the various complex rheological characteristics of concentrated suspensions, an attempt can be made to show some of the expected effects that may be produced by surfactants added to suspensions. As mentioned above, the rheology is determined by the volume fraction of the dispersed phase, the magnitude of the interaction forces and the structure of the aggregates formed, all of which are affected by the presence of adsorbed surfactant layers. This can be illustrated by considering some of the systems encountered in practice, as summarized below.

To illustrate the importance of adsorbed surfactant layers on the viscosity-volume fraction plot for suspensions, let us consider the most simple case with hard sphere interactions, i.e., the so-called neutral stability, with all potentials weak relative to Brownian motion. At low volume fraction, ϕ, the relative viscosity, η_r, of the suspension is fairly low, increasing fairly slowly with increase of ϕ, until the volume fraction approaches the so-called packing fraction, ϕ_p, where η_r increases rapidly with increase of ϕ. This is illustrated in Fig. 5 and in this case η_r is related to ϕ, by the Douherty-Krieger equation (1959; 1972)

$$\eta_r = [1-(\phi/\phi_p)]^{-[\eta]\phi_p} \tag{18}$$

where $[\eta]$ is the intrinsic viscosity (~2.5-3.0).

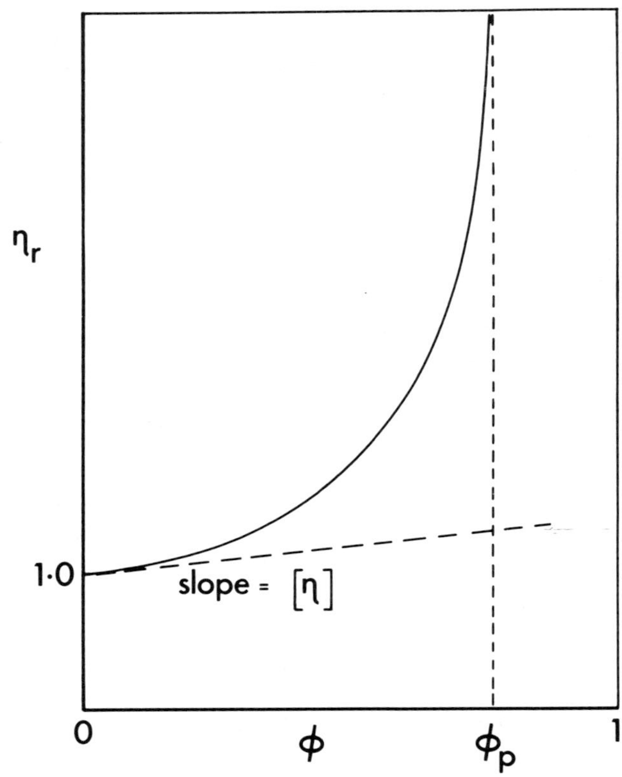

Fig. 5. Viscosity - volume fraction curve for suspensions.

Assuming now that a suspension with a particle radius of 100 nm and ϕ_p = 0.45 can be prepared in a stable form without adsorbed layers of surfactant, then such a suspension will have a low η_r and hence will be fluid. If a surfactant is now added to give an adsorbed layer thickness of 10 nm, the hydrodynamic volume fraction of the suspension now becomes $\phi_h = \phi(1 + \frac{\delta}{a})^3 = 0.45 \times (1.1)^3 = 0.6$, i.e., it reaches the so-called packing fraction and if the system still behaves as hard spheres the viscosity of the system now goes to very high values. Thus, from a simple geometrical effect, adsorbed surfactant layers cause an increase in hydrodynamic volume and hence in viscosity.

The above picture is an oversimplification of the actual situation since even at $\phi < \phi_p$ the adsorbed surfactant layers are able to interpenetrate and the suspension shows pronounced viscoelastic behaviour. This can, for example, be illustrated if an oscillating deformation (or stress) is applied to the system. By following the variation of stress (or strain) with time, it is possible to characterize the viscoelastic behaviour in terms of a storage modulus (G') and a loss modulus (G"). G' gives a measure of the elastic component, whereas G" is related to the viscous component. Measurements of this kind have been carried out on various suspensions of the aqueous and non-aqueous type (Tadros, in preparation) and they showed that at high ϕ, G' indeed exceeds G" at all frequencies, indicating the dominance of the elastic contribution due to the interaction of the steric barriers.

Ionic surfactants also affect the rheology of suspensions, due to the long range electrostatic repulsion. At low electrolyte concentrations, these systems are also expected to show viscoelastic behaviour, which in view of the long range of the interaction may be manifest at low ϕ values, particularly with small particles.

Finally, surfactants both of the ionic and nonionic type will influence the rheology of unstable suspensions, whether these are weakly or strongly flocculated. In the first case, weak flocculation, which may be produced by control of the adsorbed layer thickness, will be reflected in the formation of a pseudoplastic system with a measurable yield value and the appearance in some cases of hysteresis in the flow curves (due to thixotropy). In the second case, strong flocculation, in which strong attraction between the particles dominates the interaction, leads also to the formation of a pseudoplastic system with a pronounced yield value. This, for example, is produced by addition of a cationic surfactant to a suspension with negatively charged particles. At the isoelectric point of the suspension, strong coagulation takes place leading to the formation of a pseudoplastic system. Alternatively, a system stabilized by a nonionic surfactant may be coagulated by addition of sufficient quantity of an electrolyte. The flow behaviour of such coagulated systems (Tadros, 1980) is illustrated in Fig. 6, which is characterized by three parameters:

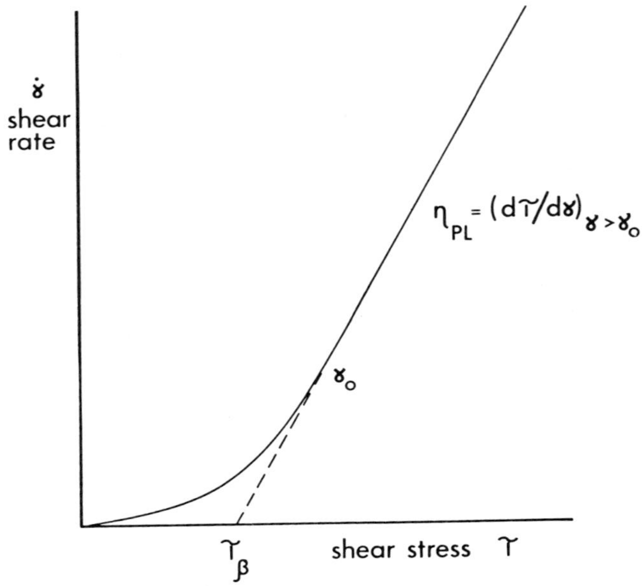

Fig. 6. Flow curve of a coagulated suspension.

(a) γ_{crit}, i.e., the shear rate above which the flow curve becomes linear. γ_{crit} is the value above which viscous stress ruptures all doublets formed; (b) the plastic viscosity $\eta_{PL} = (d\tau/d\gamma)_{\gamma>\gamma_{crit}}$, i.e., the gradient of the linear shear stress - shear rate flow curve. The pseudoplastic viscosity arises from purely hydrodynamic effects; (c) τ_β the value of the stress obtained by extrapolation of the linear portion of the flow curve to $\gamma = 0$. The residual stress arises from the residual effect of interaction potential.

Thus, as described above, addition of surfactants to a suspension will affect the flow behaviour (rheology) of suspensions in various manners depending on the effect they produce on the interparticle interaction. However, a full description of such effects is beyond the scope of the present chapter.

REFERENCES

Barrett, K.E.J. (1975). "Dispersion Polymerization in Organic Media". John Wiley and Sons, London.

Batchelor, G.K. (1974). *Ann. Rev. Fluid Mech.* 6, 227.

Bates, T.R., Gibaldi, M. and Konig, J.L. (1966). *Nature* 210, 1321.

Blackly, D. (1975). "Emulsion Polymerization". Applied Science Publishers.

Derjaguin, B.V. (1939). *Kolloid Z.* 69, 155; *Acta Physica Chemica USSR* 10, 333.

Douherty, J.T. (1959). Ph.D. Thesis Case Inst. of Technology.

Fischer, E.W. (1958). *Kolloid Z.* 160, 120.

Gibbs, J.W. (1906). "Scientific Papers". Longman Green, London. Vol. 1.

Hamaker, H.C. (1937). *Physica (Utrecht)* 4, 1058.

Hoffmann, H., Platz, G., Rehage, H. and Schorr, W. (1982). *Adv. Colloid Interface Sci.* 17, 275.

Kreiger, I.M. (1972). *Adv. Colloid Interface Sci.* 3, 111.

Mackor, E.L. (1951). *J. Colloid Sci.* 6, 490.

Mackor, E.L. and van der Waals, J.H. (1952). *J. Colloid Sci.* 7, 775.

Merill, R.C. and McBain, J.W. (1966). *J. Phys. Chem.* 46, 10.

Mewis, J. and Spaull, A.J.B. (1976). *Adv. Colloid Interface Sci.* 6, 173.

Nancollas, G.H. and Purdie, N. (1964). *Quant. Rev.* 18, 1.

Ostwald, W. (1900). *J. Phys. Chem.* 34, 493.

Parfitt, G.D. (1967). *J. Oil Chem. Assoc.* 50, 822.

Parfitt, G.D. (1981). "Dispersion of Powders in Liquids". Applied Science Publishers. 3rd edn.

Patton, T.C. (1964). "Paint Flow and Pigment Dispersion". Interscience, New York.

Rehbinder, P.A. and Likhtman, V.I. (1957). "Proc. 2nd Int. Conference Surface Activity". Butterworths, London. No.3.

Rehbinder, P.A. (1958). *Colloid J. USSR* 20, 493.

Rideal, E.K. (1922). *Phil. Mag.* 44, 1152.

Schukin, E.D. and Rehbinder, P.A. (1958). *Colloid J. USSR* 20, 601.

Tadros, Th.F. (1973). *In* "Particle Growth in Suspensions" (Ed. A.L. Smith), p.229. Academic Press, London.

Tadros, Th.F. (1980). *Adv. Colloid Interface Sci.* 12, 141.

Tadros, Th.F. (1982). *In* "The Effect of Polymers on Dispersion Properties" (Ed. Th.F. Tadros), p.1. Academic Press, London.

Tadros, Th.F. (1983). *In* "Thin Liquid Films" (Ed. B. Ivanov). Marcel Dekker, in press.

Tadros, Th.F. (in preparation). *In* "Suspension Concentrates" (Eds R.H. Ottewill and R. Rowell). Academic Press, London.

Tape, G. and Kahlweit, M. (1961). *J. Phys. Chem.* 30, 90.

Verwey, E.J.W. and Overbeek, J.Th.G. (1948). "Theory of the Stability of Lyophobic Colloids". Elsevier Publ. Co., Amsterdam.

Vincent, B. (1973). *J. Colloid Interface Sci.* 42, 270.

Vincent, B. and Whittington, S. (1982). *In* "Colloid and Surface Science" (Ed. E. Matijevic), Vol.12, p.1.

Vold, M.J. (1961). *J. Colloid Sci.* 16, 1.

Volmer, M. (1939). "Kinetik der Phase Bildung". Steinkopf, Dresden.
Washburn, E.D. (1921). *Phys. Rev.* 17, 273.

WETTING

T.D. Blake

*Kodak Limited Research Laboratories,
Headstone Drive, Harrow, Middlesex HA1 4TY, England*

INTRODUCTION

Wetting is a large subject in its own right. It has a long history and an extensive literature. Some of its key foundations were laid in the early 19th century by notable scientists like Thomas Young. Today the subject benefits from the insights of quantum electrodynamics and sophisticated analytical techniques, yet there is still much to be learnt by using simple apparatus to measure surface tensions and contact angles.

Wetting is important in many processes both industrial and natural. Some of these are touched on in the text. However, the main aim is to provide a basic but up-to-date account of the fundamentals of wetting, together with a selective review of the relevant literature. The influence of surfactants on wetting is emphasized wherever appropriate.

PHENOMENOLOGY - THE CONTACT ANGLE

When a drop of liquid is placed on a solid, the liquid either spreads to form a thin, more or less uniform film (Fig. 1a) or remains as a discrete drop (Fig. 1b). The former behaviour is generally described as complete wetting, the latter as incomplete or partial wetting. The liquid may also interact chemically or physically with the solid. Even if no bulk reaction occurs, some form of surface interaction (e.g., adsorption or dissociation) is probable with all but the most inert systems.

The outward measure of the degree of wetting is the

SURFACTANTS
ISBN 0-12-682180-1

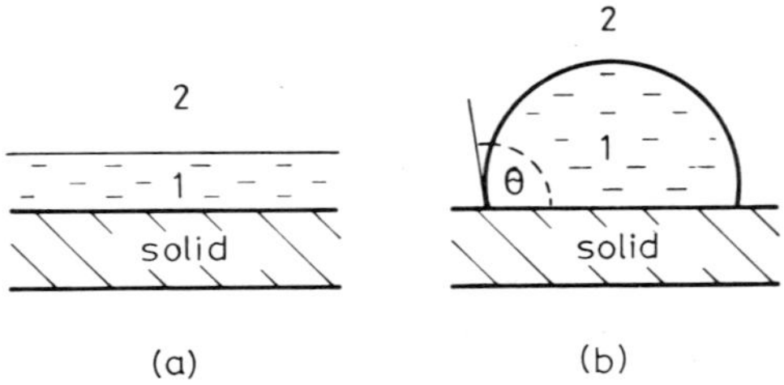

Fig. 1. 3-phase equilibrium between solid, liquid phase 1, and immiscible phase 2 (liquid or gas): (a) complete wetting; (b) formation of a contact angle, θ.

contact angle, θ, i.e., the angle formed between planes tangent to the surfaces of the solid and the liquid at the wetting perimeter. Here, solid and liquid are simultaneously in contact with each other and the surrounding phase which is typically air or some other gas, but may be another liquid more or less immiscible with the first. The wetting perimeter is therefore frequently referred to as the three-phase line, although the most common name is simply the wetting line. In using these terms, however, it is as well to remember that, at the molecular level, the three phases meet not at a line but within a zone of small but finite dimensions in which three interfacial regions merge and the structures and properties associated with one solid/fluid interface give way to those of the other. This confluent region may be called the three-phase zone.

Most wetting studies centre around measurements of the contact angle. Typical values for a few common systems are listed in Table 1. The smaller the angle, the better the liquid is said to wet the solid. Complete wetting means that the contact angle is formally zero. Conversely, complete non-wetting dictates an angle of 180°. This is only possible if the second fluid phase is a liquid.

As we shall see, much of the utility of contact angle measurements depends upon equilibrium thermodynamic arguments. Most studies are therefore carried out with more or less static systems. However, in many practical processes the aim is to cause one fluid to displace another from the surface of a solid, and to achieve this as quickly and efficiently as possible. In consequence, systematic investigations of dynamic contact angles, i.e., those associated with moving wetting lines, have become increasingly common.

Wetting measurements and, more importantly, their interpretation are complicated by the fact that even static angles are seldom single-valued. In most practical systems, the angle exhibits hysteresis, i.e., its value is dependent upon the

TABLE 1

CONTACT ANGLES OF LIQUIDS ON SOLIDS

Liquid (γ/mN m^{-1})	Solid	θ/deg
Mercury (484)	PTFE	150
Water (72.5)	PTFE paraffin wax polyethylene human skin gold glass	112 110 103 75 - 90 0 0
Methylene iodide (67)	PTFE paraffin wax polyethylene	85 61 46
Benzene (28)	PTFE graphite	46 0
n-decane (23) n-octane (21.6)	PTFE PTFE	40 30
Tetradecane/water (50.2)	PTFE	170[a]

Values drawn from Table X-2 of Adamson (1982), except (a) from Bargeman and van voorst Vader (1972).

history of the system and varies according to whether the given fluid is tending to advance across or recede from the solid surface. The limiting angles achieved just prior to movement of the wetting line (or just after movement ceases) are known as the advancing and receding contact angles, θ_A and θ_R , respectively. This choice of names is unfortunate, since they are also used more logically to describe the corresponding dynamic contact angles, but the practice is probably far too entrenched to change.* In any case, it is frequently difficult to ensure that wetting systems are truly static, so the confusion may be operational as well as semantic.

For a given system, $\theta_A \geqq \theta_R$ and θ can usually take any value between these two limits without discernible movement of the wetting line. Since smaller angles imply better wetting, we can see that the contact angle always changes in such a direction as to oppose wetting line movement.

---

* Huh and Scriven (1971) have suggested the terms advanced and receded (recessed might be better) to distinguish the static case, but the idea has not found wide acceptance.

In dynamic contact angle studies, a further complication is that once the wetting line begins to move, it often does so with a hesitant, stick-slip or "zippering" motion which is far from steady. Nevertheless, it is usually observed that advancing angles increase and receding angles decrease with increasing speed of the wetting line. The general behaviour of both static and dynamic contact angles is summarized in Fig. 2.

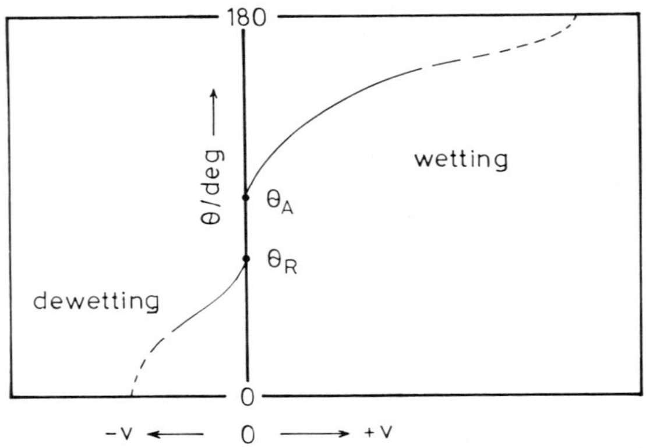

Fig. 2. Schematic variation of the contact angle, θ, *with the velocity of wetting,* v. *The behaviour shown is typical of a poorly wetting liquid with a moderate degree of contact angle hysteresis.* θ_A *and* θ_R *are the advancing and receding contact angles when* v = 0.

There are many ways of measuring the contact angle. Some are illustrated in Fig. 3. They include sessile drop (a), captive bubble (b), sliding drop (c), tilting plate (d), tensiometry (e), and various geometric methods based on measuring the shapes of menisci. They have all been adequately described elsewhere (Johnson and Dettre, 1969; Adamson, 1982). Two methods applicable to powders and other porous materials are discussed later.

THERMODYNAMICS

Thermodynamics provides the main framework for the interpretation of wetting phenomena; yet, as we shall see, the practical application of this framework is sometimes difficult.

Young's Equation

Consider the system depicted in Fig. 1b. We will call the liquid in the drop fluid 1 and the surrounding phase fluid 2. We denote the solid by the subscript S. If the system is at equilibrium, then the appropriate value of the contact angle, θ°, is given by Young's equation:

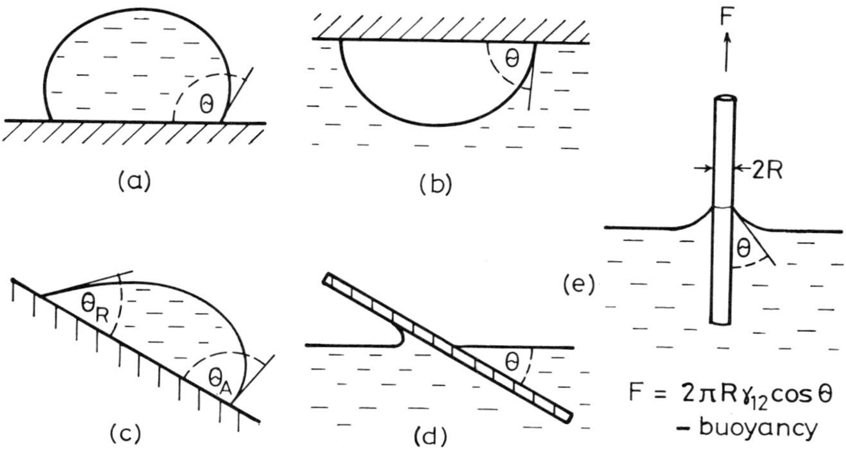

Fig. 3. Measurement of contact angles using: (a) sessile drop; (b) captive bubble; (c) sliding drop; (d) tilting plate; and (e) tensiometry.

$$\gamma_{12} \cos \theta^\circ = \gamma_{S2} - \gamma_{S1} \quad (1)$$

where γ_{12}, γ_{S1} and γ_{S2} are the interfacial tensions acting at the 12, S1 and S2 boundaries.

Equation (1) was first proposed by Young (1805) on the basis of mechanical considerations, but it may also be derived on rigorous thermodynamic grounds (Gibbs, 1928; Johnson, 1959; Buff, 1960; Everett, 1980). The variational technique generally employed shows that Young's equation is a necessary condition for hydrostatic equilibrium provided two important conditions are met.

The first condition is that the S1 and S2 interfaces must have a common tangent at the wetting line. Thus, for example, when the wetting line is located at a sharp edge, the correct three-phase equilibrium condition is given not by equation (1), but by

$$\gamma_{12} \cos \alpha \leqq \gamma_{S2} - \gamma_{S1} \quad (2a)$$

$$\gamma_{12} \cos \beta \leqq \gamma_{S1} - \gamma_{S2} \quad (2b)$$

where α and β are the angles filled by fluids 1 and 2, respectively (Gibbs, 1928). This reduces to equation (1) when $\alpha + \beta = 180°$.

The second condition requires that γ_{S1} and γ_{S2} are locally constant in the vicinity of the wetting line. This in turn requires a constant state of strain both at the surface of the solid and within the bulk (Melrose, 1965). The solid must therefore be rigid.* Notice, however, that the solid need not be inert. It might, for example, be soluble provided all the normal conditions for equilibrium are fulfilled.

Another important point that emerges from the derivation of Young's equation is that γ_{12}, γ_{S2} and γ_{S1} are all interfacial tensions and not surface free energies as is sometimes supposed. For a fluid - fluid interface γ can be regarded as the free energy change accompanying isothermal reversible creation of fresh surface area, i.e.,

$$\gamma = (\delta F/\delta A)_{T,V} \, . \tag{3}$$

Thus, for a liquid, γ can be thought of as the force required to stretch the surface. But this is not the case for a solid surface. Solids lack a mechanism enabling them to adjust to thermodynamic equilibrium and, moreover, their rigidity enables them to remain in mechanical equilibrium whilst accommodating non-uniform states of strain.

Nevertheless, an alternative definition of surface tension is available that can be applied to both solids and liquids (Gibbs, 1928; Defay *et al.*, 1966)

$$\gamma = f - \textstyle\sum_i \Gamma_i \mu_i \tag{4}$$

where f is the surface free energy per unit area, and Γ_i is the specific adsorption of the i'th component having chemical potential μ_i. This expression involves only integral quantities which are indifferent to whether or not the path taken in achieving them was isothermal and reversible. As Gibbs recognized, however, μ and f are only defined when the material and energy densities are uniform in the bulk of the solid. This is the underlying reason why Young's equation is valid only if the solid is in a uniform state of strain. If it is not, then γ_{S1} and γ_{S2} are undefined. Equation (4) shows that γ can be equated with the surface free energy only if the Gibbs dividing surface is chosen so that $\sum_i \Gamma_i \mu = 0$.

In general, Young's equation provides us with a precise thermodynamic definition of the contact angle, but it suffers from a lack of direct experimental verification because the solid surface tensions are not susceptible to direct measurement. In this sense, Young's equation differs from the analogous Neumann's triangle relationship which applies when the substrate is a liquid because, in this latter case, all the angles and tensions can be measured, and the relationship can therefore be verified (Fox, 1945).

---

* A very thorough treatment of the wetting of deformable solids has been given only recently by Rusanov (1978).

Complete Wetting

When the liquid (fluid 1 in our notation) completely wets the solid, θ° is formally zero and equation (1) reduces to

$$\gamma_{12} = \gamma_{S2} - \gamma_{S1} \quad . \tag{5}$$

For some authors this equation has presented conceptual difficulties since the wetting line for which the equation represents the hydrostatic equilibrium condition, does not exist. However, experimental studies using a liquid substrate (Johnson and Dettre, 1966) and sound thermodynamic arguments (Melrose, 1967; Rowlinson and Widom, 1982) suggest that the equation is exactly obeyed whenever spreading occurs.

The reason for this can be seen more clearly if we rewrite equation (5) to take more explicit account of the way in which the spreading liquid adsorbs at the interface between the solid and the other fluid phase. For simplicity, we consider the case in which fluid 2 is gaseous and contains no components that adsorb on the solid other than vapour from the liquid. Thus,

$$\gamma_{12} = \gamma_S - \pi_{S2,1} - \gamma_{S1}$$

where γ_S is the surface tension of the bare solid, and $\pi_{S2,1}$ is the surface pressure of fluid 1 adsorbed at the S2 interface such that

$$\gamma_S - \pi_{S2,1} = \gamma_{S2} \quad . \tag{6}$$

Under spreading conditions, $\pi_{S2,1}$ will grow reducing γ_{S2} until it exactly equals $\gamma_{12} + \gamma_{S1}$. This will be achieved when the thickness of the adsorbed film is such that those parts of the film in contact with phase 2 and the solid correspond to the 12 and S1 interfaces.

Although the surface pressure has a special relevance under spreading conditions, significant surface pressures may also occur in systems that exhibit non-zero contact angles (Derjaguin and Zorin, 1957; Melrose, 1964; Hauxwell and Ottewill, 1970; Blake *et al.*, 1971; Huh and Adamson, 1977; Tse and Adamson, 1979; Sullivan, 1981).* Nevertheless, it is only when equation (5) pertains that the adsorbed film can grow continuously until it becomes sufficiently thick to take on the properties of the bulk liquid. If θ > 0, there is no continuous transition from adsorbed phase to bulk phase through a series of increasing film thicknesses corresponding to

---

* This view has been hotly disputed in the literature, particularly for polar liquids on non-polar substrates; see for example Good (1975) and Fowkes (1980).

thermodynamically stable states (Frumkin, 1938; Dzyaloshinskii *et al.*, 1960). These differences in behaviour have been explained by Derjaguin (Derjaguin and Kusakov, 1939; Derjaguin, 1940; Derjaguin and Zorin, 1957) in terms of a disjoining pressure acting normally to the plane of the film and preventing thinning. The magnitude of the disjoining pressure varies with film thickness. In systems exhibiting non-zero contact angles, the disjoining pressure becomes negative and thus destabilizes a range of film thicknesses with respect to thinner films. In these circumstances, a draining film on reaching a certain thickness will spontaneously separate into bulk liquid droplets and adsorbed multilayers.

Line Tensions

One further complication that may be important for very small systems is the possibility that the wetting line itself may be in tension (Gibbs, 1928). If this is so, then for the liquid drop in Fig. 1b, equation (1) must be rewritten as

$$\gamma_{12} \cos \theta^\circ = \gamma_{S2} - \gamma_{S1} - \lambda/r \tag{7}$$

where λ is the line tension and r is the radius of curvature of the wetting line in the plane of the solid. Note that, in principle, λ may be either positive or negative. Various estimates (Harkins, 1937; Buff and Saltsburg, 1957; Evans and Lane, 1973; Starov and Churaev, 1980; Tarazona and Navascués, 1981) suggest that λ is of order 10^{-11} N. Hence, the correction to Young's equation should only become significant when r is of order 10 nm or less (i.e., $\lambda/r \geqq 1$ mN m^{-1}). Phenomena for which this could be important include heterogeneous nucleation (Gretz, 1966; Evans and Lane, 1973), the coalescence of liquid droplets (Torza and Mason, 1971) and capillary condensation in porous solids (Lane, 1975). Note that line tensions may also occur when the substrate is a liquid (Defay *et al.*, 1966; Pujado and Scriven, 1972), and at the perimeter of a free liquid film (Ivanov *et al.*, 1978; Platikanov *et al.*, 1980).

Spreading Coefficient, Work of Adhesion, Work of Cohesion and Adhesion Tension

For most purposes, Young's equation provides a perfectly adequate description of wetting equilibria. For the reversible wetting of planar solids, the description is complete. Nonetheless, many authors have found it convenient to define a number of additional terms and expressions.

The initial spreading coefficient (Cooper and Nuttall, 1915; Harkins, 1952) for the spreading of a liquid (fluid 1) on a solid is given by

$$S_{S12} = \gamma_S - \gamma_{S1} - \gamma_{12} \quad . \tag{8a}$$

Since spontaneous processes involve a negative free energy change, S_{S12} is defined as the work required to destroy unit areas of, respectively, S1 and 12 interfaces and leave unit area of bare solid (Fig. 4a). If S_{S12} is positive, fluid 1 will usually spread until it completely wets the solid. If S_{S12} is negative, fluid 1 will form a non-zero contact angle. Exceptions to these rules occur with <u>autophobic</u> liquids (Hare and Zisman, 1955). These are typically amphipathic polar liquids that adsorb strongly onto the solid surface to form an oriented monolayer on which the bulk liquid will no longer spread. From a thermodynamic viewpoint, the adsorbed layer lowers γ_S to such an extent that $\gamma_{S2} < (\gamma_{S1} + \gamma_{12})$ and a non-zero contact angle is formed. If equation (6) applies, then such behaviour may be exemplified by writing

$$S_{S12} = \gamma_{S2} + \pi_{S2,1} - \gamma_{S1} - \gamma_{12} \quad . \tag{8b}$$

Alternatively, we can use the <u>equilibrium</u> or <u>final spreading coefficient</u>

$$S^\circ_{S12} = \gamma_{S2} - \gamma_{S1} - \gamma_{12} \quad . \tag{9}$$

The <u>work of adhesion</u> is usually defined (Dupré, 1869) as

$$Wa_{S12} = \gamma_S + \gamma_{12} - \gamma_{S1} \tag{10}$$

which is the work required to separate unit area of S1 interface to leave unit areas of 12 interface and bare solid surface (Fig. 4b). If we use equation (6), then equation (10) may

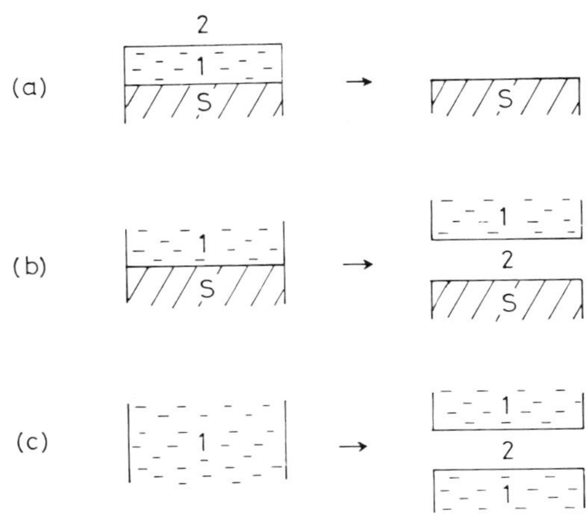

Fig. 4. (a) Initial spreading coefficient; (b) work of adhesion; (c) work of cohesion.

also be written as

$$Wa_{S12} = \gamma_{S2} + \pi_{S2,1} + \gamma_{12} - \gamma_{S1} \quad . \tag{11}$$

However, another common definition (Bangham and Razouk, 1937) is the <u>reversible work of adhesion</u>

$$Wa^{\circ}_{S12} = \gamma_{S2} + \gamma_{12} - \gamma_{S1} \quad . \tag{12}$$

$Wa^{\circ}_{S12} = Wa_{S12}$ only if $\pi_{S2,1} = 0$. In strongly interacting systems, Wa°_{S12} is $<<$ Wa_{S12}. Note that we can use equations (9) and (12) when fluid 2 contains additional components that adsorb on the solid and/or if fluids 1 and 2 are both liquids (competitive wetting).

The <u>work of cohesion</u> is the work of adhesion when both phases are the same (Fig. 4c). For a liquid, fluid 1, in the presence of fluid 2,

$$Wc_{12} = 2\gamma_{12} \quad . \tag{13}$$

Hence,

$$S_{S12} = Wa_{S12} - Wc_{12} \tag{14}$$

and

$$S^{\circ}_{S12} = Wa^{\circ}_{S12} - Wc_{12} \quad . \tag{15}$$

Finally, the <u>adhesion tension</u> is defined as

$$\tau_{S12} = \gamma_{S2} - \gamma_{S1} \tag{16}$$

and is the net force exerted by the solid on the liquid at the wetting line. Although neither γ_{S2} nor γ_{S1} can be measured directly, their difference can be determined, since from Young's equation

$$\tau_{S12} = \gamma_{12} \cos \theta^{\circ} . \tag{17}$$

Similarly, the utility of equations (8) to (12) can be much improved:

$$S_{S12} = \gamma_{12} (\cos \theta^{\circ} - 1) + \pi_{S2,1} \tag{18}$$

$$S^{\circ}_{S12} = \gamma_{12} (\cos \theta^{\circ} - 1) \tag{19}$$

$$Wa_{S12} = \gamma_{12} (1 + \cos \theta^{\circ}) + \pi_{S2,1} \tag{20}$$

$$Wa^{\circ}_{S12} = \gamma_{12} (1 + \cos \theta^{\circ}) . \tag{21*}$$

Each of these terms is therefore experimentally accessible. Studies of wetting have a long history, and it is perhaps not surprising that the subject had produced such a luxuriant

---

* Equation (21) is usually known as the Dupré equation (Dupré, 1869).

growth of expressions from such a small number of essential parameters. One interesting variation is obtained by combining equations (13) and (21):

$$\cos \theta^\circ = -1 + 2(Wa^\circ_{S12}/Wc_{12}). \tag{22}$$

This shows that when $Wc_{12} = Wa^\circ_{S12}$, $\theta^\circ = 0$, and when $Wc_{12} = 2\ Wa^\circ_{S12}$, $\theta^\circ = 90^\circ$. According to Bashforth and Adams (1883), Clairault had appreciated as long ago as 1743 that the height to which a liquid will rise in a capillary is determined by the relative attraction of the liquid particles for themselves and for the walls of the tube. He concluded that even if the former attractions were of greater magnitude than the latter, provided they were not twice as great, the liquid would still rise. Nevertheless, whilst equation (22) is clearly foreshadowed by Clairault's conclusion, the idea expressed, that competition between the cohesion of a liquid to itself and its adhesion to a solid should give rise to an angle of contact both constant and specific to a given system at equilibrium, was expounded for the first time by Young in the same paper in which he proposed equation (1).

THE MOLECULAR BASIS FOR WETTING

A proper treatment of this topic is well beyond the scope of the present introductory text. Nevertheless, it would be wrong to omit what has become one of the most important areas of study. A readable account of intermolecular forces is given by Israelachvili and Tabor (1973). For a more detailed review of dispersion forces see Mahanty and Ninham (1976). For the statistical thermodynamics of interfaces see Rowlinson and Widom (1982).

Young's Equation and Intermolecular Forces

As noted in the last section, the general idea that wetting has its origin in what we would now call forces of intermolecular attraction was well appreciated at the beginning of the nineteenth century and possibly earlier, even though the nature and range of these forces was not understood. The situation had improved only slightly by the time that Rayleigh (1890a) gave the first "modern" analysis of wetting in terms of a non-specific but fairly short-range attractive intermolecular potential and a very short-range repulsion. Nevertheless, the analysis was sufficiently complete to show that Young's equation provides the correct macroscopic equilibrium condition, and that the contact angle given by Young's equation remains correct to within a microscopically small distance of the wetting line. Interactions with the solid cause the fluid/fluid interfacial tension and the contact angle to depart from their macroscopic values only at still smaller distances. Recent analyses (Berry, 1974; White, 1977) strongly support these conclusions and tend to refute contrary suggestions (Jameson and del Cerro, 1976; Saville, 1977). Our present knowledge suggests that macroscopic values should hold to within about 1 nm of the wetting line, except for very small contact angles. With the latter, the film of liquid at the S2 interface may be relatively thick and the transition between bulk liquid and film relatively gradual. In either case, deviations from macroscopic values may be accounted for in terms of a line tension.

Since Rayleigh's time there obviously have been major improvements in our understanding of intermolecular forces. For our purposes, it is sufficient to consider the various van der Waals forces. These include dipole-dipole (Keesom, 1921), dipole-induced dipole (Debye, 1920), and dispersion forces (London, 1930). For a definitive review see Margenau and Kestner (1969).

Dispersion forces are especially important since they occur between all molecules and are additive. They are the principal attractive intermolecular forces in non-polar liquids like alkanes. London's expression for the dispersion interaction between two molecules separated by a distance r is

$$u = -\beta r^{-6} \tag{23}$$

where the constant β depends on the electric polarizability of the molecules. Because of the additive nature of dispersion forces, pair-wise interactions may be summed to give the total interaction between condensed media (Hamaker, 1937). Thus, for example, the attractive force per unit area between two semi-infinite flat plates separated by a distance d is

$$F = A/6\pi d^3 \tag{24}$$

where A is the Hamaker constant

$$A = \pi^2 n^2 \beta \tag{25}$$

and n is the number of interacting dispersion centres per unit volume. Due to the finite speed of propagation of electromagnetic radiation, as r increases beyond about 10 nm, u begins to fall off more rapidly than predicted by equation (23) (Casimir and Polder, 1948). For $d > 50$ nm, these retarded interactions give a relationship of the form $F = B/d^4$ instead of equation (24). However, the main point to note is that dispersion forces between condensed media are of much longer range than those between individual molecules. One result of this is that the former can be measured with only moderate difficulty, either directly (e.g., Derjaguin and Abrikossova, 1954; 1957; Kitchener and Prosser, 1957; Sparnaay, 1958; Van Silfhout, 1966; Tabor and Winterton, 1969; Israelachvili and Tabor, 1972; Rabinovich *et al.*, 1982), or indirectly from the stability of thin liquid films on solids (e.g., Sabisky and Anderson, 1973; Blake, 1975). Typical values of the Hamaker constant in equation (24) are of order 10^{-19}J.

An important consequence of the long range of dispersion forces is that they play a major role in the stability of colloidal dispersions. The balance between attractive dispersion forces and electrostatic double-layer repulsion is the basis of the celebrated DLVO theory.* Indeed, this topic has provided the major theme for the study of forces between condensed media. In contrast, the application of intermolecular force theory to wetting has received rather less attention.

Calculation of Surface Tensions and Contact Angles

Fowler (1937) was one of the first to use the improved understanding of intermolecular forces to calculate the surface tension of a simple liquid. He determined the energy required to break a column of liquid of unit cross-section and remove the two halves to infinite separation. This gave the work of cohesion equal to twice the surface tension. Fowler used statistical thermodynamic methods to integrate the intermolecular forces. These were calculated from the Lennard-Jones equation, which is like the London expression, equation (23), but contains an additional term which varies as r^{-12} to account for very short-range intermolecular repulsion. In a later statistical treatment, Kirkwood and Buff (1949) obtained the surface tension directly from the variation in the local intermolecular force field as the interface is traversed. Using molecular distribution functions estimated from experimental data, they found good agreement with experiment for the surface tension of Argon. More recent publications (e.g., Navascués and Berry, 1977; Sullivan, 1981; Rowlinson and Widom, 1982) suggest that we are still some way from being able to use these exact methods to calculate surface tensions and contact angles for systems of practical interest.

Girifalco and Good (1957) have proposed a rather more empirical approach to the problem. According to the Berthelot principle, the interaction constant for two different particles equals the geometric mean of the interaction constant for the individual particles. Applied to London constants,

* For a fairly recent account see Napper and Hunter (1975).

this gives

$$\beta_{ij} = (\beta_i \beta_j)^{\frac{1}{2}} \tag{26}$$

which leads to

$$A_{ij} = (A_i A_j)^{\frac{1}{2}} \tag{27}$$

for Hamaker constants. By analogy, Girifalco and Good suggested that the work of adhesion between two different liquids can be expressed in terms of the geometric mean of their respective works of cohesion

$$Wa_{12} = \Phi(Wc_1 Wc_2) \tag{28}$$

hence,

$$\gamma_{12} = \gamma_1 + \gamma_2 - 2\Phi(\gamma_1\gamma_2)^{\frac{1}{2}} \tag{29}$$

where the constant, Φ, depends on the relative molecular size and polar content of the interacting media. For nonpolar media, equation (29) was found to work quite well with $\Phi \cong 1$, though values of Φ from 0.32 to 1.15 were found for dissimilar substances, such as water and alkanes or water and mercury. Good and Girifalco (1960) and Good (1964) subsequently extended their theory to solid/liquid interfaces and contact angles. Using Young's equation for the case when fluid 2 is a gas (air) and can be ignored, they obtained

$$\cos\theta° = -1 + 2\Phi(\gamma_S/\gamma_1)^{\frac{1}{2}} - \pi_{S,1}/\gamma_1 \; . \tag{30}$$

where $\pi_{S,1}$ is the surface pressure of fluid 1 adsorbed at the solid/gas interface. This expression gave reasonable values of γ_S for nonpolar substrates.

The next step was taken by Fowkes (1962; 1964) who proposed that surface and interfacial tensions can be subdivided into independent, additive terms arising from different types of intermolecular interactions. For water, in which both hydrogen bonding (h) and dispersion forces (d) are active,

$$\gamma = \gamma^h + \gamma^d$$

whereas for nonpolar substances like the alkanes, γ is simply equal to γ^d. By applying the geometric mean relationship to only the γ^d term Fowkes obtained

$$Wa_{12} = 2(\gamma_1^d\gamma_2^d)^{\frac{1}{2}} \tag{31}$$

hence,

$$\gamma_{12} = \gamma_1 + \gamma_2 - 2(\gamma_1^d\gamma_2^d)^{\frac{1}{2}} \tag{32}$$

where the non-dispersive contributions to γ_1 and γ_2 are supposed to be unaltered at the 12 interface. Similar equations can be written for the solid/liquid (S1) interface, and using Young's equation to eliminate γ_S, we can obtain an equation for the contact angle

$$\cos\theta^\circ = -1 + 2(\gamma_S{}^d\gamma_l{}^d)^{\frac{1}{2}}/\gamma_l - \pi_{S,l}/\gamma_l \qquad (33)$$

Various studies (e.g., Fowkes, 1964; Tamai *et al.*, 1967; 1977; Bargeman, 1972; Hamilton, 1972; 1974; Schultz *et al.*, 1977; Pashley and Israelachvili, 1981) have shown that equations (32) and (33) are quite effective for materials that interact only through dispersion forces, and give reasonable predictions of γ^d for liquids and solids in which other forces are active. The value of γ^d for water appears to be about 21.8 mN m^{-1}, leaving $\gamma^h \cong 51$ mN m^{-1}. Much of the more recent work has been carried out using two immiscible liquids and a solid, as pioneered by Tamai *et al.* (1967) and Bargeman (1972). More complex equations have also been proposed to account for polar intereactions (Kaelble, 1970; Kaelble and Uy, 1970; Wu, 1971). These equations have been used to investigate biological surfaces (El-Shimi and Goddard, 1974) and hydrophilic polymer surfaces (Andrade *et al.*, 1979; Ko *et al.*, 1981). For an alternative approach using solubility parameters see Panzer (1973).

In his later papers, Fowkes (1964a; 1967) supplements his theory by showing that for nonpolar liquids equation (32) can be obtained by summation of pair-wise interactions. Padday and Uffindell (1968) have integrated the London equation to similar purpose. The procedure is instructive and may be illustrated as follows.

The work of cohesion of phase i is the work done per unit area in breaking a slab of i and removing to two halves to infinite separation. If the average distance between neighbouring molecules or molecular groups is r°_i, then this work can be calculated by integrating equation (24) from r°_i to infinity and changing the sign. Thus,

$$Wc_i{}^d = 2\gamma_i{}^d = A_{ii}/12\pi(r^\circ_i)^2 \qquad (34)$$

and

$$\gamma_i{}^d = A_{ii}/24\pi(r^\circ_i)^2. \qquad (35)$$

Likewise, using equation (27), the work of adhesion between two dissimilar phases, i and j, is

$$W_{ij}{}^d = \gamma_i{}^d + \gamma_j{}^d - \gamma_{ij}{}^d = A_{ij}/12\pi(r^\circ_{ij})^2. \qquad (36)$$

Hence

$$\gamma_{ij}{}^d = \gamma_i{}^d + \gamma_j{}^d - [A_{ij}/12\pi(r^\circ_{ij})^2]. \qquad (37)$$

If

$$r^\circ_{ij} = (r^\circ_i r^\circ_j)^{\frac{1}{2}} \tag{38}$$

then

$$\gamma_{12}{}^d = \gamma_1{}^d + \gamma_2{}^d - 2(\gamma_1{}^d\gamma_2{}^d)^{\frac{1}{2}}. \tag{39}$$

If i and j are solid and liquid, then combination with Young's equation gives

$$\cos\theta^\circ = -1 + [A_{Sl}/12\pi(r^\circ_{Sl})^2\gamma_l] - \pi_{S,l}/\gamma_l \tag{40}$$

which is identical with equation (33).

In principle, at least, we can obtain r°_i and r°_j from molecular volumes or X-ray data, and we can calculate A_{ii} and A_{ij} by a variety of methods (see, for example, Gregory, 1969). The methods used by Padday and Uffindell gave good values for the n-hexane/water and n-hexane/mercury interfacial tensions, and for the contact angle of water on wax. However, it is important to remember, that in all these calculations, the values we obtain are strongly dependent upon the choice of r°. A 0.1 nm change in r° can change γ^d by 50%.

The Macroscopic Approach

Despite its apparent success, the Hamaker summation method suffers from a number of limitations:

a) The summation procedures ignore the influence of neighbouring molecules on pair-wise interactions.
b) Hamaker constants are usually evaluated at a single dominant interaction frequency in the ultraviolet, whereas molecular interactions often involve a large part of the electromagnetic spectrum. This can become important in wetting systems if the dispersion properties of the various media differ only slightly in the u.v. (Richmond *et al.*, 1973). Longer wavelength interactions also remain unretarded at greater separations.
c) With the summation procedure it is difficult to account for interactions across a third medium other than a vacuum.
d) With polar media, temperature-dependent Debye and Keesom forces may become significant.

However, all these problems are avoided by the macroscopic theory of dispersion forces formulated by Dzyaloshinskii *et al.* (1960; 1961). The DLP theory treats the interacting bodies as continuous and ascribes the interactions to fluctuating electromagnetic fields arising from spontaneous electric and magnetic polarizations within the various media. One important result is that the interactions are described completely in terms of the complex dielectric constants of the media. The equations are complex and will not be given here. The fact that the theory has become increasingly accessible to surface and colloid chemists is due largely to the efforts of Parsegian, Ninham, Langbein, Richmond and others who have shown how the theory can be employed using readily obtainable spectroscopic data (for details see Mahanty and Ninham, 1976). The predictions have usually been more consistent with experiment than those of the microscopic Hamaker approach (e.g., Richmond *et al.*, 1973; Donners *et al.*, 1977).

The significance of the macroscopic theory to the interpretation of wetting phenomena was considered by Dzyaloshinskii *et al.*, in their original papers (1960; 1961). For example, they showed how dispersion forces can give rise to liquid films that, dependent upon thickness, are stable, metastable or unstable. The problem has been pursued in greater detail by Israelachvili (1973). Again, the treatment is based on calculating the work of cohesion and the work of adhesion, and still relies on Young's equation for the contact angle. Nevertheless, Israelachvili gives the full equations for non-polar media in a variety of wetting configurations involving one or two liquids with or without adsorbed layers. He also shows what approximations and combining laws are required to produce the simpler Hamaker expressions. In sample calculations, good agreement with experiment is shown for the surface tensions of various saturated hydrocarbons and the dispersion component for water. Israelachvili stresses that his analysis is expected to hold only where adsorbed films are thick enough to be treated as continuous media. He points out that strictly, neither the macroscopic nor the microscopic theories of dispersion forces may be used at distances approaching interatomic spacing, but that given a sensible choice for r° (D in his notation), both theories are found to predict reasonable values for surface energies, and it is only this that justifies their use at short range. In order to be consistent with his macroscopic approach, Israelachvili uses a value of D of the order of atomic spacing ($\sim$ 0.2 nm) rather than the molecular spacing appropriate to the summation method.

The macroscopic theory has also been applied to wetting by Owens *et al.* (1978). Using an expression for the interfacial tension given by Mitchell and Richmond (1974), they have derived an equation similar to equation (40), but with r°_{Sl} replaced by a "cut-off wavelength", k^{-1}. In physical terms, k^{-1} is proportional to the mean diameter or spacing of the molecular groups at the 12 interface. Comparison with experimental data for n-alkanes on PTFE shows that k^{-1} is sensibly constant (∿0.16 nm) for octane and higher alkanes, but increases slightly at lower chain lengths. Similar values are found for n-alcohols on PTFE. Owens *et al.* also consider the effects of surfactant adsorption (see below).

Future Developments

Much of the interest in dispersion forces centres around their relatively long range. However, wetting behaviour is frequently dominated by short-range effects. This is illustrated by the integration of equation (24) to give γ_i^d, where it can be seen that the short-range part contributes most of the integral (equation (35)). The molecular structure and organization at the solid/liquid interface are especially important. This is shown by the general ability of deposited monolayers to mask the underlying properties of the substrate (Langmuir, 1920), and by phenomena such as autophobism. Nevertheless, provided the properties of the surface layers are known, it should be possible to include them within the macroscopic theory. This has already been done for soap films (where the presence of electrolyte must be considered) and for other layered structures (e.g., Mahanty and Ninham, 1976; Donners *et al.*, 1977). Modern, highly sensitive surface analytical methods such as ESCA should prove helpful in providing structural data for solid surfaces (e.g., Phillips and Dettre, 1976).

A more difficult task for quantitative molecular theories is to accommodate the structural (entropic) effects of short-range polar forces such as hydrogen bonding. Some progress has already been made for colloidal systems (Grimson *et al.*, 1980), and recent measurements of forces between crossed mica cylinders have given very direct evidence of the nature and range of hydration forces and other structure-dependent effects close to interfaces (Horn and Israelachvili, 1980; Pashley, 1981; Pashley and Israelachvili, 1981).

Finally, we note that computer-based molecular dynamics techniques previously applied to interfaces are now being used to study wetting phenomena both static (Saville, 1977) and dynamic (Schukin and Yushchenko, 1977).

THE CRITICAL SURFACE TENSION OF WETTING

Background

For many practical purposes, it is helpful to have some systematic way of characterizing the "wettability" of a substrate. Zisman and co-workers (Fox and Zisman, 1950; Zisman, 1964) have found that the contact angle exhibited by a liquid on a low energy surface* is largely dependent on the surface tension of the liquid, γ_{12} (where phase 2 is air plus vapour). In particular, they found that for a given substance and a series of related liquids (such as n-alkanes, siloxanes or dialkyl ethers) $\cos\theta$ is a reasonably linear function of γ_{12}. This is illustrated in Fig. 5. For unrelated liquids of widely ranging surface tension, the line broadens into a band which tends to be curved for high surface tension polar liquids. The surface tension at the point where the line cuts the $\cos\theta = 1$ axis is known as the critical surface tension of wetting, γ_C. It is the surface tension of a liquid that would just spread on the substrate to give complete wetting.

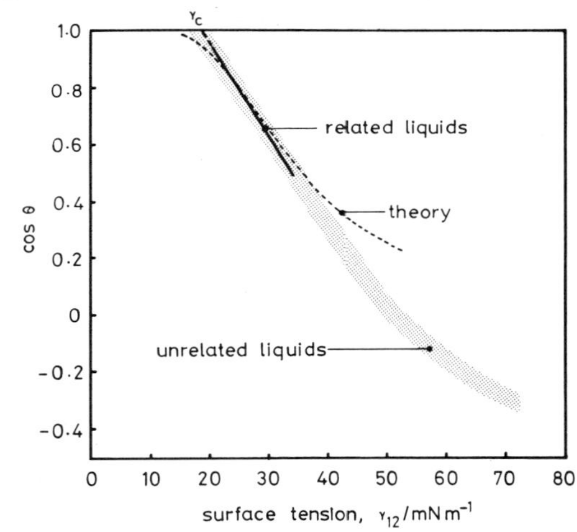

Fig. 5. Schematic Zisman plot for related liquids (solid line) and unrelated liquids of more widely ranging surface tensions (shaded band). The intercept on the $\cos\theta = 1$ axis is the critical surface tension of wetting, γ_C. The figure is based on data reported by Zisman (1964) for PTFE. The dotted curve illustrates the theoretical relationship between $\cos\theta^0$ and γ for a non polar solid having $\gamma_S{}^d = 20$ mN m^{-1}. The curve was obtained using the Fowkes equation [Equation (33)] with a sensible allowance $\pi_{S,1}$ as $\cos\theta^0 \rightarrow 1$.

---

* Defined as having a surface free energy of less than 100 mJ m^{-2} (Fox and Zisman, 1950). The description is usually applied to the surfaces of materials like hydrocarbons, fluorocarbons and polymers, i.e., materials with low melting points. High-melting solids, such as metals or ionic solids, usually have surface free energies ranging from hundreds to several thousand mJ m^{-2}. However, as normally encountered, these surfaces are usually covered with adsorbed layers of low energy substances such as water.

If $\gamma_{12} \leqq \gamma_C$, the liquid will spread. If $\gamma_{12} > \gamma_C$, the liquid will form a non-zero contact angle. Typical values are 43, 39, 31 and 18 mN m^{-1} for, respectively, polyethyleneterephthalate, PVC, polyethylene, and PTFE. The lowest known value is about 6 mN m^{-1}. This was found for condensed monolayers of perfluorolauric acid (Hare *et al.*, 1954).

Theoretical Basis

To some extent, the value of γ_C must depend on the set of liquids used to measure it. This is especially true if there are polar interactions. Zisman describes γ_C as "a useful empirical parameter whose relative values act as one would expect of $\gamma_S°$, the specific surface free energy of the solid". But other authors have been tempted to identify γ_C with γ_S or γ_S^d. According to the Good and Girifalco equation (equation (30)) with $\pi_{S,1} = 0$ and $\cos\theta° = 1$

$$\gamma_{12} = \gamma_1 = \Phi^2\gamma_S = \gamma_C \; . \qquad (41)$$

For nonpolar liquids and solids, $\Phi \cong 1$, hence $\gamma_C \cong \gamma_S$. Similarly, using the Fowkes equation (equation (33)),

$$\gamma_{12} = \gamma_1 = (\gamma_1^d\gamma_S^d)^{\frac{1}{2}} = \gamma_C . \qquad (42)$$

Hence, if $\gamma_1^d = \gamma_{12}$, then $\gamma_C = \gamma_S^d$. Equations (30) and (33) show that if $\pi_{S,1} = 0$, a plot of $\cos\theta°$ versus $(\gamma_{12})^{-\frac{1}{2}}$ should be linear with intercept $(\gamma_C)^{-\frac{1}{2}}$ on the $\cos\theta° = 1$ axis. The experimental results appear to support this prediction (Good and Girifalco, 1960; Fowkes, 1962). We might therefore expect that for nonpolar solids, $\gamma_C = \gamma_S$ provided there is no significant adsorption at the solid/air interface, i.e., provided $\pi_{S,1} = 0$. However, this last condition is most unlikely to be fulfilled when $\theta° = 0$.

The significance of γ_C can also be examined using Young's equation. If $\cos\theta° = 1$, then

$$\gamma_C = \gamma_{12} = \gamma_S - \pi_{S,1} - \gamma_{S1} \qquad (43)$$

Thus, $\gamma_C = \gamma_S$ only if <u>both</u> γ_{S1} and $\pi_{S,1}$ are zero. Evidently, γ_C is not a property of the solid alone. This important point also emerges from the theoretical interpretations of γ_C given by Owens *et al.* (1978) and Sullivan (1981).

The Zisman relationship can be expressed as

$$\cos\theta° = 1 - C\,(\gamma_{12} - \gamma_C) \qquad (44)$$

where the constant C usually has a value between 0.03 and 0.04. Equation (44) is incompatible with the Girifalco and

Good, and Fowkes equations except when $\cos\theta° = 1$ (and $\pi_{S,1} = 0$). A plot of $\cos\theta°$ versus γ_1 calculated using the Fowkes equation with a sensible allowance for $\pi_{S,1}$ as $\cos\theta° \rightarrow 1$, will be S-shaped as shown schematically by the dotted line in Fig. 5. Similar curve shapes are predicted by Sullivan and should also follow from the treatment given by Owens *et al.* Interestingly, Zisman plots for the n-alkanes appear to fit an S-shaped curve at least as well as a straight line (e.g., Fig. 3 of Zisman (1964)). As yet, we do not know whether the curve should become asymptotic to the $\cos\theta° = 1$ axis or should intersect it at a finite angle. The first option would indicate a second order transition between partial and complete wetting as predicted by Sullivan, whereas the second option would require the transition to be first order as concluded by Cahn (1977). Similar comments apply to the shape of the line obtained by plotting $\cos\theta°$ versus $(\gamma_{12})^{-\frac{1}{2}}$. If $\pi_{S,1} \neq 0$, then the line will either become asymptotic to the $\cos\theta° = 1$ axis or simply intersect at a lower value of γ_{12} than that given by linear extrapolation.

The fact that the various experimental plots are not markedly curved as $\cos\theta \rightarrow 1$ suggests either than $\pi_{S,1}$ is not usually very great or that it is not very great under the conditions obtaining during typical measurements of θ. In general, however, we cannot expect γ_c to provide more than a lower bound for $\gamma_S{}^d$ or γ_S.

Measurement with Solutions

The measurement of γ_c requires a range of liquids of widely varying surface tension $\gamma_{12} > \gamma_c$. This is difficult to achieve with pure liquids. An alternative approach is to use a series of aqueous surfactant solutions of varying concentrations or binary mixtures such as alcohols or glycols in water.

Bernett and Zisman (1959a) found that for hydrocarbon wetting agents on polyethylene or PTFE substrates, the solution method gave values of γ_c in reasonable agreement with those determined with pure liquids. They supposed that roughly equal adsorption of the solute at the 12 and S1 interfaces caused the solutions to behave as though they were hydrocarbon liquids with surfaces composed of oriented, packed hydrophobic groups. However, in subsequent experiments, Bernett and Zisman (1959b) found that fluorinated surfactants gave normal values of γ_c for PTFE, but values that were about 11 mN m^{-1} too low for polyethylene. They concluded that the effect was due to adsorption of the fluorinated compounds on the polyethylene surface, thereby altering its characteristics to those of a less wettable, fluorinated surface.

Further problems with the solution method were encountered

by Dann (1970) and Murphy *et al.* (1972). They found that even with hydrocarbon wetting aids, the solution method gave values of γ_c that were usually significantly lower than those obtained with pure liquids. Dann attributed this to polar interactions, while Murphy *et al.* considered it was due to solute adsorption at both the S1 and S2 interfaces. Very recently, Pyter *et al.* (1982) have persuasively argued that the low values of γ_c result from low adsorption at the S1 interface causing γ_{S1} to be higher than if the wetting liquid were a pure, nonpolar liquid. They find that with hydrocarbon surfactants, the effect is more significant on semi-polar solids, but with fluorocarbon surfactants like perfluoro-octanoic acid, the effect is important on both semi-polar solids and nonpolar hydrocarbon solids such as paraffin. Reference to Young's equation shows that larger values of γ_{S1} will cause larger contact angles and hence lower value of γ_c. There is no need to postulate adsorption at the S2 interface. We will return to the question of surfactant adsorption at low energy surfaces in the next section.

THE EFFECT OF SURFACTANT ADSORPTION

In this section we discuss the way in which surfactants can change the contact angle by adsorbing at the various interfaces and modifying the local interfacial tensions.

General Comments

For an ideal dilute solution of surface active solute, the Gibbs equation (Defay *et al.*, 1966) is

$$(d\gamma/d\ln c)_{T,P} = -RT\Gamma \quad (45)$$

where Γ is the specific adsorption at concentration c, and R, T and P have their usual significance. The effect on the contact angle can be seen by writing Young's equation as

$$\cos\theta° = (\gamma_{S2} - \gamma_{S1})/\gamma_{12} \; . \quad (46)$$

Addition of surfactant will reduce $\theta°$ if either γ_{S1} or γ_{12} (or both) are reduced and γ_{S2} remains more or less unchanged. The reduction will be more significant when γ_{S2} is large, e.g., if phase 2 is a gas rather than a liquid. Addition of surfactant will usually increase $\theta°$ only if the solute adsorbs at the S2 interface. This requires some transport mechanism either via phase 2 or by prior contact with the solution (as, for example, in mineral flotation). Alternatively, the same result can be achieved by adding a suitable surfactant to a liquid phase 2, or by transfer of an insoluble monolayer from the 12 interface to the S2 interface.

Following Smolders (1961), the general relationship between the change in contact angle with surfactant concentration and the adsorption at the three interfaces is obtained by differentiating Young's equation with respect to ln c at constant T and P,

$$(d\gamma_{12}\cos\theta°/d\ln c) = (d\gamma_{S2}/d\ln c) - (d\gamma_{S1}/d\ln c) \qquad (47)$$

and using equation (45). Thus

$$\sin\theta°(d\theta°/d\ln c) = RT(\Gamma_{S2} - \Gamma_{S1} - \Gamma_{12}\cos\theta°). \qquad (48)$$

Since $\gamma_{12}\sin\theta°$ is always positive, $(d\theta°/d\ln c)$ will always have the same sign as the RHS of this equation. We may therefore distinguish three modes of behaviour:

a) $(d\theta°/d\ln c) < 0$, $\Gamma_{S2} < \Gamma_{S1} + \Gamma_{12}\cos\theta°$;

b) $(d\theta°/d\ln c) = 0$, $\Gamma_{S2} = \Gamma_{S1} + \Gamma_{12}\cos\theta°$;

c) $(d\theta°/d\ln c) > 0$, $\Gamma_{S2} > \Gamma_{S1} + \Gamma_{12}\cos\theta°$.

In case (a), addition of surfactant leads to improved wetting, while in case (c) the surfactant causes de-wetting. In many practical situations, the mode of behaviour may vary with surfactant concentration, e.g., from de-wetting to wetting. However, one mode often dominates, and this obviously has important implications for the underlying pattern of adsorption. Cases (a) and (b) above are common with predominantly nonpolar low-energy surfaces, whereas case (c) is usually observed only with more polar substrates.

Surfactants do not always adsorb at the 12 interface. For example, many polymeric materials adsorb only at the solid/liquid interface, and oil-soluble surfactants frequently adsorb strongly at oil/solid or oil/water interfaces, but have negligible adsorption at an oil/air interface. Nevertheless, we are often concerned with water- or oil-soluble surfactants that <u>do</u> adsorb at the 12 interface and lower γ_{12}. In such cases it is sometimes helpful to distinguish the results of surfactant addition according to the way in which θ, $\gamma_{12}\cos\theta$ or some other thermodynamic quantity, such as the work of adhesion, varies with γ_{12}.

Nonpolar Surfaces

These surfaces are poorly wetted by water. If fluid 2 is a gas (air), the degree of wetting can usually be improved by adding a suitable surfactant. If fluid 2 is an oil, the situation is less straightforward. We consider the simpler case first.

Wetting by Aqueous Solutions when Fluid 2 is Air If the adsorption of surfactant at the solid/air interface is small, i.e., $\Gamma_{S2} \cong 0$, then from equations (45) and (47)

$$(d\gamma_{12}\cos\theta°/d\ln c) = -(d\gamma_{S1}/d\ln c) = RT\Gamma_{S1}. \quad (49)$$

Hence, we can determine the adsorption at the S1 interface from the variation of $\gamma_{12}\cos\theta°$ with lnc. Fowkes and Harkins (1940) used this method to show that the adsorption of low molecular weight amphiphiles such as n-butyl alcohol is almost the same at the paraffin wax/aqueous solution interface as at the solution/air interface, i.e., $\Gamma_{S1} \cong \Gamma_{12}$. Hence, from equation (45), $(d\gamma_{S1}/d\gamma_{12}) \cong 1$. Moreover, if $\Gamma_{S2} \cong 0$, then $(d\gamma_{S2}/d\gamma_{12}) \cong 0$. Subsequent studies have suggested that this pattern of adsorption is quite common for solutions of hydrocarbon surfactants like Aerosol OT (AOT) and sodium dodecylsulphate (SDS) on completely nonpolar surfaces such as paraffin wax, glyceryl tristearate and PTFE (Lucassen-Reynders, 1963; Wolfram, 1966; Bargeman and van Voorst Vader, 1973). Negligible adsorption at the paraffin wax/air interface has also been reported by Padday (1967).

These observations can be placed on a more quantitative footing as follows. The reversible work of adhesion between solution and solid can be written as (equation (12))

$$Wa°_{S12} = \gamma_{S2} + \gamma_{12} - \gamma_{S1} \quad (50)$$

Differentiating this with respect to γ_{12} gives

$$(dWa°_{S12}/d\gamma_{12}) = (d\gamma_{S2}/d\gamma_{12}) + 1 - (d\gamma_{S1}/d\gamma_{12}) \quad (51)$$

and setting $(d\gamma_{S2}/d\gamma_{12}) = 0$, and $(d\gamma_{S1}/d\gamma_{12}) = 1$, we obtain

$$(dWa°_{S12}/d\gamma_{12}) = 0. \quad (52)$$

This means that for systems in which $\Gamma_{S1} = \Gamma_{12}$ and $\Gamma_{S2} = 0$, $Wa°_{S12}$ is independent of surfactant concentration. According to equation (21)

$$Wa°_{S12} = \gamma_{12}(1 + \cos\theta°).$$

But, at the concentration at which $\cos\theta° = 1$, $\gamma_{12} = \gamma_c$. Hence, at <u>all</u> concentrations, $Wa°_{S12} = 2\gamma_c$, and

$$\cos\theta° = -1 + Wa°_{S12}/\gamma_{12} = -1 + 2\gamma_c/\gamma_{12}. \quad (53)$$

Johnson and Dettre (1969) used a slightly different route to derive this equation, and showed that it effectively accounts for the wettability of polyethylene by aqueous solutions of hydrocarbon surfactants (Bernett and Zisman, 1959a). The equation is less successful with the results obtained with fluorinated surfactants (Bernett and Zisman, 1959b).

We can derive a more general expression than equation (53), if we suppose that $(\Gamma_{S1}/\Gamma_{12}) = (d\gamma_{S1}/d\gamma_{12})$ is constant but not necessarily equal to one. If $(d\gamma_{S1}/d\gamma_{12}) = \xi$, then from equation (51)

$$(dWa^\circ_{S12}/d\gamma_{12}) = 1 - \xi. \tag{54}$$

Integrating this and using the subscript W to denote the aqueous phase with c = 0

$$Wa^\circ_{S12} - Wa^\circ_{SW2} = (1 - \xi)(\gamma_{12} - \gamma_{W2}) \tag{55}$$

whence, on eliminating Wa°_{S12} using equation (21), we obtain

$$\cos \theta^\circ = -\xi + [\gamma_{W2}(\xi - 1) + Wa^\circ_{SW2}]/\gamma_{12}. \tag{56}$$

This reduces to equation (53) when $\xi = 1$. If we use equation (21) once more to obtain Wa°_{SW2} from θ°_W, the contact angle when c = 0, then

$$\cos \theta^\circ = -\xi + \gamma_{W2}(\cos \theta^\circ_W + \xi)/\gamma_{12}. \tag{57}$$

A plot of $\cos \theta^\circ$ versus $1/\gamma_{12}$ should therefore yield a straight line of slope $\gamma_{W2}(\cos \theta^\circ_W + \xi)$ and intercept $\cos \theta^\circ = -\xi$. Notice that for a given solid, all surfactants with the same ξ should give the same line.

Equation (56) was derived by Owens *et al.* (1978) with the work of adhesion expressed in terms of a Hamaker function which they obtained via the DLP macroscopic theory (see above). Their data for aqueous solutions of anionic and non-ionic surfactants on polyethylene and PTFE give excellent straight lines when $\cos \theta$ is plotted against $1/\gamma_{12}$. From the intercept, $\xi \cong 1.2$, i.e., $\Gamma_{S1} \cong 1.2\ \Gamma_{12}$. The rationale for expecting surfactants to adsorb equally at the solid/solution and solution/air interfaces is that the driving force is the same, i.e., the free energy lost by excluding the hydrophobic surfactant tails from the water. The results of Owens *et al.* suggest that an additional force is at work at the solid/solution interface, presumably the dispersion attraction between the surfactant tails and the solid, which should be slightly greater than that between water and polyethylene or PTFE.

If we ignore the possibility of short-range entropic effects (hydrophobic bonding) at the solid/water interface, we can calculate Wa°_{SW2} in equation (56) via the Fowkes formula, equation (31). This is permissible since $Wa°_{SW2}$ refers only to water and the nonpolar solid substrate, and $\Gamma_{S2} = 0$. Thus,

$$\cos\theta° = -\xi + [\gamma_{W2}(\xi - 1) + 2(\gamma_W{}^d\gamma_S{}^d)^{\frac{1}{2}}]/\gamma_{12}. \quad (58)$$

If $\xi = 1$, then

$$\cos\theta° = -1 + 2(\gamma_W{}^d\gamma_S{}^d)^{\frac{1}{2}}/\gamma_{12} \quad (59)$$

which is equivalent to equation (33). This is an interesting result, because it implies that $\gamma_1{}^d = \gamma_W{}^d$ at all surfactant concentrations, even though γ_{12} may change by 50% or more.

Wolfram (1966) and Bargeman and van Voorst Vader (1973) obtained equation (59) in the rearranged from

$$\gamma_{12}\cos\theta° = -\gamma_{12} + 2(\gamma_W{}^d\gamma_S{}^d)^{\frac{1}{2}}. \quad (60)$$

Wolfram found several systems (solutions of alcohols and cationic surfactants on paraffin wax and PTFE) that gave the predicted linear relationship between $\gamma_{12}\cos\theta$ and γ_{12} with a slope of approximately -1. For water and paraffin wax, the intercept at $\gamma_{12}\cos\theta = 0$ gave $Wa°_{SW2} = 48$ mJ m^{-2}. The calculated value using literature values for $\gamma_W{}^d$ and $\gamma_S{}^d$ (21.8 and 25.5 mN m^{-1}, respectively) was 47.3 mJ m^{-2}. For solutions of anionic surfactants on flat polycrystalline layers of glyceryl tristearate, Bargeman and van Voorst Vader found

$$\gamma_{12}\cos\theta = -1.00\gamma_{12} + 51.8. \quad (61)$$

Using $\gamma_S{}^d = 28.5$ mN m^{-1}, they calculated $Wa°_{SW2} = 50$ mJ m^{-2}, which is also in good agreement with experiment. Bargeman and van Voorst Vader made other comparisons using published data for various surfactants on PTFE, paraffin wax and beeswax. In each case, the data were consistent with equation (60) and, where $\gamma_S{}^d$ values were available, the agreement between calculated and measured values of $Wa°_{SW2}$ was very good (47.5 and 38.5 mJ m^{-2} measured, against 47.3 and 41.2 mJ m^{-2} calculated, for paraffin wax and PTFE, respectively).

Not all the literature data conform to equation (60) in giving $(d\gamma_{12}\cos\theta/d\gamma_{12}) = -1$. For example, Lucassen-Reynders (1963) found the empirical relationship

$$\gamma_{12}\cos\theta = -0.56\gamma_{12} + 21 \quad (62)$$

for AOT solutions on glyceryl tristearate crystals. Bargeman and van Voorst Vader attributed this disparity with their own findings to the method used by Lucassen-Reynders to determine θ (measurement of the capillary pressure within a porous plug of crystals). However, if we suppose that $\xi = 0.56$ for this system then, taking $Wa°_{SW_2} = 51.8$ mJ m^{-2} from equation (61), we find that equation (56) predicts

$$\gamma_{12}\cos\theta = -0.56\gamma_{12} + 20.1 \qquad (63)$$

which is experimentally indistinguishable from equation (62). Similarly, Finch and Smith (1981) have used the results of Murphy *et al.* (1972) for aqueous solutions on polystyrene and found

$$\gamma_{12}\cos\theta = -0.42\gamma_{12} + 42.1. \qquad (64)$$

This is consistent with $\xi = 0.42$ if $Wa°_{SW_2} = 84.1$ mJ m^{-2}. From the original data, $\cos\theta_W \cong 0.1$. Inserting this in equation (21) gives $Wa°_{SW2} = 79.8$ mJ m^{-2}.

Murphy *et al.* have suggested that it may be wrong to assume that $\Gamma_{S2} = 0$ even for nonpolar solids. If both Γ_{S1} and Γ_{S2} are significant, then from equations (45) and (47)

$$(d\gamma_{12}\cos\theta°/d\ln c) = RT(\Gamma_{S1} - \Gamma_{S2}). \qquad (65)$$

Murphy *et al.* investigated the wetting of polystyrene, polyfluorobutene, polymethylmethacrylate (PMMA) and polyvinylidene fluoride by aqueous solutions of methanol, butanol and 2,2,2-trifluoroethanol. Their data showed that the variation of $(\Gamma_{S1} - \Gamma_{S2})$ with concentration was rather complex and not always consistent with $\Gamma_{S2} = 0$. However, this may have been due to the fact that the solutes were fairly volatile and at least three of the solids had some polar content. Some of the systems investigated by Wolfram (1966) also showed complex wetting behaviour and again this was especially marked when the solids were semi-polar (PMMA and polyethyleneterephthalate (PET)).

It is important to recognize that if the adsorption at both the S1 and S2 interfaces is proportional to Γ_{12}, then we may still expect a linear relationship between $\gamma_{12}\cos\theta$ and γ_{12}, and between $\cos\theta$ and $1/\gamma_{12}$. If $(d\gamma_{S2}/d\gamma_{12}) = \omega$ and, as before, $(d\gamma_{S1}/d\gamma_{12}) = \xi$, then from equation (51)

$$(dWa°_{S12}/d\gamma_{12}) = 1 + \omega - \xi \qquad (66)$$

whence,

$$\cos\theta° = (\omega - \xi) + [\gamma_{W2}(\xi - \omega - 1) + Wa°_{SW2}]/\gamma_{12}$$

$$= (\omega - \xi) + \gamma_{W2}(\cos\theta°_W + \xi - \omega)/\gamma_{12}. \qquad (67)$$

This reduces to equation (57) when $\omega = 0$. Owens *et al.* (1978) report the effect of insoluble monolayers of octadecanol, dipalmitoyl lecithin and sodium octadecylsulphate on the wetting of Nylon 6-6 by water. The data give a reasonably good straight line when plotted as cos θ versus $1/\gamma_{12}$. The intercept is at cos θ = 0.33, which implies that $\omega > \xi$. If $\omega = 1$, i.e., direct transfer from the 12 interface to the S2 interface, then $\xi = 0.67$.

The present author has applied equation (67) to data obtained by Ottewill and Vincent (1972) for aqueous solutions of ethanol on polystyrene. A good fit is obtained with $Wa°_{SW2} = 76.1$ mJ m^{-2} and $(\omega - \xi) = -0.6$. Independent measurements of ethanol adsorption on polystyrene latex particles are consistent with $\xi = 1$. In which case, $\omega = 0.4$ for this system.

From all these studies, it is clear that the variation of γ_{12} and θ with surfactant concentration provides only part of the information necessary to identify the complete pattern of adsorption at the three interfaces. For an unambiguous assessment we need additional information, e.g., the adsorption at the S1 or S2 interfaces, or the relationship between them.

This need has been highlighted in the recent paper by Pyter *et al.* (1982). Following Lucassen-Reynders (1963) and Wolfram (1966), we may differentiate Young's equation with respect to γ_{12}, and use the Gibbs equation (equation 45) to obtain

$$(d\gamma_{12}\cos\theta°/d\gamma_{12}) = (\Gamma_{S2} - \Gamma_{S1})/\Gamma_{12}. \qquad (68)^*$$

If $\Gamma_{S2} = 0$, then we can see that a plot of $\gamma_{12}\cos\theta°$ versus γ_{12} must have a slope of $-(\Gamma_{S1}/\Gamma_{12})$. Pyter *et al.* use this result to illustrate the wide range of wetting behaviour possible when surfactants and other wetting agents are added to a solvent. They consider the particular case of water on paraffin wax with $\cos\theta°_W = -0.276$ as shown in Fig. 6. Note that in this type of plot a straight line through the origin is the locus for a particular value of $\theta°$ and has slope cos $\theta°$. Three possibilities for $(\Gamma_{S1}/\Gamma_{12})$ are shown in the figure.

a) $(\Gamma_{S1}/\Gamma_{12}) = 1$. This gives the relationship predicted by the Johnson and Wolfram equations (equations (53) and (60)). The line intersects that for cos $\theta° = 1$ at γ_c = 26 mN m^{-1}. Hence, for complete wetting, the surfactant must be capable of lowering γ_{12} to 26 mN m^{-1} or less.
b) $(\Gamma_{S1}/\Gamma_{12}) = 0$. Complete wetting cannot be achieved at any value of γ_{12}.

---

* The same result is obtained by differentiating equation (67), since $\omega = (\Gamma_{S2}/\Gamma_{12})$ and $\xi = (\Gamma_{S1}/\Gamma_{12})$. Conversely, integration of equation (68) offers an alternative route to equations (67) etc.

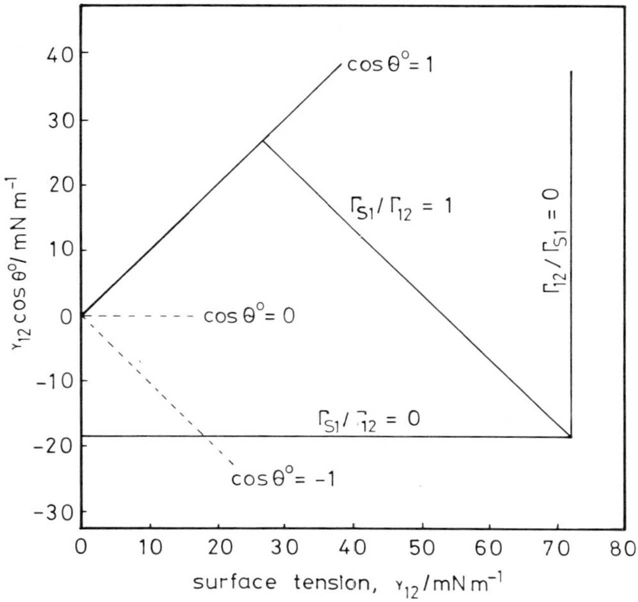

Fig. 6. Theoretical plots of γ_{12}, $\cos\theta^0$ versus surface tension, γ_{12}, for aqueous surfactant solutions on paraffin wax. The adsorption at the S2 interface is assumed to be zero. The figure is based on that of Pyter, et al. (1982), but this type of plot has also been used by Wolfram (1966) and Lucassen-Reynders (1983). Note that a straight line through the origin is the locus for a particular value of θ^0 and has slope $\cos\theta^0$.

c) $(\Gamma_{12}/\Gamma_{S1}) = 0$. Complete wetting can occur without any change in γ_{12}.

Clearly, adsorption at the S1 interface is crucial for improved wetting.

Pyter *et al.* present data for aqueous solutions of AOT and perfluorooctanoic acid on Nylon 11, PMMA, paraffin wax (Parafilm M) and polyethylene. Only the data for AOT on paraffin wax has a constant slope of about -1. The data for AOT on Nylon and PMMA have an apparently constant but very shallow slope of about -0.1. The remaining systems show very complex behaviour. Pyter *et al.* show how independent evidence of the magnitude of Γ_{S1} can be used to clarify the underlying adsorption pattern. One important point to emerge is that fluorocarbon surfactants appear to adsorb less strongly than hydrocarbon surfactants on nonpolar hydrocarbon solids. This results in larger values of γ_{S1} and θ° at any particular γ_{12} and, in consequence, a lower value of γ_c (see above). Note, however, that fluorocarbon surfactants can still be good wetting agents since they usually give lower values of γ_{12} than their hydrocarbon counterparts.

Competitive Wetting by Aqueous Solutions and Oil Lucassen-Reynders (1963) investigated the competitive wetting of glyceryl tristearate crystals by aqueous solutions of AOT and paraffin oil. She found the remarkable result that θ was almost constant at 108° over the entire AOT concentration range. A plot of $\gamma_{12}\cos\theta$ versus γ_{12} gave the relationship

$$\gamma_{12}\cos\theta = -0.31\gamma_{12} - 0.2 \tag{69}$$

i.e., a line passing almost exactly through the origin in Fig. 6. Comparable results were obtained with either cetyl alcohol in the oil or SDS in the water. With sodium dodecylsulphonate and flat polycrystalline layers of glyceryl tristearate, Bargeman and van Voorst Vader (1973) found

$$\gamma_{12}\cos\theta = -0.98\gamma_{12} + 0.2 \tag{70}$$

which corresponds to a nearly constant contact angle of about 170°. Almost exactly the same relationship was found on changing the substrate to PTFE or the oil to tetradecane. Again, it did not matter whether oil-soluble or water-soluble surfactants were used.

Application of equation (67) to Bargeman and van Voorst Vader's results shows that for their systems, $Wa°_{SW2}$ is very small: about 2 mJ m^{-2}. Lucassen-Reynders' results give a larger value (∿50 mJ m^{-2}) if the coefficient of γ_{12} is correct. However, there is every reason to expect that the work of adhesion between water and a nonpolar solid will be small in the presence of a nonpolar oil. This is because the only attractive forces between each phase are dispersion forces of similar magnitude. Moreover, all these forces are weak relative to the inwardly directed cohesive forces of water due to hydrogen bonding. Hence, both $\theta°_W$ and $W°_{S2W}$, the reversible work of adhesion between oil and solid in the presence of water, will be substantial.

We can give these conclusions a more quantitative basis by using the Fowkes equation to calculate $W°_{SW2}$ and $W°_{S2W}$ directly (Bargeman, 1972). Starting with equation (12) and substituting for each surface tension in turn using equation (32) yields

$$Wa°_{SW2} = 2[\gamma_2 - (\gamma_W{}^d\gamma_2)^{\frac{1}{2}} - (\gamma_S{}^d\gamma_2)^{\frac{1}{2}} + (\gamma_S{}^d\gamma_W{}^d)^{\frac{1}{2}}]. \tag{71}$$

Bargeman and Van Voorst Vader have evaluated this equation for paraffin oil and tetradecane on PTFE, and paraffin oil on glyceryl tristearate. The values they obtained were 2.0, 0.4 and 0.6 mJ m^{-2}, respectively. In contrast, if we carry out the same procedure for $Wa°_{S2W}$, we find corresponding values of 101.6, 99.8 and 103.4 mJ m^{-2}.

Nearly constant contact angles have also been observed on more polar substrates. Smolders (1967) investigated the relative wetting of Nylon-6 and PET by n-heptane and aqueous solutions of lauroyldiethanolamide and dodecylpyridinium bromide. Although there was considerable contact angle hysteresis, θ_A and θ_R were more or less independent of surfactant concentration. Both solids were more wettable by water than were glyceryl tristearate and PTFE. Nylon-6 was more wettable than PET. Application of equation (67) suggests adsorption of surfactant at both S1 and S2 interfaces.

Smolders also studied solutions of DPB against chloroform on Nylon-6. This system showed a steady <u>increase</u> in the water contact angle with increasing concentration of surfactant. The slope of the $\gamma_{12}\cos\theta_A$ versus γ_{12} plot is about +0.53, which implies strong adsorption at the Nylon-6/chloroform interface.

All these results have important implications for detergency. One of the main functions of a detergent is to reduce θ_W and thereby "roll up" spots of oily contaminant for easy removal by mechanical agitation. Evidently, this function is likely to be severely curtailed on nonpolar surfaces. Equation (67) shows that if $\omega = 0$ and $\xi = 1$, the most effective surfactants are likely to be those which lower γ_{12} by the greatest amount. The equation also shows that poor adsorption at the solid/water interface ($\xi < 1$) or any adsorption at the solid/oil interface ($\omega > 0$) will be highly detrimental to the roll-up mechanism.

Polar Solids

This section is chiefly concerned with ionic solids such as minerals and ceramics, but the field also includes the more polar organic solids.

In many industrial processes, surfactants are used to modify the wetting characteristics of polar solids. Typical applications include: mineral flotation; detergency; the wetting of powders prior to dispersion; the use of cationics to displace water and promote adhesion between bitumen and stone chippings in road making; and promoting dropwise condensation in distillation. Each of these topics has its own wide literature (e.g., see the recent extensive review of wetting and flotation given by Finch and Smith (1981)). Here, we consider only a few principles of general application.

Polar solids usually have high surface free energies and can therefore be wetted by high surface tension liquids. However, many of these liquids are themselves polar and frequently interact with the solid in some specific way. Water is a classic example. In many cases, complete wetting by water is possible only if the surface can form sufficient hydrogen bonds to overcome the internal cohesion of water. Autophobic liquids constitute another special case (see above).

The situation becomes even more complex when one considers wetting by surfactant solutions. This is because surfactant adsorption and its degree of reversibility are strongly influenced by a large number of variables. These include:

a) Surface polarity (dipole moment, etc.).
b) Surface charge and the nature of the charge-determining ions.
c) Solution pH and ionic strength. This affects both the surface charge and the degree of ionization of an ionic surfactant.
d) The degree of surface hydration.
e) The presence of specific ions, such as Cu^{++} or Ca^{++}, with which chelation complexes or insoluble surface compounds may be formed, e.g., calcium oleate on the surface of calcium carbonate.

As we saw in the last section, surfactant adsorption on weakly polar surfaces (e.g., polymers such as PMMA and PET) appears to be sufficiently non-specific to allow us to apply predictive models developed primarily for nonpolar surfaces. With ionizable surfaces, however, it is necessary to fall back to the more general wetting criteria proposed by Smolders (1961) and discussed above. We consider two cases involving the wetting of ionic solids by aqueous surfactant solutions.

In the first case, the surfactant ions and the solid surface have like charge. Because of this, there is little adsorption at the solid surface and little change in either γ_{S1} or γ_{S2}. Any reduction in θ is caused by adsorption at the 12 interface and the resulting reduction in γ_{12}. Once complete wetting is achieved, the spreading film will be stabilized against rupture by electrostatic repulsion between the ionic double layers associated with the S1 and 12 interfaces.

In the second case, the surfactant ions and the solid surface have charges of opposite sign. Examples are dodecylammonium chloride and silica between pH 3 and 12, and SDS and alumina (Al_2O_3) below pH 9. The variation of θ and γ_{12} with surfactant concentration is illustrated schematically in Fig. 7. We suppose that the solid is initially wetted by water.

At low concentrations, the surfactant ions are attracted by the oppositely charged sites on the solid and adsorb there more strongly than at the 12 interface. Adsorption at the solid/solution interface reduces the interaction between the water and the solid and thus produces a net increase in

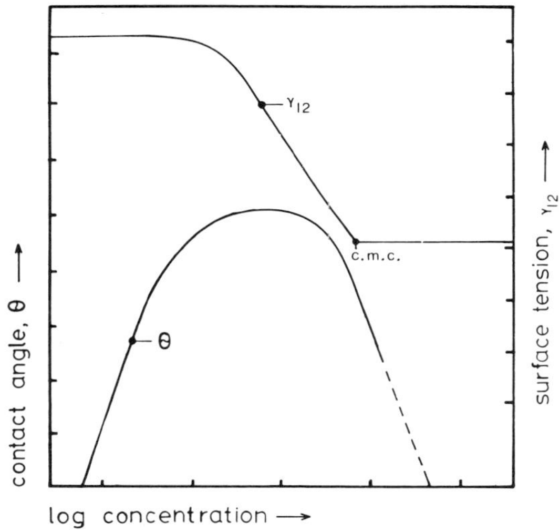

Fig. 7. Surfactant adsorption on polar surfaces. Schematic plots of θ and γ_{12} versus log concentration, for aqueous solutions of ionic surfactants when the surfactant ions and solid surface have charges of opposite sign. After Johnson and Dettre (1969).

γ_{S1}.* This, in turn, causes an increase in θ and retraction of the wetting line to leave surfactant adsorbed at the solid/air boundary, which now has a lower surface tension than before. There is some debate in the literature as to whether the adsorption is greatest at the solid/solution or the solid/air interface (Finch and Smith, 1981). However, provided some mechanism exists by which the surfactant ions can reach the S2 interface (e.g., transfer from the 12 interface during retraction) adsorption there would appear to be energetically more favourable.

As the surfactant concentration is increased, monolayer adsorption at the S1 interface approaches completion, whilst adsorption at the 12 interface becomes increasingly significant, thus reducing γ_{12} and $(d\theta/d\ln c)$ until θ reaches either a maximum or a plateau region. At this point the charge on the solid surface is usually close to zero, and conditions also correlate roughly with those at which mineral flotation occurs most readily (Fuerstenau and Healy, 1972).

Finally, at sufficiently high concentrations, which are generally just below the critical micelle concentration

---

* Although γ_{S1} is increased, the specific surface free energy is reduced, else the surfactant would not adsorb (equation (4)). This highlights the risks of confusing surface tensions with surface free energies.

(c.m.c.), θ, may begin to decline. The main reason for this is the adsorption of a second monolayer of surfactant at the solid/solution interface. The second layer is adsorbed with its ionic groups oriented towards the solution, thereby increasing the interaction with the water and reducing γ_{12}. The process has been termed hemi-micelle formation (Gaudin and Fuerstenau, 1955) and appears to occur only with long-chain surfactants. In some systems, e.g., decylamine on platinum (Ruch and Bartell, 1960), the contact angle eventually falls to zero, i.e., there is complete rewetting.

Very recently, Pashley and Israelachvili (1981) have obtained convincing proof as to the reality of both the first and second layers, by directly measuring their thicknesses for cetyltrimethylammonium bromide (CTAB) adsorbed at the surface of mica. In this system, the first layer appears to adsorb by ion exchange. The second layer is more weakly bound. It can be removed by rinsing in water and does not remain at the solid/air interface. First layer adsorption at the solid/air interface is sufficient to prevent rewetting by CTAB solutions, for which $\gamma_{12} \cong 37.5$ mN m^{-1} at the c.m.c. Pashley and Israelachvili found that above the c.m.c., θ_A was roughly constant at about 46°, though θ_R was smaller and continued to fall with increasing concentration. They estimated that $\gamma_{S2} \cong 27$ mN m^{-1}, which is close to the value expected for a hydrophobic surface made up of roughly equal areas of exposed CH_2 and CH_3 groups.

The adsorption of the second layer at the solid/solution interface increases the net surface charge once more, and gives the S1 and 12 interfaces the same polarity. Electrostatic repulsion between the S1 and 12 interfaces can lead to metastable films on hydrophobic surfaces (Blake and Kitchener, 1972), and this may have an influence on the kinetics of mineral flotation at high surfactant concentrations (Kitchener, 1975).

CONTACT ANGLE HYSTERESIS

In previous sections we have seen how the contact angle provides a quantitative measure of wetting and, through Young's equation, enables us to express useful quantities, such as the work of adhesion, in terms of measurable parameters, namely γ_{12} and $\theta°$. Unfortunately, as discussed above, the contact angle often exhibits hysteresis so that $\theta°$ cannot be defined unambiguously by experiment.

Contact angle hysteresis must have been obvious to the earliest experimenters. It was certainly known to Rayleigh (1890b). The term hysteresis was first used in this context by Sulman (1919). Adam and Jessop (1925) suggested that the phenomenon was due to a frictional force acting at the wetting line and tending to prevent displacement. More recent explanations have been concerned with the more specific effects of surface roughness, surface heterogeneity and kinetic factors. Fairly recent reviews have been given by Johnson and Dettre (1969) and Blake and Haynes (1973).

Surface Roughness, Surface Heterogeneity and Metastable Configurations

In order to explain the effects of roughness on the observed contact angle, θ, Wenzel (1936) proposed that

$$\gamma_{12} \cos\theta = R_W(\gamma_{S2} - \gamma_{S1}) \tag{72}$$

where R_W is the roughness factor defined as the ratio of real and apparent surface areas. Substitution for $(\gamma_{S2} - \gamma_{S1})$ from equation (1) yields

$$\cos\theta = R_W \cos\theta°. \tag{73}$$

Cassie and Baxter (1944; 1948) extended Wenzel's treatment to heterogeneous surfaces and derived the equation

$$\cos\theta = a' \cos\theta' + a'' \cos\theta'' \tag{74}$$

where a' and a" are the fractions of the solid surface with different intrinsic contact angles θ' and θ".

Although both equations (73) and (74) predict that θ will differ from θ°, neither offers a satisfactory explanation of hysteresis, since each predicts that θ will be single-valued unles R_W, a' and a" vary locally at the wetting line. However, R_W, a' and a" are all defined as average quantities, not local ones.

Derjaguin (1946) appears to have been the first to suggest that contact angle hysteresis may result from the fact that a liquid interface in contact with a rough surface can adopt any one of a series of metastable configurations each with a different apparent contact angle. He showed that if the interface were displaced, the wetting line would move in a series of thermodynamically irreversible jumps from one metastable configuration to the next. Similar ideas were put forward independently by Shuttleworth and Bailey (1948), but there was no quantitative treatment until Johnson and Dettre (1964) carried out a detailed analysis of the configurations open to a liquid drop resting on idealized rough and heterogeneous surfaces.

The idealized rough surface comprised a concentric pattern of sinusoidal corrugations. A drop of liquid placed axisymmetrically on such a surface will exhibit an apparent contact angle θ that is related to θ° by

$$\theta = \theta^\circ + \alpha \tag{75}$$

where α is the slope of the solid surface at the wetting line (Fig. 8a). The value of θ is therefore dependent on the location of the wetting line and, hence, upon such factors as drop volume and gravitational forces. The model heterogeneous surface comprised a series of concentric bands having alternate characteristic contact angles θ' and θ'', such that $\theta' > \theta''$. The boundaries were supposed to be infinitely sharp and to contribute nothing to the free energy of the system. A drop of liquid placed on this type of surface will spread or retract until the wetting line rests at a boundary and the drop assumes some configuration such that $\theta' \geqq \theta \geqq \theta''$ (Fig. 8b).

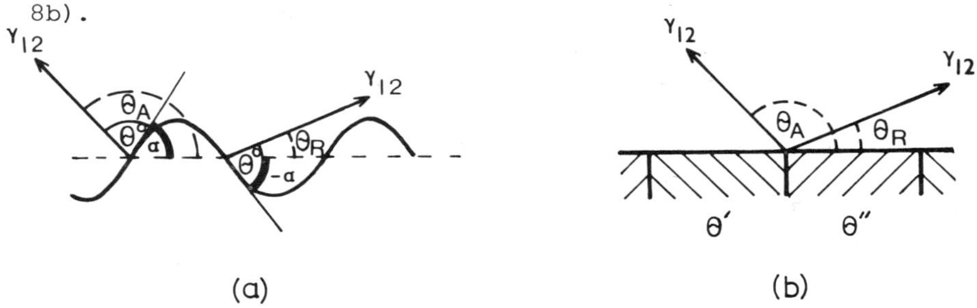

Fig. 8. Origin of contact angle hysteresis on model surfaces: (a) rough surfaces, where $\theta = \theta^0 + \alpha$; (b) heterogeneous surfaces, where $\theta' \geqslant \theta \geqslant \theta''$.

Johnson and Dettre showed that, for each model system, there is a family of metastable configurations of differing θ. Each configuration is separated from its neighbours by energy barriers, E_θ, that hinder movement of the wetting line but tend to zero as θ approaches its theoretical limits. On the rough surface, these limits are equal to $(\theta^\circ + \alpha_{max})$ and $(\theta^\circ + \alpha_{min})$, where α_{max} and α_{min} are the maximum and minimum slope angles. On the heterogeneous surface, the limits are simply θ' and θ''. However, Johnson and Dettre proposed that the drop will have some vibrational energy, E_D, such that the energy barriers hinder the wetting line only if $E_D < E_\theta$. Thus, the amount of hysteresis actually observed is dependent on the level of vibration. The configuration for which E_θ is greatest has the lowest free energy. On a rough surface, this configuration approximates to that predicted by Wenzel, equation (73), whereas on a heterogeneous surface it tends to that predicted by Cassie and Baxter, equation (74).

Dettre and Johnson (1964; 1965; 1967) subsequently reported a series of ingenious experiments on artificially roughened and specially constructed surfaces. Their results confirm qualitatively their theoretical predictions, as do the results of earlier studies by Bartell and Shepard (1953).

One prediction is that small patches of low contact-angle material in a predominantly high contact-angle surface will markedly decrease the receding contact angle, θ_R, and leave the advancing angle, θ_A, relatively unaffected. Thus, θ_A is more characteristic of the surface that θ_R. This effect probably accounts for the widespread observation that θ_A is more reproducible than θ_R on poorly wetted surfaces. However, the converse also applies, so that for a predominantly low contact angle surface, θ_R is more representative than θ_A.

Another prediction experimentally confirmed is that when a liquid spreads on a sufficiently rough, poorly wetted surface, some of the displaced phase is trapped in the valleys between asperities. The result is a composite solid/fluid interface. A composite interface is also formed on a poorly wetted, porous surface (Cassie and Baxter, 1944). In either case, θ_R tends to increase causing a marked reduction in hysteresis. A practical consequence of this is the excellent water-shedding properties of woven waterproof fabrics and the leaves of plants.

Real Surfaces

Models such as those used by Johnson and Dettre are necessarily highly idealized in terms of their geometry. A recent study by Hitchcock *et al.* (1981) illustrates some of the problems of interpreting contact angle hysteresis on real surfaces in terms of practical quantities such as the degree of roughness as measured by surface profilometers.

Observations of liquids spreading on real surfaces show that often only quite short lengths of the wetting line are in motion at any one time (Huh and Scriven, 1971). The wetting line typically lies across a more or less random array of heterogeneities or asperities and thus provides a highly irregular boundary condition for the adjacent fluid/fluid interface. In general, there will only be a limited number of configurations of constant mean curvature open to the interface, and spontaneous transitions between them are presumably responsible for the locally unsteady movement of the wetting line (Blake and Haynes, 1973). Support for this view may be drawn from a theoretical perturbation analysis of the stability of non-axisymmetric drops on complex model surfaces and a randomly rough surface (Huh and Mason, 1977a). This study also emphasizes the fact that it is not just the degree of roughness that is important but also the nature of the roughness, i.e., the "surface texture". It is well known that on some rough surfaces wicking occurs, i.e., the liquid spreads between the asperities rather than over them. A system of microscratches, for example, can considerably aid spreading provided $\theta° < 90°$. Conversely, if $\theta° > 90°$ then spreading will be inhibited.

Wicking and related behaviour have been demonstrated using scanning electron microscopy in an elegant study of spreading on various surfaces including cleaved mica, nickel grids, hexagonally close-packed latex particles and machined surfaces (Oliver and Mason, 1977). This work also highlights the problem of setting practical limits to the size of the surface irregularities likely to cause hysteresis. The experiments show that sharp-edged steps of less than 50 nm can be sufficient to hinder movement of the wetting line and cause variations in θ. Further studies (Oliver *et al.*, 1977) suggest that the range of angles open to a meniscus pinned at such a step is correctly specified by inequalities (2a) and (2b). On the other hand, steps as large as 800 nm apparently present no barrier if they have ragged geometry and unsharp edges (Mori *et al.*, 1982). Presumably, for an observable effect, some minimum length of the wetting line must be arrested simultaneously with a configuration that allows the meniscus to retain its constant curvature. A theoretical analysis of strip and patch-wise heterogeneous surfaces (Neumann and Good, 1972) suggests that this minimum length could be about 100 nm. Boruvka and Neumann (1978) propose that line tension may play a role in increasing the number of permissible configurations on more highly differentiated surfaces. In the sub-microscopic limit, however, the influence of irregularities of little more than molecular size must be smoothed out by thermal fluctuations in much the same way that larger barriers may be swamped by the vibrational disturbances and capillary waves considered by Johnson and Dettre (1964) and, more recently, by Bayramli *et al.* (1981).

Kinetic Factors

According to strict convention, the term hysteresis should not be applied to contact angle variation arising from non-equilibrium, kinetic sources (Blake and Haynes, 1973). In many experimental situations, a contact angle that changes with time leads inevitably to movement of the wetting line. The result is a dynamic contact angle, which we will discuss in more detail below.

However, in preparing for contact angle measurement, some movement of the wetting line is unavoidable, and this must cause at least a temporary disturbance of equilibrium. For example, the state of adsorption at the S1 and S2 interfaces will not, in general, be the same, and displacement of, say, fluid 2 by fluid 1 will involve destruction of S2 interface and creation of fresh S1 interface. If the sorption-desorption processes or the accompanying transport processes to and from the various interfaces are slow compared with the rate of displacement, then a non-equilibrium contact angle will result. If the external driving force is now removed, the contact angle will relax towards its equilibrium value (see, for example, Phillips and Riddiford, 1967). But if the relaxation processes are sufficiently slow relative to experimental time scales, then a non-equilibrium angle may persist in an apparently stable system, and it will be difficult to distinguish this effect from true hysteresis. Hansen and Miotto (1957) have suggested that this type of behaviour would also result from slow molecular reorientation following movement of the wetting line.

Clearly, if contact angle measurements are to be given equilibrium significance, great care must be taken to ensure that equilibrium has been achieved. This caution is especially pertinent to measurements with surfactant solutions, where there may be no ready mechanism by which a non-volatile solute can reach the solid/vapour interface, except by prior contact with the solution. For a recent review of time-dependent contact angles in systems containing surfactants, see Finch and Smith (1981).

Another example of how kinetic factors may cause contact angle variation arises from the possibility that the liquid penetrates the solid surface. The penetration of nonpolar surfaces by water is commonly cited (Edser, 1922; Kawasaki, 1962; Zisman, 1964), and Timmons and Zisman (1966) have reported a general relationship between molecular size and the extent of contact angle hysteresis. The evidence suggests that the observed behaviour simply reflects the non-attainment of a uniformly penetrated surface during the period of observation. In extreme cases, penetration may cause swelling. If so, then the wetting line will rest along a labile ridge, and contact angle variation should be similar to that observed when the wetting line is pinned at an edge.

In summary, therefore, contact angle variability can be caused by a number of factors, but unless proper attention is paid to the attainment of equilibrium, we should be cautious in ascribing the phenomenon to permanent features of the system such as surface roughness or intrinsic heterogeneity. Indeed, in many practical systems all the factors we have discussed may operate together.

WETTING KINETICS

The Dynamic Contact Angle

Contact angles associated with moving wetting lines are described as dynamic contact angles. The same name is sometimes given to contact angles that change with time (Wilkinson and Elliott, 1974). As noted above, one phenomenon frequently leads to the other, and there are quite a number of mechanisms which could cause the dynamic contact angle to differ from its equilibrium value even at very slow rates of wetting line displacement.

Although speed-independent regimes have sometimes been reported (Elliott and Riddiford, 1967; Johnson *et al.*, 1977) contact angles are usually dependent on both speed and direction of wetting line displacement, i.e., they are velocity-dependent. Experiment shows that advancing angles increase and receding angles decrease with increasing rate of displacement (Fig. 2). This behaviour is clearly related to contact angle hysteresis in that both imply thermodynamic irreversibility (Blake and Haynes, 1973; Everett, 1980). In the static case, this is attributed to spontaneous transitions between metastable equilibrium states. In the dynamic case, however, the interface may fail to attain any kind of equilibrium in the time available. Irreversibility will then result from the spontaneous progress towards equilibrium. In both cases, the contact angle changes in such a way as to oppose movement of the wetting line.

The first experimental study of dynamic contact angles appears to have been that of Ablett (1923). Notable recent studies include those of Inverarity (1969), Schwartz and Tejada (1970; 1972), Hoffman (1975), Burley and Kennedy (1976; 1978), Gribanova and Mulchanova (1978), and Petrov and Radoev (1981). Velocities investigated range from a few micrometres per second to metres per second.

Figure 9 shows fairly typical data. This particular set were obtained by Burley and Kennedy for a Mylar polyester tape entering a pool of 0.456 poise aqueous glycerine solution. At low rates of wetting, the advancing contact angle appears to be a steep function of the wetting velocity, v. As v increases, the slope first decreases and then increases again as θ approaches 180° at v_{180}. At still higher velocities, the wetting line develops a sawtooth shape (Fig. 10a) and air is entrained from the trailing vertices (Derjaguin and Levi, 1964; Wilkinson, 1975; Burley and Kennedy, 1976a; Blake and Ruschak, 1979).

Much of the interest in dynamic contact angles lies in maximizing the wetting velocity at the onset of entrainment. For example, in liquid coating operations, entrainment of air

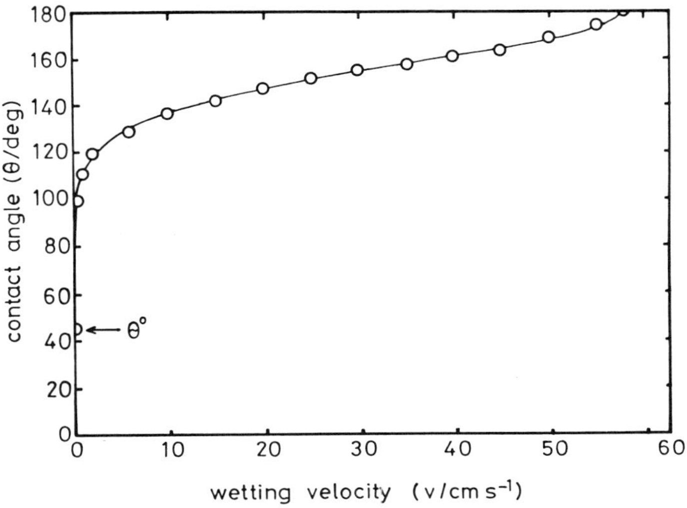

Fig. 9. Velocity dependence of the contact angle: data of Burley and Kennedy (1976) for 0.456 poise aqueous glycerine solutions on Mylar polyester tape.

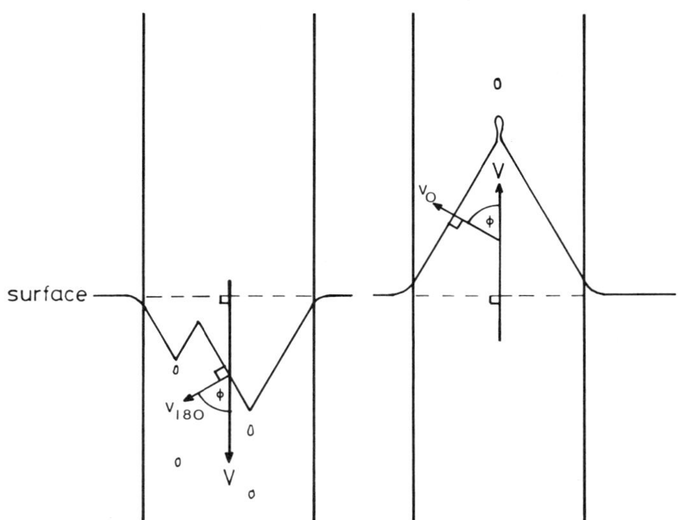

Fig. 10. Formation of sawtooth wetting lines showing entrainment from trailing vertices: (a) wetting; (b) dewetting (Blake and Ruschak, 1979).

leads to patchy or uneven coatings; in petroleum recovery, entrainment of crude oil by gas or water flood may reduce the efficiency of recovery and the ultimate yield of a reservoir. Blake and Ruschak (1979) have argued that the occurrence of the sawtooth wetting line shows that for any given system, v_{180} is the maximum velocity at which a wetting line can advance normal to itself. If an attempt is made to wet a solid at some velocity $V > v_{180}$, the wetting line must lengthen and slant at some angle ϕ relative to its orientation at velocities below v_{180}, such that

$$\cos \phi = v_{180}/V. \quad (76)$$

Similarly for dewetting; if an attempt is made to dewet a solid (for which $\theta_R > 0$) at a velocity greater than that at which θ becomes zero, v_0, then the wetting line again consists of at least two straight segments slanted at an angle ϕ relative to the normal orientation (Fig. 10b). In this case

$$\cos \phi = v_0/V \quad (77)$$

and drops of liquid or a continuous rivulet may be entrained from the trailing vertices. Blake and Ruschak report data in good agreement with these simple relationships.

Of perhaps more academic interest: the very steep velocity-dependence shown in Fig. 9 as $v \to 0$ strongly suggests the kinetic origin of at least some instances of apparent contact angle hysteresis.

Effects of Viscosity and Surface Tension

The available evidence indicates that the velocity-dependence of the contact angle increases with increasing liquid viscosity, η, and decreasing surface tension, γ. As an indication of these trends, we may note that Hoffman (1975), Jiang *et al.* (1979), and Rillaerts and Joos (1980) have all found empirical correlations between θ and the capillary number, $Ca = (\eta v/\gamma)$, such that θ always increases with Ca.

The effect of viscosity is especially well illustrated by the results of Inverarity (1969) and Burley and Kennedy (1976), who carried out systematic studies using solutions of progressively higher viscosity on glass and polymer fibres, and on polyester tapes. It is noticeable that although v_{180} is fairly sharply defined for low viscosity solutions, the approach to 180° becomes almost asymptotic at high viscosities (> 1 poise, say). However, in general, the results suggest that higher viscosities lead to greater velocity-dependence of θ and lower values of v_{180}; i.e., viscous forces tend to oppose wetting. Studies of the rate of spreading of polymer droplets have shown that viscosity has a similar, inhibiting effect on the rate of change of θ with time (Arslanov and Ogarev, 1977).

The effect of surface tension is less clear cut. Burley and Kennedy (1976a) have investigated the wetting of polyester tape by a series of solutions having the same viscosity, but with surface tensions ranging from 62.3 to 28.2 mN m^{-1}. They confirmed that lower surface tensions resulted in increased velocity-dependence of θ and therefore lower values of v_{180}. The change was roughly in proportion to the change in γ. On the other hand, practical experience suggests that the addition of surfactants can improve the rate of wetting (see below).

Quite a number of authors (e.g., Cherry and Holmes, 1969; Blake and Haynes, 1969; Ruckenstein, 1969; Lau and Burns, 1973; and Pier *et al.*, 1977) have proposed that the driving force for wetting due to surface forces may be represented by

$$F_W = \gamma_{12}(\cos\theta^\circ - \cos\theta). \tag{78}$$

Here, F_W is effectively the out-of-balance surface tension acting at the wetting line in the direction of wetting. The work done by F_W is the irreversible work expended by surface forces in promoting wetting at some non-zero velocity rather than infinitely slowly. Clearly, if γ_{12} is reduced, θ must increase to keep F_W constant. But, in reducing γ_{12}, we may also change θ°. If these changes are brought about by adding a surfactant, then the net effect on θ will depend on the relative adsorption at the 12, S1 and S2 boundaries. Unfortunately, there appear to have been no reported systematic studies of the effect of surfactants on the dynamic contact angle.

Theoretical Interpretation

Despite increasing attention, our understanding of the kinetics of wetting is far from complete. This is partly because of a shortage of experimental evidence, but also because the theoretical problem is highly complex. We can only hint at it here.

One major difficulty concerns the significance of the dynamic contact angle and how its experimentally determined value can be related not only to the velocity of wetting, but also to the material properties of the system. The problem is compounded by our uncertainty of the true nature of the contact angle very close to the wetting line even under equilibrium conditions. Recent hydrodynamic studies of wetting (Lacey, 1982; Hocking and Rivers, 1982) have treated this local angle as fixed, i.e., velocity-independent. White (1977) has suggested that at the molecular level equilibrium contact angles are always zero (or 180°). This could also be true in the dynamic case, but for any other angle it appears highly unlikely that the balance of forces at the wetting line would be so totally unaffected by the motion of that line that the angle could remain constant.

With respect to the hydrodynamics of wetting, there seems to be fairly good agreement that in order to avoid a mathematical singularity to the equations of flow, it is necessary to relax classical boundary conditions and allow limited slip between fluid and solid in the immediate vicinity of the wetting line (Huh and Scriven, 1971; Hocking, 1976; 1977; 1982; Dunsan, 1976; Huh and Mason, 1977b; Greenspan, 1978; and Lacey, 1982). Yet the physical basis for slip and the appropriate form of the slip condition remain a matter for speculation.

One approach is to devise a molecular model of the moving wetting line. This has been attempted by Blake and Haynes (1969), and independently by Cherry and Holmes (1969), and also, more recently, by Ruckenstein and Dunn (1977). Blake and Haynes proposed that for displacement of the wetting line, irreversible work must be expended in overcoming energy barriers to molecular displacement within the three-phase zone. They developed their theory around a simple adsorption/desorption model of surface flow. In contrast, Cherry and Holmes were concerned with the spreading of viscous polymers and assumed that the energy barriers were associated with viscous flow. Both analyses were based on the Eyring theory of absolute reaction rates as applied to transport processes (Glasstone *et al.*, 1941), and in both cases it was assumed that the driving force for wetting was given by equation (78). The analysis adopted by Blake and Haynes led directly to an equation linking v and θ. Comparison with experiment has shown good agreement over a wide range of velocities (e.g., Blake and Haynes, 1969; Schwartz and Tejada, 1972; Gribanova and Mulchanova, 1978). The curve through the data in Fig. 9 was obtained by fitting the same equation. Petrov and Radoev (1981) have modified the equation by adding a term to account for viscous losses near the wetting line. They show that the resulting expression accurately describes the velocity-dependence of the contact angle found during the deposition of Langmuir-Blodgett multilayers.

The Blake and Haynes equation contains two adjustable parameters that have to be found by experiment. The Petrov and Radoev version contains three. The experimental values are thoroughly consistent with the physical model, but are system-dependent, and no means has been found to determine them with sufficient accuracy *a priori*. A more complete solution to the theoretical problems posed by wetting kinetics probably awaits an analysis that combines hydrodynamics with a slip condition based on a molecular mechanism of greater precision than those described above, and a proper treatment of the intermolecular forces operating close to the wetting line. Until all three elements are in place, we are unlikely to be able to make completely *a priori* predictions for even simple systems. In the meantime, dynamic contact angles remain a fertile area for the experimentalist.

SOME PRACTICAL ASPECTS OF WETTING

Porous Materials

In practice, we are frequently faced with solid surfaces that are not only rough and heterogeneous but also porous. Contact angle measurements on such surfaces (e.g., measurement of the angle between a ceramic casting slip and a plaster mould (Taylor and Howlett, 1974)), though far from ideal, can still be useful in assessing practical wettability and for comparing one surface with another. The apparent contact angle may have no clear significance within the context of Young's equation; its value lies in its function as a boundary condition for the free liquid surface.

Similarly, once the liquid has penetrated within a porous material, the geometrical complexity of the pore spaces may preclude any exact knowledge of the contact angle, but an apparent angle, θ_{app}, can still be inferred from measurements of the static capillary pressure (Bartell and co-workers, 1927; 1942) or the rate of liquid penetration (Eley and Pepper, 1946; Crowl and Wooldridge, 1967; Bruil and van Aartsen, 1974).

One simple approach is to treat the porous material as a bundle of uniform cylindrical capillary tubes. Application of the Laplace equation for the pressure drop across a hemispherical meniscus within a tube of radius R gives

$$\Delta p = (2\gamma_{12} \cos \theta)/R. \qquad (79)$$

For the porous material, an equivalent cylindrical radius, R_c, can be obtained either by measuring Δp for a completely wetting liquid, or from single-phase permeability experiments combined with measurement of the total pore volume or surface area. Substitution in equation (79) yields θ_{app}. Since the value of Δp will depend on whether equilibrium is approached by pore filling or drainage, θ_{app} will exhibit hysteresis (Everett, 1980).

An approximate expression for the rate of penetration of a liquid into a cylindrical capillary can be obtained by combining equation (79) with the Poiseuille formula for the average velocity of the liquid under conditions of fully developed laminar flow. Thus,

$$(dx/dt) = (\gamma_{12} R\cos\theta)/(4\eta x) \tag{80}$$

where the liquid is considered to penetrate a distance x in time t. Equation (80) is usually known as the Washburn equation (Washburn, 1921). Integration gives

$$x^2 = (\gamma_{12} Rt\cos\theta)/(2\eta). \tag{81}$$

This equation has recently been verified for capillaries as small as 0.2 μm radius (Fisher and Lark, 1979).

For porous materials it is customary to multiply x by a constant, κ, to account for the tortuous path of the capillaires, whence

$$x^2 = (R_c/\kappa^2)(\gamma_{12} t\cos\theta_{app})/(2\eta). \tag{82}$$

A plot of x^2 versus t usually gives a good straight line. The value of (R_c/κ^2) can be found from a separate experiment with a completely wetting liquid, hence θ_{app} can be determined. The method is particularly convenient for powders where θ would be difficult to measure by more conventional means. A suitable package can be constructed by filling a tube closed at one end by a glass frit (Bruil and van Aartsen, 1974). A similar approach has been employed for filamentary materials (Chwastiac, 1973). Note, however, that the contact angle determined in this way will be a dynamic angle.

From equation (82) we can see that the driving force for capillary wetting is proportional to $\gamma_{12}\cos\theta$. Thus, the addition of a surfactant will not necessarily improve the rate of wetting. It will only be successful if θ is initially nonzero and the surfactant adsorbs at the solid/liquid interface, so reducing γ_{S1}. This can be seen on reference to Young's equation. If $\theta = 0$ initially, a reduction in γ_{12} will reduce the driving force. If $\theta > 0$ initially and γ_{S1} remains constant, then (assuming γ_{S2} to be unaffected) a reduction in γ_{12} will reduce the contact angle but leave $\gamma_{12}\cos\theta$ unchanged. If the porous material has a large surface to volume ratio, then adsorption on the capillary walls may leave the solution depleted near the wetting front. To be effective, therefore, the surfactant may have to be present in high concentration.

Draves Test

In textile processing, dyeing, etc., surfactants are used to increase the rate of wetting by aqueous solutions. The Draves test (Draves and Clarkson, 1931; Draves, 1939) provides a practical means of assessing the performance of these solutions since it approximates the conditions of use. In the standard test* a 5 g skein of naturally waxed cotton yarn, comprising a 54 inch (137 cm) loop of 120 threads, is attached to a 3 g hook and totally immersed in the surfactant solution by means of an anchor. The solution gradually penetrates the skein which then suddenly sinks. The better the wetting, the shorter the time required for sinking. Comparative measurements are made at a fixed temperature and concentration: usually 25°C and 0.1%, respectively. Sinking times, t, can vary from < 1 s to > 5 minutes. For a given surfactant, it is usually found that log t is a linear function of log c. This relationship has been explained by Fowkes (1953). The current status of the Draves test has been reviewed by Rosen (1978), who also tabulates results obtained for a wide variety of common surfactants.

A variation of the Draves test has been used to compare the performance of surfactant solutions in wetting coal dust (Walker *et al.*, 1952; Glanville and Haley, 1982). The latter authors find that the time taken for the dust to sink correlates better with practical experience than do predictions based on capillary penetration experiments using a packed column of dust. Garhsva *et al.* (1978) have used the sinking test to determine the wettability of hydrophobic powders by aqueous solutions of various alcohols.

Practical Effects of Contact Angle Hysteresis

The fact that stable contact angles can be multi-valued often has important practical consequences. One example relates to the attachment of bubbles to particles in mineral flotation, where it is believed that the stability of attachment is enhanced by the configurational flexibility afforded by contact angle hysteresis (Blake, 1973).

Another example relates to the retention of agricultural sprays on the surfaces of leaves. Furmidge (1962) found that for a liquid drop on an inclined surface, there is a linear relationship between the width of the largest drop that can be formed on the surface without run-off, and the component of the weight of the drop acting down the incline. It is easy to show that the restraining force is due to the difference between the work of wetting and dewetting as the drop slides (Rosano, 1951). If the drop is rectangular, then the difference is $\gamma_{12}(\cos\theta_R - \cos\theta_A)$ per unit width of the drop.**

---

* Standard Method D-2281, *Amer. Soc. Test. Mater.*, Philadelphia, 1980.

** Accurate solutions for more realistic shapes have been obtained by Brown *et al.* (1980) using the finite-element method to integrate the Laplace equation.

Furmidge defined the retention factor

$$F = \theta_m[\gamma_{12}(\cos\theta_R - \cos\theta_A)/\rho]^{\frac{1}{2}} \tag{83}$$

where $\theta_m = (\theta_R + \theta_A)/2$, and ρ is the density of the liquid. For a given angle of inclination, F is roughly proportional to the volume of liquid retained on the surface at the moment of incipient run-off. The factor provides a useful means of comparing the efficacy of surfactant formulations for use in sprays, where a balance must be achieved between spreading and retention (Tadros, 1978).

The phenomenon of drop retention is clearly related to the frequently troublesome Jamin effect (Jamin, 1860). This is the resistance to motion exhibited by incompletely wetting liquids in capillary tubes. Consider a liquid index within a cylindrical tube of small radius R. If we attempt to move the index with an applied pressure, the contact angles at the leading and trailing meniscii will adjust until they equal θ_A and θ_R, respectively. Any further increase in pressure will cause the index to move. The pressure drop across the index just prior to movement will therefore be

$$\Delta p = 2\gamma_{12}(\cos\theta_R - \cos\theta_A)/R. \tag{84}$$

If R is very small and there are many separate indices, the overall pressure drop can be very large and can seriously impede flow. The Jamin effect has been extensively investigated by Schwartz and co-workers (1964).

REFERENCES

Ablett, R. (1923). *Phil. Mag.* 46, 244.

Adam, N.K. and Jessop, G. (1925). *J. Chem. Soc.* 127, 1863.

Adamson, A.W. (1982). "Physical Chemistry of Surfaces". 4th edn. Wiley, New York.

Andrade, J.D., Ma, S.M., King, R.N. and Gregonis, D.E. (1979). *J. Colloid Interface Sci.* 72, 488.

Arslanov, V.V. and Ogarev, V.A. (1977). *Kolloidn. Zh.* 39, 934.

Bangham, D.H. and Razouk, R.I. (1937). *Trans. Faraday Soc.* 33, 1459.

Bargeman, D. (1972). *J. Colloid Interface Sci.* 40, 344.

Bargeman, D. and van Voorst Vader, F. (1973). *J. Colloid Interface Sci.* 42, 467.

Bartell, F.E. and Osterhof, H.J. (1927). *Ind. Engng Chem.* 19, 1277.

Bartell, F.E. and Benner, F.C. (1942). *J. Phys. Chem.* 46, 847.

Bartell, F.E. and Shepard, J.W. (1953). *J. Phys. Chem.* 57, 211.

Bashforth, F. and Adams, J.C. (1883). "An Attempt to Test the

Theories of Capillary Action". Camb. Univ. Press, Cambridge.
Bayramli, E., van de Ven, T.G.M. and Mason, S.G. (1981). *Colloids Surfaces* 3, 279.
Bernett, M.K. and Zisman, W.A. (1959a). *J. Phys. Chem.* 63, 1241.
Bernett, M.K. and Zisman, W.A. (1959b). *J. Phys. Chem.* 63, 1911.
Berry, M.V. (1974). *J. Phys. A: Math., Nucl. Gen.* 7, 231.
Blake, T.D. and Haynes, J.M. (1969). *J. Colloid Interface Sci.* 30, 421.
Blake, T.D., Cayias, J.L., Wade, W.H. and Zerdecki, J.A. (1971). *J. Colloid Interface Sci.* 37, 678.
Blake, T.D. and Kitchener, J.A. (1972). *J. Chem. Soc. Faraday 1* 68, 1435.
Blake, T.D. (1973). *Ver. Deut. Ing. Berlin* 182, 117.
Blake, T.D. and Haynes, J.M. (1973). *In* "Progress in Surface and Membrane Science" (Eds J.F. Danielli, M.D. Rosenberg and D.A. Cadenhead), Vol.6, pp.125-138. Academic Press, New York.
Blake, T.D. (1975). *J. Chem. Soc. Faraday 1* 71, 192.
Blake, T.D. and Ruschak, K.J. (1979). *Nature* 282, 489.
Boruvka, L. and Neumann, A.W. (1978). *J. Colloid Interface Sci.* 65, 315.
Brown, R.A., Orr, F.M. Jr and Scriven, L.E. (1980). *J. Colloid Interface Sci.* 73, 76.
Bruil, H.G. and van Aartsen, J.J. (1974). *Colloid Polym. Sci.* 252, 32.
Buff, F.P. (1960). *In* "Handbuch der Physik" (Ed. S. Flugge), Vol.X, pp.281-304. Springer-Verlag, Berlin and New York.
Buff, F.P. and Saltsburg, H. (1957). *J. Chem. Phys.* 26, 23.
Burley, R. and Kennedy, B.S. (1976a). *Chem. Engng Sci.* 31, 901.
Burley, R. and Kennedy, B.S. (1976b). *Brit. Polymer J.* 8, 140.
Burley, R. and Kennedy, B.S. (1978). *In* "Wetting, Spreading, and Adhesion" (Ed. J.F. Padday), pp.327-360. Academic Press, London and New York.
Cahn, J.W. (1977). *J. Chem. Phys.* 66, 3667.
Casimir, H.B.D. and Polder, D. (1948). *Phys. Rev.* 73, 360.
Cassie, A.B.D. and Baxter, S. (1944). *Trans. Faraday Soc.* 40, 546.
Cassie, A.B.D. (1948). *Disc. Faraday Soc.* 3, 361.
Cherry, B.W. and Holmes, C.M. (1969). *J. Colloid Interface Sci.* 29, 174.
Chwastiak, S. (1973). *J. Colloid Interface Sci.* 42, 298.
Cooper, W.A. and Nuttall, W.A. (1915). *J. Agr. Sci.* 7, 219.
Crowl, V.T. and Wooldridge, W.D.S. (1967). *SCI (Soc. Chem. Ind., London) Monogr.* 25, 200. See also subsequent discussion, especially the comments of N.R. Morrow, pp.210-212.
Dann, J.R. (1970). *J. Colloid Interface Sci.* 32, 302.
Debye, P. (1920). *Phys. Z.* 21, 178.

Defay, R., Prigogine, I., Bellemans, A. and Everett, D.H. (1966). "Surface Tension and Adsorption". Longmans, London.

Derjaguin, B.V. and Kusakov, M. (1939). *Acta Physicochim.* 10, 25, 153.

Derjaguin, B.V. (1940). *Trans. Faraday Soc.* 36, 212.

Derjaguin, B.V. (1946). *C.R. Acad. Sci. USSR* 51, 361.

Derjaguin, B.V. and Abrikossova, I.I. (1954). *Disc. Faraday Soc.* 18, 24.

Derjaguin, B.V. and Abrikossova, I.I. (1957). *In* "Proc. 2nd Int. Congr. Surface Activity" Vol.3, p.398. Butterworths, London.

Derjaguin, B.V. and Zorin, Z.M. (1957). *In* "Proc. 2nd Int. Congr. Surface Activity" Vol.2, p.145. Butterworths, London.

Derjaguin, B.V. and Levi, S.M. (1964). "Film Coating Theory", pp.140-143. Focal Press, London.

Dettre, R.H. and Johnson, R.E. Jr (1964). *Adv. Chem. Ser.* 43, 136.

Dettre, R.H. and Johnson, R.E. Jr (1965). *J. Phys. Chem.* 69, 1507.

Dettre, R.H. and Johnson, R.E. Jr (1967). *SCI (Soc. Chem. Ind., London) Monogr.* 25, 144.

Donners, W.A.B., Rijnbout, J.B. and Vrij, A. (1977). *J. Colloid Interface Sci.* 60, 540.

Draves, C.Z. and Clarkson, R.G. (1931). *Amer. Dyestuff Rep.* 20, 201.

Draves, C.Z. (1939). *Amer. Dyestuff Rep.* 28, 425.

Dupré, A. (1869). "Théorie mécanique de la chaleur", p.201. Paris.

Dussan V., E.B. (1976). *J. Fluid Mech.* 77, 665.

Dzyaloshinskii, I.E., Lifshitz, E.M. and Pitaevskii, L.P. (1960). *Soviet Phys. J.E.T.P.* 37, 161.

Dzyaloshinskii, I.E., Lifshitz, E.M. and Pitaevskii, L.P. (1961). *Adv. Phys.* 10, 165.

Edser, E. (1922). *Brit. Ass. Adv. Sci. 4th Collect. Rep.*, 291.

Eley, D.D. and Pepper, D.C. (1946). *Trans. Faraday Soc.* 42, 697.

Elliott, G.E.P. and Riddiford, A.C. (1967). *J. Colloid Interface Sci.* 23, 389.

El-Shimi, A. and Goddard, E.D. (1974). *J. Colloid Interface Sci.* 48, 242, 249.

Evans, L.F. and Lane, J.E. (1973). *J. Atmos. Sci.* 30, 326.

Everett, D.H. (1980). *Pure and Appl. Chem.* 52, 1279.

Finch, J.A. and Smith, G.W. (1981). *In* "Anionic Surfactants: Physical Chemistry of Surfactant Action" (Ed. E.H. Lucassen-Reynders), pp.317-384. Marcel Dekker, New York.

Fisher, L.R. and Lark, P.D. (1979). *J. Colloid Interface Sci.* 69, 486.

Fowkes, F.M. and Harkins, W.D. (1940). *J. Am. Chem. Soc.* 62, 3377.
Fowkes, F.M. (1953). *J. Phys. Chem.* 57, 98
Fowkes, F.M. (1962). *J. Phys. Chem.* 66, 382.
Fowkes, F.M. (1964a). *Ind. Engng Chem.* 56, 40.
Fowkes, F.M. (1964b). *Adv. Chem. Ser.* 43, 99.
Fowkes, F.M. (1967). *SCI (Soc. Chem. Ind., London) Monogr.* 25, 3.
Fowkes, F.M., McCarthy, D.C. and Mostafa, M.A. (1980). *J. Colloid Interface Sci.* 78, 200.
Fowler, R.H. (1937). *Proc. Royal Soc. Ser. A* 159, 229.
Fox, W. (1945). *J. Am. Chem. Soc.* 67, 700.
Fox. H.W. and Zisman, W.A. (1950). *J. Colloid Sci.* 5, 514.
Frumkin, A.N. (1938). *Acta Physicochim.* 9, 313; *Zhur. fiz. Khim.* 12, 337.
Fuerstenau, D.W. and Healy, T.W. (1972). *In* "Adsorptive Bubble Separation Techniques" (Ed. R. Lemlich), pp.91-131. Academic Press, New York and London.
Furmidge, C.G.L. (1962). *J. Colloid Sci.* 17, 309.
Garhsva, S., Contreras, S. and Goldfarb, J. (1978). *Colloid Polymer Sci.* 246, 241.
Gaudin, A.M. and Fuerstenau, D.W. (1955). *Trans. A.I.M.E.* 202, 958.
Gibbs, J.W. (1928). "The Collected Works of J. Willard Gibbs", Vol.1. Longmans-Green, New York.
Girifalco, L.A. and Good, R.J. (1957). *J. Phys. Chem.* 61, 904.
Glanville, J.O. and Haley, L.H. (1982). *Colloids Surfaces* 4, 213.
Glasstone, S., Laidler, K.J. and Eyring, H.J. (1941). "The Theory of Rate Processes". McGraw-Hill, New York.
Good, R.J. and Girifalco, L.A. (1960). *J. Phys. Chem.* 64, 561.
Good, R.J. (1964). *Adv. Chem. Ser.* 43, 74.
Good, R.J. (1975). *J. Colloid Interface Sci.* 52, 308.
Greenspan, H.P. (1978). *J. Fluid Mech.* 84, 125.
Gregory, J. (1969). *Adv. Colloid Interface Sci.* 2, 396.
Gretz, R.D. (1966). *J. Chem. Phys.* 45, 3160.
Gribanova, E.V. and Mulchanova, L.I. (1978). *Kolloidn. Zh.* 40, 30, 217.
Grimson, M.J., Richayzen, G. and Richmond, P. (1980). *Mol. Phys.* 39, 61.
Hamaker, H.C. (1937). *Physica (Utrecht)* 4, 1058.
Hamilton, W.C. (1972). *J. Colloid Interface Sci.* 40, 219.
Hamilton, W.C. (1974). *J. Colloid Interface Sci.* 47, 672.
Hansen, R.S. and Miotto, M. (1957). *J. Am. Chem. Soc.* 79, 1765.
Hare, E.F., Shafrin, E.G. and Zisman, W.A. (1954). *J. Phys. Chem.* 58, 236.
Hare, E.F. and Zisman, W.A. (1955). *J. Phys. Chem.* 59, 335.

Harkins, W.D. (1937). *J. Phys. Chem.* 5, 135.
Harkins, W.D. (1952). "The Physical Chemistry of Surface Films". Reinhold, New York.
Hauxwell, F. and Ottewill, R.H. (1970). *J. Colloid Interface Sci.* 34, 473.
Hitchcock, S.J., Carroll, N.T. and Nicholas, M.G. (1981). *J. Materials Sci.* 16, 714.
Hocking, L.M. (1976). *J. Fluid Mech.* 76, 801.
Hocking, L.M. (1977). *J. Fluid Mech.* 79, 209.
Hocking, L.M. (1982). *In* "Proc. 2nd Int. Coll. Drops and Bubbles", pp.315-321. JPL Publication No.82-7, NASA-JPL, Pasadena.
Hocking, L.M. and Rivers, A.D. (1982). *J. Fluid Mech.* 121, 425.
Hoffman, R.L. (1975). *J. Colloid Interface Sci.* 50, 228.
Horn, R.G. and Israelachvili, J.N. (1980). *Chem. Phys. Letts* 71, 192.
Hu, P. and Adamson, A.W. (1977). *J. Colloid Interface Sci.* 59, 605.
Huh, C. and Scriven, L.E. (1971). *J. Colloid Interface Sci.* 35, 85.
Huh, C. and Mason, S.G. (1977a). *J. Colloid Interface Sci.* 60, 11.
Huh, C. and Mason, S.G. (1977b). *J. Fluid Mech.* 81, 401.
Inverarity, G. (1969). *Brit. Polymer J.* 1, 245.
Israelachvili, J.N. and Tabor, D. (1972). *Proc. Royal Soc. Ser. A* 331, 19.
Israelachvili, J.N. and Tabor, D. (1973). *In* "Progress in Surface and Membrane Science" (Eds J.F. Danielli, M.D. Rosenberg and D.A. Cadenhead), Vol.7, pp.1-55. Academic Press, New York and London.
Israelachvili, J.N. (1973). *J. Chem. Soc. Faraday 2* 69, 1729.
Ivanov, I.B., Toshev, B.V. and Radoev, B.P. (1978). *In* "Wetting, Spreading and Adhesion" (Ed. J.F. Padday), pp.37-60. Academic Press, London and New York.
Jameson, G. and del Cerro, M.C.G. (1976). *J. Chem. Soc. Faraday 1* 72, 883.
Jamin, J. (1860). *C.r.h. Acad. Sci.* 50, 172.
Jiang, T.S., Oh, S.G. and Slattery, J.C. (1979). *J. Colloid Interface Sci.* 69, 74.
Johnson, R.E. Jr (1959). *J. Phys. Chem.* 63, 1655.
Johnson, R.E. Jr and Dettre, R.H. (1964). *Adv. Chem. Ser.* 43, 112; *J. Phys. Chem.* 68, 1744.
Johnson, R.E. Jr and Dettre, R.H. (1966). *J. Colloid Interface Sci.* 21, 610.
Johnson, R.E. Jr and Dettre, R.H. (1969). *In* "Surface and Colloid Science" (Ed. E. Matijevic), Vol.2, pp.85-153. Interscience-Wiley, New York.

Johnson, R.E. Jr, Dettre, R.H. and Brandreth, D.A. (1977). *J. Colloid Interface Sci.* 62, 205.
Kaelble, D.H. (1970). *J. Adhesion* 2, 66.
Kaelble, D.H. and Uy, K.C. (1970). *J. Adhesion* 2, 50.
Kawasaki, K. (1962). *J. Colloid Sci.* 17, 169.
Keesom, W.H. (1921). *Phys. Z.* 22, 129.
Kirkwood, J.G. and Buff, F.B. (1949). *J. Chem. Phys.* 17, 338.
Kitchener, J.A. and Prosser, A.P. (1957). *Proc. Royal Soc. Ser. A* 242, 403.
Kitchener, J.A. (1975). *Chem. Ind.* 18 Jan., p.54.
Ko, Y.C., Ratner, B.D. and Hoffman, A.S. (1981). *J. Colloid Interface Sci.* 82, 25.
Lacey, A.A. (1982). *Stud. Appl. Math.* 67, 217.
Lane, J.E. (1975). *J. Colloid Interface Sci.* 52, 155.
Langmuir, I. (1920). *Trans. Faraday Soc.* 15, 62.
Lau, W.W.Y. and Burns, C.M. (1973). *J. Colloid Interface Sci.* 45, 295.
London, F. (1930). *Z. Phys.* 63, 245; *Z. Phys. Chem. Abt. B* 11, 222.
Lucassen-Reynders, E.H. (1963). *J. Phys. Chem.* 67, 969.
Mahanty, J. and Ninham, B.W. (1976). "Dispersion Forces". Academic Press, London and New York.
Margenau, H. and Kestner, N.R. (1969). "Theory of Intermolecular Forces". Pergamon, Oxford.
Melrose, J.C. (1964). *Adv. Chem. Ser.* 43, 158.
Melrose, J.C. (1965). *J. Colloid Sci.* 20, 801.
Melrose, J.C. (1967). *SCI (Soc. Chem. Ind., London) Monogr.* 25, 123.
Mitchell, D.J. and Richmond, P. (1974). *J. Colloid Interface Sci.* 46, 118.
Mori, Y.H., van de Ven, T.G.M. and Mason, S.G. (1982). *Colloids and Surfaces* 4, 1.
Murphy, W.J., Roberts, M.W. and Ross, J.R.H. (1972). *J. Chem. Soc. Faraday 1* 68, 1190.
Napper, D.H. and Hunter, R.J. (1975). *In* "Surface Chemistry and Colloids" (Ed. M. Kerker), Vol.7, p.161. Butterworths, London.
Navascués, G. and Berry, M.V. (1977). *Mol. Phys.* 34, 649.
Neumann, A.W. and Good, R.J. (1972). *J. Colloid Interface Sci.* 38, 341.
Oliver, J.F., Huh, C. and Mason, S.G. (1977). *J. Colloid Interface Sci.* 59, 568.
Oliver, J.F. and Mason, S.G. (1977). *J. Colloid Interface Sci.* 60, 480.
Ottewill, R.H. and Vincent, B. (1972). *J. Chem. Soc. Faraday 1* 68, 1533.
Owens, N.F., Richmond, P., Gregory, D., Mingins, J. and Chan, D. (1978). *In* "Wetting, Spreading and Adhesion" (Ed. J.F.

Padday), pp.127-145. Academic Press, London and New York.
Padday, J.F. (1967). *In* "Proc. 4th Int. Congr. Surface Activity, Brussels, 1964", pp.299-306. Gordon and Breach, New York.
Padday, J.F. and Uffindell, N.D. (1968). *J. Phys. Chem.* 72, 1407.
Panzer, J. (1973). *J. Colloid Interface Sci.* 44, 142.
Pashley, R.M. (1981). *J. Colloid Interface Sci.* 80, 153; 83, 531.
Pashley, R.M. and Israelachvili, J.N. (1981). *Colloids Surfaces* 2, 169.
Petrov, J.G. and Radoev, B.P. (1981). *Colloid Polymer Sci.* 259, 753.
Phillips, M.C. and Riddiford, A.C. (1967). *J. Colloid Interface Sci.* 41, 77.
Phillips, R.W. and Dettre, R.H. (1976). *J. Colloid Interface Sci.* 56, 251.
Pier, W., Pointer, A.B. and Fabre, S. (1977). *Chem. Engng Sci.* 32, 1491.
Platikanov, D., Nedyalkov, M. and Scheludko, A. (1980a). *J. Colloid Interface Sci.* 75, 612.
Platikanov, D., Nedyalkov, M. and Nasteva, V. (1980b). *J. Colloid Interface Sci.* 75, 620.
Pujado, P.R. and Scriven, L.E. (1972). *J. Colloid Interface Sci.* 40, 82.
Pyter, R.A., Zografi, G. and Mukerjee, P. (1982). *J. Colloid Interface Sci.* 89, 144.
Rabinovich, Ya.I., Derjaguin, B.V. and Churaev, N.V. (1982). *Adv. Colloid Interface Sci.* 16, 63.
Rayleigh, Lord (1890a). *Phil. Mag.* 30, 285, 456.
Rayleigh, Lord (1890b). *Phil. Mag.* 30, 397.
Richmond, P., Ninham, B.W. and Ottewill, R.H. (1973). *J. Colloid Interface Sci.* 45, 69.
Rillaerts, E. and Joos, P. (1980). *Chem. Engng Sci.* 35, 883.
Rosano, H.L. (1951). *Mém. serv. chim. état (Paris)* 36, 437.
Rosen, M.J. (1978). "Surfactants and Interfacial Phenomena", pp.188-196. Wiley, New York.
Rowlinson, J.S. and Widom, B. (1982). "Molecular Theory of Capillarity". Clarendon Press, Oxford.
Ruch, R.J. and Bartell, L.S. (1960). *J. Phys. Chem.* 64, 513.
Ruckenstein, E. (1969). *Chem. Engng Sci.* 24, 1223.
Ruckenstein, E. and Dunn, C.S. (1977). *J. Colloid Interface Sci.* 59, 135.
Rusanov, A.I. (1978). *J. Colloid Interface Sci.* 63, 330.
Sabisky, E.S. and Anderson, C.H. (1973). *Phys. Rev. A* 7, 790.
Saville, G. (1977). *J. Chem. Soc. Faraday 2* 73, 1122.
Schukin, E.D. and Yushchenko, V.S. (1977). *Kolloidn. Zh.* 39, 331.
Schultz, J., Tsutsumi, K. and Donnet, J.B. (1977). *J. Colloid*

Interface Sci. 59, 277.
Schwartz, A.M., Rader, C.A. and Huey, E. (1964). *Adv. Chem. Ser.* 43, 251.
Schwartz, A.M. and Tejada, S.B. (1970). NASA Contract Rep. CR-72728.
Schwartz, A.M. and Tejada, S.B. (1972). *J. Colloid Interface Sci.* 38, 359.
Shuttleworth, R. and Bailey, G.L.J. (1948). *Disc. Faraday Soc.* 3, 16.
Smolders, C.A. (1961). *Rec. Trav. Chim.* 80, 650.
Smolders, C.A. (1967). *SCI (Soc. Chem. Ind., London) Monogr.* 25, 319.
Sparnaay, M.J. (1958). *Physica (Utrecht)* 24, 751.
Starov, V.M. and Churaev, N.V. (1980). *Kolloidn. Zh.* 42, 703.
Sullivan, D.E. (1981). *J. Chem. Phys.* 74, 2604.
Sulman, H.L. (1919). *Inst. Mining Met. Bull.* 29, 88.
Tabor, D. and Winterton, R.H.S. (1969). *Proc. Royal Soc. Ser. A* 312, 435.
Tadros, Th.F. (1978). *In* "Wetting, Spreading and Adhesion" (Ed. J.F. Padday), pp.424-437. Academic Press, London. and New York.
Tamai, Y., Makuuchi, K. and Suzuki, M. (1967). *J. Phys. Chem.* 71, 4176.
Tamai, Y., Matsunaga, T. and Horiuchi, K. (1977). *J. Colloid Interface Sci.* 60, 112.
Tarazona, P. and Navascués, G. (1981). *J. Chem. Phys.* 75, 3114.
Taylor, D. and Howlett, D.J. (1974). *Trans. Brit. Ceramic Soc.* 73, 19.
Timmons, C.O. and Zisman, W.A. (1966). *J. Colloid Interface Sci.* 22, 165.
Torza, S. and Mason, S.G. (1971). *Kolloid Z.* 246, 593.
Tse, J. and Adamson, A.W. (1979). *J. Colloid Interface Sci.* 72, 515.
Van Silfhout, A. (1966). *Proc. Kon. Akad. Wetensch. Ser. B* 69, 501.
Walker, P.L. Jr, Peterson, E.E. and Wright, C.C. (1952). *Ind. Engng Chem.* 44, 2389.
Washburn, E.W. (1921). *Phys. Rev.* 17, 273.
Wenzel, R.N. (1936). *Ind. Engng Chem.* 28, 988.
White, L.R. (1977). *J. Chem. Soc. Faraday 1* 73, 390.
Wilkinson, M.C. and Elliott, T.A. (1974). *J. Colloid Interface Sci.* 48, 187.
Wilkinson, W.L. (1975). *Chem. Engng Sci.* 30, 1227.
Wolfram, E. (1966). *Kolloid Z.* 211, 84.
Wu, S. (1971). *J. Polymer Sci. C* 34, 19.
Young, T. (1805). *Phil. Trans. Royal Soc. (London)* 95, 65.
Zisman, W.A. (1964). *Adv. Chem. Ser.* 43, 1.

SURFACTANTS IN ENHANCED OIL RECOVERY

E.L. Neustadter

BP Research Centre,
Sunbury-on-Thames, Middlesex TW16 7LN, England

INTRODUCTION

Interfacial effects play an important part in any consideration of oil recovery because the oil and connate water are retained in a porous rock matrix of enormous solid/liquid and liquid/fluid interfacial areas (cf the order of 10^9 m^2/acre ft of reservoir). Any recovery process must be able to displace oil and water from this high surface area matrix, where pore sizes can be in the micron region. We believe that an understanding of the microscopic displacement process is essential if a logical attempt is to be made to modify these processes favourably.

Oil can be recovered initially by primary production using the natural energy of the reservoir. This is then followed by water injection into the reservoir to maintain reservoir pressure and assist in oil displacement. Generally about 40% of the oil can be recovered economically up to this stage. Some 20-30% of the residual oil will be trapped in pore spaces of the water-swept regions of the reservoir and it is this oil we seek to recover. The rest of the unrecovered oil is located in unswept regions of the reservoir. Recovery of this oil can be improved by use of polymers, polymer blocking agents or foams.

The majority of surfactant floods carried out to date, principally in the USA, have used concentrated (∿10%) surfactant microemulsion systems containing petroleum sulphonates and solubilizing short chain alcohols. These have been injected as a low pore volume slug, usually followed by aqueous polymer solution, to prevent 'fingering' of the drive

SURFACTANTS
ISBN 0-12-682180-1

water into the viscous surfactant slug.

However in the North Sea such systems are not applicable. The petroleum sulphonates are not soluble in sea water which has to be used to inject the surfactant. It would also be logistically difficult to handle the large amounts of surfactant required for concentrated slug injection on a North Sea platform. Furthermore, the well spacing in the North Sea (approximately 1 km) is much larger than for the on land reservoirs where slug injection has been practised. A consequence of this would be that the low volume concentrated slug would be dispersed. For these reasons a continuous injection of a low concentration surfactant system (approximately ½%) is preferable for North Sea application. Alkyl ethoxylated sulphonates and alkyl ethoxylate methyl carboxylates are surfactant classes which could be considered for the highly saline, high temperature North Sea environment.

This chapter is mainly concerned with the processes that employ aqueous surfactant and/or polymer solutions to improve oil recovery. We discuss the factors influencing oil mobilization, surfactant screening tests, adsorption effects, polymer surfactant interactions, some effects exerted by the reservoir and conclude with a discussion of areas requiring further study.

SCREENING TESTS FOR SURFACTANT FLOODS

A number of laboratory tests are generally carried out to assess the suitability of any surfactant system.

a) Low interfacial tension measurements of specific crudes with the reservoir fluids.
b) Compatibility of the surfactant system with the reservoir fluid. This is particularly important if the reservoir fluid is highly saline.
c) Stability tests of the surfactant in the reservoir fluid at reservoir temperatures. In the North Sea temperatures as high as 100°C may be experienced and the surfactant systems require to be stable over a period of years.
d) Sand pack and sand column tests are carried out to determine oil recovery from residual oil saturation, which is the oil remaining in place after water flooding. Sand packs are particularly useful for the visualization of the macroscopic displacement process. Oil bank formation can be observed and this is particularly important for an effective oil displacement process. An oil bank in this context is defined as a region of high oil saturation generated by the surfactant system.
e) Phase behaviour - the presence of liquid crystalline phases needs to be established, because these could lead

to undesirable pore blocking effects. For this reason Krafft and cloud point boundaries need to be determined. In the presence of oil, solubilization of oil into the aqueous phase can occur and partitioning of the surfactant into the oil phase can cause surfactant loss. The occurrence of a middle phase between oil and water is often associated with low interfacial tension systems. Highly viscous oil water interfaces may also lead to stable oil-in-water emulsion inhibiting the production of an oil bank.

Core displacement tests carried out by reservoir engineers are the final test carried out before a surfactant system can be considered for field test application. In these tests oil displacement by the surfactant solution is carried out in a core of the actual reservoir material with oil under reservoir conditions of high pressure (in the region of 3000 psi) and temperature ($\sim$100°C).

MECHANISMS OF OIL MOBILIZATION

Four mechanisms will be discussed.

Low Surface Tension Flood

This relies on reducing the oil/water interfacial tensions to very low levels ($\sim 10^{-3}$ mN m^{-1}) so that oil ganglia trapped in a water wet pore by interfacial tension forces can be mobilized. The capillary pressure resistance to flow is proportional to the oil/water interfacial tension divided by the diameter of the constriction. Pore throat diameters, through which the oil ganglias must pass, are in the region of 10 μ for North Sea sand formations and equilibrium interfacial tensions are typically $\sim$30 mN m^{-1}. Pressure gradients in the reservoir are of the order of 1 psi/ft and are much too low to overcome the capillary pressure forces. Mobilization could be achieved if the capillary number, the ratio of surface tension to viscous forces, could be reduced by a factor of 10^4. Viscous forces cannot be increased greatly because of the limited pressure resistance of the reservoir. Reduction of the interfacial tension to about 10^{-3} mN m^{-1} could produce the desired effect.

Measurements of interfacial tension to about 1 mN m^{-1} can be carried out with the pendant-drop technique, however, below this range only the spinning-drop and micro-sessile-drop techniques operate. The latter technique has been used principally by Professor L.E. Scriven of Minnesota University, but more commonly the spinning-drop method is employed.

A drop of the less dense phase is introduced into a

horizontal capillary containing the other liquid phase and the tube rotated at high speeds (approximately 10,000 rpm). It is important that the rate of rotation is constant and able to be determined accurately. The interfacial tension can be calculated from the rate of rotation, the density of the fluids and the shape of the drop. A spinning-drop apparatus is supplied by the University of Texas. Equipment is also now available from the University of Clausthal which enables interfacial tension measurements to be carried out under reservoir conditions. This can be important because the light ends present in the "live" crude could modify the interfacial tension behaviour of the system.

Spontaneous Emulsification

Spontaneous emulsification can arise due to the diffusion of surfactants across the oil/water interface caused by chemical potential gradients. This could result in the mobilization of trapped oil with a minimum energy input. One mechanism of spontaneous emulsification is diffusion stranding. This is exemplified by a drop of toluene containing ethanol suspended in water. The ethanol diffuses into the aqueous phase creating interfacial turbulence at the same time dragging fine droplets of toluene into the aqueous phase.

Interfacial Rheology

Movement of oil/water interfaces through pores of varying geometry will involve compression and expansion of the oil/water interface. If this interface is highly elastic or viscous this could present a resistance to oil mobilization superimposed on capillary pressure.

Highly incompressible interfacial films can form at crude oil/water interfaces. It has also been shown that such interfacial films between crude oil and water can exist under reservoir conditions. Such films stabilize oil drops released from pore surfaces, resisting the formation of an oil bank essential for efficient oil displacement.

We have constructed an interfacial film balance based on the design by Brookes and Pethica to study the compressibility characteristics of these interfacial films. A biconical pendulum viscometer has been built to study the shear viscosity of oil/water interfaces. We have been able to show that the oil/water interfacial rheology governs the stability of water in crude oil emulsions.

Compression of the oil/water interface shows that for a 20% compression the interfacial viscosity of a crude oil/water interface can increase by three orders of magnitude. Interfacial area changes of this order could occur on passage of an interface through a pore throat.

We are currently using these techniques together with surface wetting and interfacial tension studies to assess the influence of these interfacial properties on immiscible displacement behaviour.

Surface Wetting

It has long been known that the wettability of a surface can greatly influence oil displacement. However, the origin of a particular wettability and the economic control of wettability are still research goals.

We may consider a pore doublet consisting of two pores of unequal diameter connected in parallel, from which oil must be displaced at flow rates (approximately 1 mm/s) which give equivalent reservoir capillary numbers. This is the unit system and not the single capillary.

If the system is water wet, water will imbibe into the narrow pore, leaving oil trapped in the wider pore. The resultant system has a low permeability with the possibility of mobilizing the oil in the wider pore channel.

In the oil wet case, oil will be trapped in the narrow pore, with water in the wider pore. The system has a high permeability and attempts at oil mobilization lead to water breakthrough in the wider pore.

However, an actual surface may have regions of mixed wettability owing to surface heterogeneities. This can lead to oil trapping, e.g. when an oil/water interface advances through a water wet capillary and meets an oil wet section. In this case an oil blob would be retained on the oil wet section.

The properties of the crude oil surfactants and their adsorption at the solid/liquid interface govern the wetting behaviour. These surfactants can be considered as amphoteric and possibly nonionic species. Typical surface active species are naphthenic and aromatic carboxylic acids, and substituted pyridines and quinolines which have a sufficiently large hydrocarbon moiety and polar substituent group. The equilibrium interfacial tension versus pH graph for all crude oils studied exhibits the same characteristics. The interfacial tension is low (approximately 10 mN m^{-1}) at low pH values (approximately pH = 2) and high pH values (approximately pH = 10). At intermediate pH values (approximately pH = 6) the interfacial tension is about 20 mN m^{-1}. Oil/water displacement studies (as a function of aqueous phase pH) have been carried out in 1 mm bore silica capillaries in the velocity range of 1 mm s^{-1} approximating to reservoir capillary numbers. At high pH values (approximately pH = 11) the system exhibits water wet behaviour, a water film being present between the silica surface and the oil phase. The thickness of this film varies with the displacement rate. At acid pH values (pH = 3.5) the silica surface is rendered oil wetted. At intermediate pH values mixed wetting behaviour is observed.

We have been able to rationalize those wetting effects on the basis of charge effects.

Silica has an iep of about 2 and North Sea oil about 5. The silica surface is thus negative over the pH range 3.5 - 11 whilst the oil changes from positive to negative at high pH. Charge repulsion at high pH values stabilizes a water film between the oil and the surface. At low pH, crude oil can contact the surface putting down a film of positively charged crude oil surfactants. This surfactant adsorption also leads to contact angle hysteresis.

In the presence of salts charge effects are damped out and the surface will become oil wetted. We have found that polyvalent metal ions at low concentrations < 1 ppm can render a surface oil wet at alkaline pH values. At this level charge effects are not damped so we are dealing with specific interactions.

Surface roughness also greatly influences the wetting behaviour. Thus under conditions in which a smooth silica surface exhibits oil wet characteristics in the presence of crude oil and water, irregularly textured sand grains prewetted by water appear water wet. This apparently arises because oil displacing water from such a surface cannot displace the water from the depressions in such a surface as it advances over it, being able to adhere only to the asperities. Subsequent displacement of the oil by water then creates a predominantly water wet surface with only very small amounts of oil adhering to the previously oil wetted asperities.

It is important to note that studies with model hydrocarbon oils cannot accurately predict the behaviour of crude oil systems. The interfacial effects observed with crude oils are governed by the adsorption of the crude oil surfactants at the interface. With model oil systems neither the variation in wetting behaviour with pH nor the interfacial rheological effects obtained with crude oils are observed.

ADSORPTION AT SOLID/LIQUID AND LIQUID/LIQUID INTERFACES

For low concentration surfactant floods surfactant adsorption is probably the most important single parameter affecting the efficiency of the operation. With the wide well spacing of North Sea wells it can take in the order of four years to pass one pore volume of liquid through the reservoir. Until the adsorptive capacity of the reservoir for the surfactant is satisfied no surfactant will be available at the flood front to release oil. Hence with a strongly adsorbing surfactant in dilute aqueous solution an intolerably long time lag would occur. Furthermore, if surfactant adsorption rendered the formation oil wet this would further delay oil production. Rapid oil production is associated with the release of oil ganglia from water wet pores. If the pores are oil wet, oil is released slowly over a long period of time, associated with large volumes of water, both of these factors would be unacceptable in a North Sea context.

The effect of oil/water wettability on solid surfaces, simulating reservoir materials, are hence usually carried out in conjunction with adsorption measurements. Surfactant adsorption can also affect:- 1) chemical balance of EOR chemicals; 2) permeability of rock matrix particularly as affected by polymer adsorption; 3) the cost-effectiveness of EOR process chemicals.

A continuous flow chromatographic apparatus has been constructed. This enables us to study adsorption and desorption under flow conditions used at reservoir pressures and temperatures (3000 psi, 100°C). This is not possible with conventional batch techniques. It is also possible to study adsorption rate effects which are important when considering scaling from adsorption in short cores or high frontal flow rates to reservoir conditions.

Using this equipment, it has been shown that the pressure dependence of adsorption of sodium dodecylsulphate on reservoir sand is negligible in the range of 10 - 3000 psi. However, though pressure may not greatly influence adsorption *per se*, the presence of light ends in the crude could exert a considerable effect and such experiments can only be carried out under pressure.

In addition, in a waterflood process, adsorption at the liquid/liquid interface is desirable to reduce oil/water interfacial tension, but undue partitioning of the surfactant with the oil phase leading to surfactant loss must be avoided.

Partition coefficient measurements for the surfactant between oil and water are therefore of importance. Adsorption measurements of surfactant at the solid/liquid interface must therefore take into account the effect of the oil phase on surfactant adsorption.

POLYMER SURFACTANT INTERACTIONS

Polymers are added to the aqueous phase to increase its viscosity. The viscosity of the displacing phase must substantially exceed the viscosity of the oil phase to reduce fingering of the aqueous phase into the oil phase leading to trapping of the oil phase.

Polymer surfactant interactions may alter the surface activity of added surfactants and influence their adsorption characteristics. The rheological properties of the polymer and its influence on surface wettability may also be affected by polymer/surfactant interactions.

Polymer adsorption in pore channels or pore throats may also lead to blocking of such pores. It is not clear at this stage to what extent polymer adsorption contributes to the viscosifying effect observed with dilute polymer solutions.

RESERVOIR EFFECTS

Because the oil recovery is essentially a slow process (a flood front may only move at the rate of ft/day) any surfactants or polymers must be stable for a period of years at elevated temperatures in the presence of brine.

Processes of oil mobilization such as spontaneous emulsification which depend on surfactant concentration gradients and which may occur in small cores may not be relevant in the reservoirs where greater dispersion may occur.

The surface of the reservoir rock is far from uniform and it is also chemically heterogeneous. The surface of sand grains may differ greatly from the bulk. Thus ESCA studies have shown that the surface of formation sand grains consists of a high proportion of aluminium, we believe probably from kaolinite clay.

Specific ion effects and surface roughness can greatly affect wettability and therefore it is difficult to transfer model studies to the reservoir situation.

In addition, clay minerals which are often present can induce ion exchange with any added surfactant, possibly causing the latter to precipitate. As the ionic strength of the aqueous displacing fluid alters clay, deflocculation can occur and this can lead to pore blocking.

AREAS FOR FURTHER STUDY - THESE APPLY PARTICULARLY IN THE CONTEXT OF NORTH SEA OPERATIONS

1) The compatibility of any enhanced oil recovery (EOR) chemicals with other oilfield chemical additives. This is particularly important if the EOR chemicals are added at low concentrations.
2) Emulsion resolution of produced crude. This is a problem which is becoming more prevalent as more crude oil is produced wet. In North Sea fields such as Forties, water injection is practised for purposes of pressure maintenance and eventual water breakthrough is inevitable.

3) Specific surfactant and polymers will need to be developed. Conditions in North Sea fields are very severe - a high aqueous phase salinity and temperatures in the region of 100°C. Polymers currently in use in the USA are unlikely to be unsuitable. In any case most of these polymers were not developed with EOR specifically in mind.
4) A reduction in the sensitivity of surfactant systems will need to be achieved. Thus most of the systems for achieving low interfacial tensions currently available do so only across a very narrow range of salinity, oil composition and temperature.
5) Development of surfactant systems specifically for North Sea application. A low concentration continuous injection surfactant aqueous flood would be useful, because water injection is already practised. Surfactant concentration should preferably be less than 5000 ppm, both from the point of view of cost and logistics.
6) Alternative recovery mechanisms. This chapter has only dealt with aqueous surfactant systems. Other techniques will need to be carefully evaluated. Most field tests on EOR techniques to date have been carried out on shallow, low temperature, low salinity and moderate to high permeability reservoirs. This of course does not approximate to conditions pertaining in North Sea reservoirs.
7) Microscopic displacement mechanisms. Further work to elucidate these mechanisms is required. It is particularly important to be able to carry out laboratory studies of some phenomena under reservoir conditions, e.g. surface wetting, interfacial tension and adsorption.
8) At this stage laboratory experiments do not allow us to predict effects observed in the field. It is important to achieve a correlation between laboratory and field experiments.

ACKNOWLEDGEMENT

Permission to publish this review has been granted by the British Petroleum Company plc.

MACROMOLECULAR SURFACTANTS

R.I. Hancock

ICI PLC, Petrochemicals and Plastics Division,
PO Box 90, Wilton, Middlesbrough, Cleveland TS6 8JE, England

INTRODUCTION

The most common problem encountered in the "real" world of surfactants is of matching a commercially available product to a required end-effect. To many industrial users the surfactants they use have a mystique hitherto only attributed to medieval alchemy. Apparently complex problems can be readily understood and solved if practitioners have a basic systematic knowledge of the manufacture, compositions and performance trends of commodity and speciality surfactants. The objective of this chapter is to summarize the key information which the technologist needs to problem-solve and innovate.

MAIN CLASSES OF NONIONIC SURFACTANTS, THEIR RAW MATERIALS AND MANUFACTURE

Main Classes

These are:

$R(OCH_2CH_2)_nOH$	Alcohol ethoxylate
R' $(OCH_2CH_2)_nOH$	Alkylphenol ethoxylate
R" CON $[(CH_2CH_2O)_nH]_2$	Fatty acid ethanolamide

$HO(CH_2CH_2O)_p(CH_3CHCH_2O)_q(CH_2CH_2O)_pH$ — Ethylene oxide propylene oxide block copolymer

$R''N\left<\begin{matrix}(CH_2CH_2O)_nH\\(CH_2CH_2O)_nH\end{matrix}\right.$ — Fatty amine ethoxylate

$\left.\begin{matrix}R''COO(CH_2CH_2O)_nH\\+\\R''COO(CH_2CH_2O)_nOCR''\end{matrix}\right\}$ — Fatty acid ethoxylates

$R''COOCH_2[CH(OH)]_4CH_2OH$ — Fatty acid sorbitol ester

$R''COOCH_2[CH(OH)]_4CH_2(OCH_2CH_2)_nOH$ — Fatty acid sorbitol ester ethoxylate (nominal only)

Where:

R = linear or monobranched alkyl radical; $C_8 \rightarrow C_{20}$, odd and even carbon numbers.

R' = linear or highly branched alkyl radical; C_8, C_9, C_{10} or C_{12}.

R" = linear alkyl or alkenyl radical; $C_{12} \rightarrow C_{18}$, even carbon numbers only.

n = 1 → 40

p = 2 - 100

q = 15 - 50

It is clear that each molecule has separate polar and non-polar functional groups. The polar groups confer some degree of water-solubility on the molecule and are referred to *en masse* as its hydrophilic, lyophobic or oleophobic portion(s). The nonpolar part of the molecule is water-insoluble and usually oil-soluble and is referred to as the hydrophobe, lyophile or oleophile. Polypropylene oxide is not immediately obvious as a nonpolar species and indeed is insoluble in non-polar hydrocarbons but in the molecular weight range (1000-4000) used for this class of surfactant it is water-insoluble and rejected to phase boundaries where it is interfacially active.

Water-solubility is conferred by the ether oxygens of poly-ethylene oxide chains; hydroxyl groups and amine (or amine oxide) groups. Of these the first two are more important. In certain environments, especially at high temperatures or high electrolyte concentrations only hydroxyl groups are solvated by water. This causes the solution properties of a nonionic surfactant to vary depending on their numbers and ratios. To aid the selection of a nonionic surfactant Griffin (1949) devised the Hydrophobe Lyophobe Balance (HLB) concept. This assigns dimensionless numbers to surfactants on the basis of their oil-water-solubility such that those with HLB of <10 are water-insoluble and >12 water-soluble. Its main use is in selecting emulsifiers where small changes in molecular polarity are often critical. For a simple nonionic such as an alcohol ethoxylate, its HLB number is one fifth of its weight percentage of polyethylene oxide.

Surfactant Raw Materials

The two basic raw materials for surfactant production are natural oils and fats (especially beef and mutton tallows and coconut oil) and crude oil. The routes via which they are subsequently converted through key intermediates to the products described above are outlined in Fig. 1.

The synthetic route chosen to a specific surfactant influences its availability, economics, performance, carbon chain distribution, hydrophobe branching, chemical and oxidative stability and its environmental acceptability. Perhaps because of the oil price induced inflation in petrochemicals during the last decade and the wild swings in vegetable oil and fat prices, surfactants from natural and synthetic sources continue to find their own places in the market. In the last decade the cost sensitivity of crude oil based surfactants to the actions of the OPEC cartel has and is inducing producers to buffer their effects by additional investment in fatty chemical manufacturing plant. New petrochemical investment has concentrated on polymerizing the most readily available and flexible raw material (ethylene) in the most efficient way (oligomerization) to produce the most environmentally acceptable surfactants of widest applicability (alcohol ethoxylates). Older synthetic routes using n-paraffins, cracked wax olefins, propylene trimer/tetramer and di-isobutylene are either declining in volume or at best static. In the area of natural products coconut oil is the best contender as most widely useful hydrophobe source, whilst increased plantings of new "Mawa" hybrid palms in the Philippines should ensure its availability.

Approximate USA and WE production of the major nonionics is shown in Table 1.

Ethoxylate Manufacture

Ethylene oxide reacts with substrates containing "active" hydrogen under either acidic or basic conditions. Although (Lewis) acid (e.g. BF_3) catalysts give more specific products

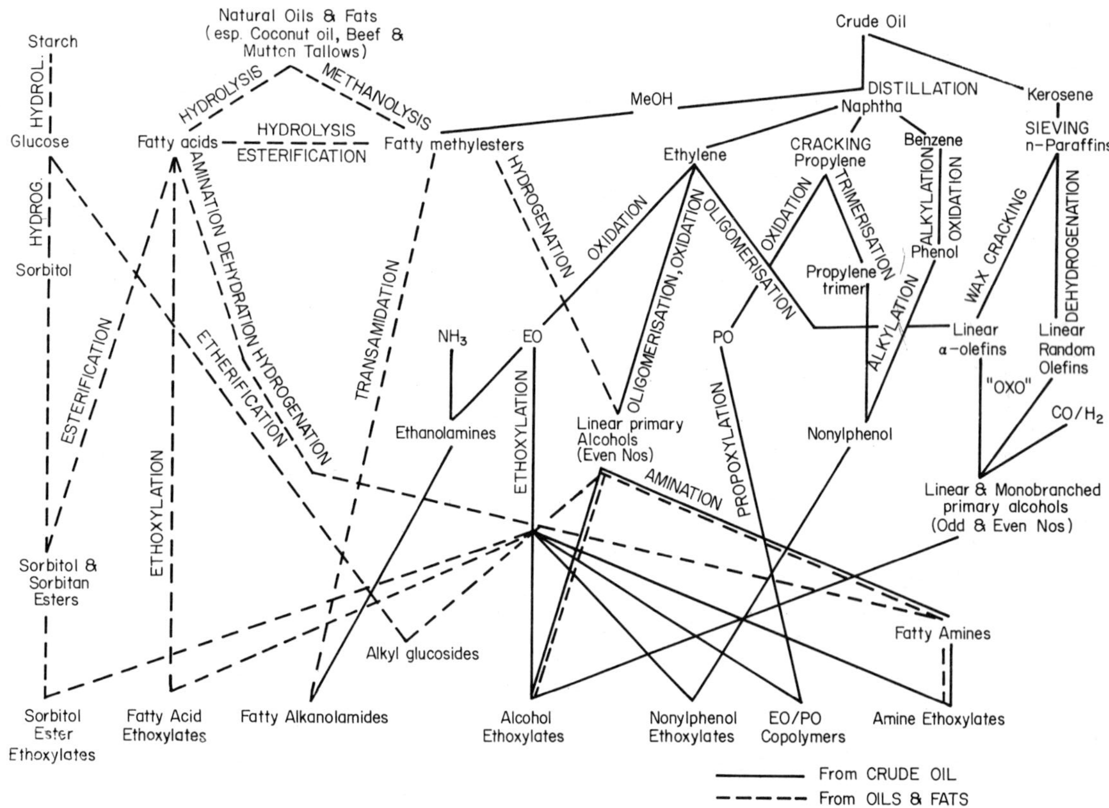

Fig. 1. Schematic diagram of the routes used for preparation of surfactants.

TABLE 1

Major nonionics; current production (tonnes), USA and WE

	USA	WE
Alcohol Ethoxylates	230,000	225,000
Alkylphenol Ethoxylates	140,000	100,000
Fatty acid Ethoxylates	25,000	20,000
Fatty acid Alkanolamides	30,000	40,000
Fatty amine Ethoxylates	5,000	5,000
Poly Propylene oxide/Poly Ethylene oxide copolymers	N.K.	5,000

(especially with secondary alcohols) they are less pure, containing aldehydes, dioxan and glycols. Thus virtually all commercial production uses alkali metal hydroxides (or methoxides), especially NaOH and KOH at concentrations between 0.1% and 2% w/w on hydrophobe.

Reaction proceeds by nucleophilic attack by the hydrophobe anion (e.g. alkoxide or phenate) on one of the ethylene oxide carbon atoms. The monoethoxylate anion thus generated can then exchange with a molecule of hydrophobe or condense with another mole of ethylene oxide:

$$(1)\quad RXH + KOH \xrightarrow[\text{Vacuum}]{120^\circ C} RX^{\ominus}K^{\oplus} + H_2O$$

$$(2)\quad RX^{\ominus}K^{\oplus} + \underset{\diagdown O \diagup}{CH_2 - CH_2} \xrightarrow[\text{1-6 BARS}]{120^\circ\text{-}160^\circ C} RXCH_2CH_2O^{\ominus}K^{\oplus}$$

$$(3)\quad RXCH_2CH_2O^{\ominus}K^{\oplus} + n(\underset{\diagdown O \diagup}{CH_2 - CH_2}) \longrightarrow RX(CH_2CH_2O)^{\ominus}_{n+1}K^{\oplus}$$

$$(4)\quad RXCH_2CH_2O^{\ominus}K^{\oplus} + RXH \longrightarrow RX^{\ominus}K^{\oplus} + RXCH_2CH_2OH$$

R = Alkyl, aryl or alkyl-aryl

$$X = \text{-O-},\ \text{-NH-},\ \text{S-},\ -C(=O)O-,\ -C(=O)NH-$$

If reaction (3) predominates the ethoxylate has a wide homologue distribution (e.g. alcohol ethoxylates) and if reaction (4) is dominant the homologue distribution is narrow (e.g. phenolic ethoxylates). This is further discussed below.

The reaction rate is therefore proportional to hydrophobe, catalyst, and ethylene oxide concentrations (or partial pressure). Good mixing is very important both to dissipate the heat generated (22 k.cal. mole^{-1}) and to ensure anion exchange is kinetically and not diffusion controlled (the latter complication widens the homologue distribution). Reactor temperature is from 120 to 160°C. Between these limits rates are rapid and byproduct formation (especially polyethylene glycol content) is acceptable. The gas phase above the reaction is normally at least 55% v/v nitrogen to prevent explosive ethylene oxide polymerization. Reactors are invariably of stainless steel (carbon steel gives poor coloured products), batch rather than continuous and the more modern use a recirculating loop design which gives complete mixing and efficient cooling.

When the stoichiometric amount of ethylene oxide has been added the reactor is "cooked" down prior to neutralization with an (organic) acid (e.g. adipic, acetic, phosphoric), vacuum applied to remove traces of untreated ethylene oxide (which is very toxic and irritant) and the product discharged. Contact between hot reactants or products and air has to be avoided or poor colour results.

COMPOSITIONS AND IMPURITIES

Alkyl Phenol and Alcohol Ethoxylates

Due to the acidity of the phenolic proton the addition of one mole of ethylene oxide to a mole of an alkyl-phenol forms a

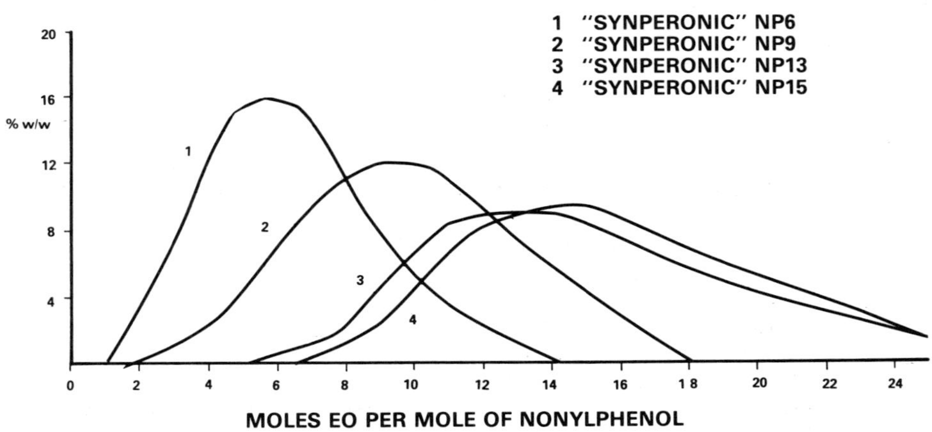

Fig. 2. Homologue distribution of nonylphenol ethoxylates.

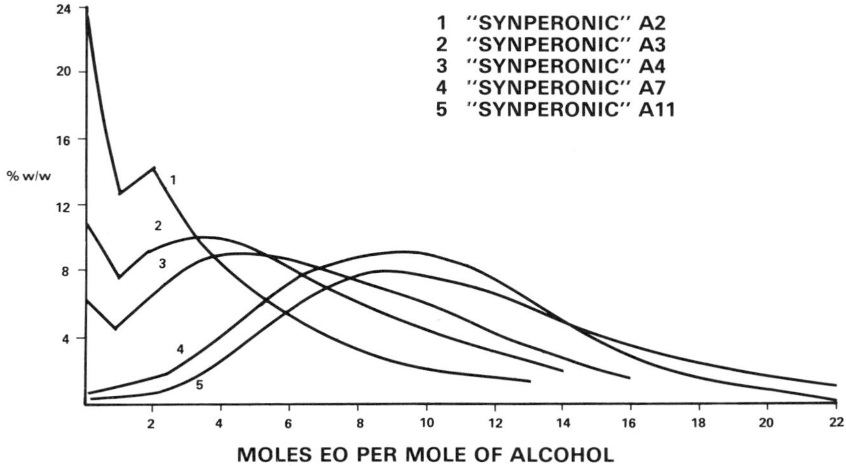

Fig. 3. Homologue distribution of "Synprol" alcohol ethoxylates.

specific monoethoxylate before any higher homologues are formed. Alkylphenol ethoxylates therefore have a relatively narrow homologue distribution which is effectively insensitive to variation in reaction pressure or temperature and catalyst type or concentration (within normal limits). Some typical homologue distributions are shown in Fig. 2.

The alcohol proton is slightly less acidic than that of its connate ethoxylates. This results in the formation of 2-mole and higher ethoxylates whilst substantial amounts of starting alcohol remain unreacted. Homologue distributions are therefore broad (Fig. 3) which gives these products both advantages (e.g. wide spectrum detergency and emulsification performance) and disadvantages (such as higher volatility, melting points and lower effective activity) when compared with alkylphenol ethoxylates. For most purposes unreacted alcohol and homologues with more than 20 moles ethylene oxide per mole of alcohol are not surface active.

Both groups of ethoxylates contain common impurities. These are:

1) Polyethylene glycols (PEG); produced by condensation of ethylene oxide with water which can arise from numerous sources (e.g. catalyst, substrate, reactor coolant leaks or by reaction of aldehydic impurities with ethoxylates). As a PEG contains two active protons of similar reactivity to ethoxylate protons more PEGs are produced the higher the molecular weight of the required ethoxylate. Thus a

three mole alcohol ethoxylate will typically contain 0.2-0.5% PEG, a seven-mole 1-2% and an eleven-mole 2-3%.

2) Aldehydes are present in both starting materials. Ethylene oxide (EO) fed to ethoxylation plants typically contains 10-30 ppm acetaldehyde, fatty alcohols 50-200 ppm of C_{10}-C_{20} aldehydes and alkyl phenols 100-250 ppm of aromatic aldehydes. There is also some evidence that aldehydes are formed during ethoxylation if reactor temperatures are too high and when making products with more than 20 moles EO. Colour (and sometimes odour) problems in ethoxylates are almost invariably caused by aldehydes. The best way of handling such problems is by pretreatment of the feed-stock with sodium borohydride ($NaBH_4$).

3) Inorganic catalyst residues at levels of up to 0.1% w/w are present if the ethoxylate is not filtered. These are usually sodium/potassium acetate/phosphate. They rarely interfere with subsequent use but are sometimes seen as a slight haze in liquid products.

Fatty Amine Ethoxylates

Fatty amines react readily with ethylene oxide either auto-catalytically or with addition of acids or bases. The most interesting and important phase of this process is the addition of the first two moles of ethylene oxide. Both amino protons can be replaced by ethylene oxide to give a specific two-mole ethoxylate before any further chain extension. In the absence of catalyst the reaction slows thereafter but the rate up to

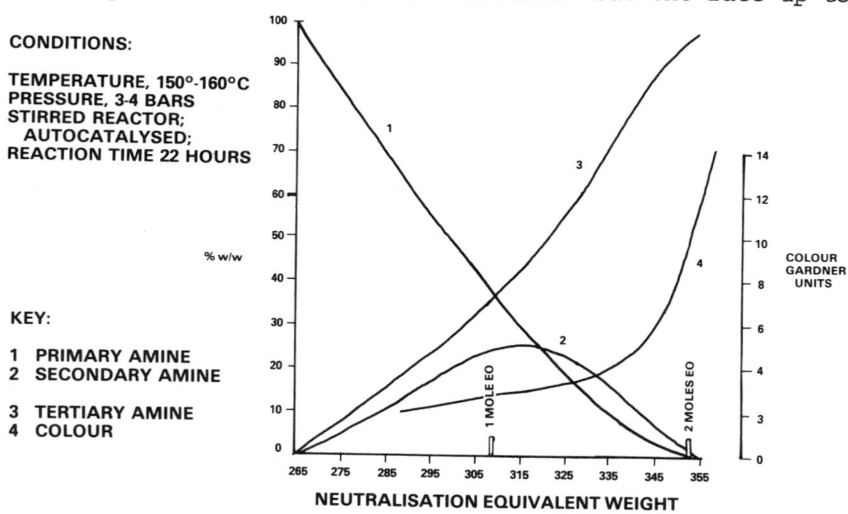

Fig. 4. Ethoxylation of hydrogenated tallow amine.

five moles EO is still acceptable. For higher ethoxylates it is usual to add alkali catalyst after two moles of EO have been added. As can be seen from Fig. 4, the colour of the ethoxylate intensifies near the two moles EO addition. This is a characteristic of all amine ethoxylates but most pronounced in those derived from natural oils and fats. Reaction specificity is not measurably affected by pressure or temperature within normal operating conditions.

Apart from the formation of PEG, natural fatty amine ethoxylates often have unpleasant nitrile odours and can contain small amounts (up to 2-3%) of monofatty amides (RCONHR). These compounds are less easy to ethoxylate than amines and usually adversely affect the ethoxylates' acid and alkali solubility.

Fatty Acid Ethoxylates

Fatty acids are ethoxylated using alkaline catalysts to give a complex mixture of unconverted fatty acid, monofatty PEG ester, difatty PEG esters and free polyethylene glycols. Both ratios and molecular weights of these products vary as reaction proceeds. This is illustrated in Figs 5 and 6.

These ester ethoxylates are particularly useful as emulsifiers with a wide spectrum of activity under neutral conditions. If the fatty acid is first converted to a sorbitol ester then the derived ethoxylates are even more complex mixtures with very wide application as emulsifiers.

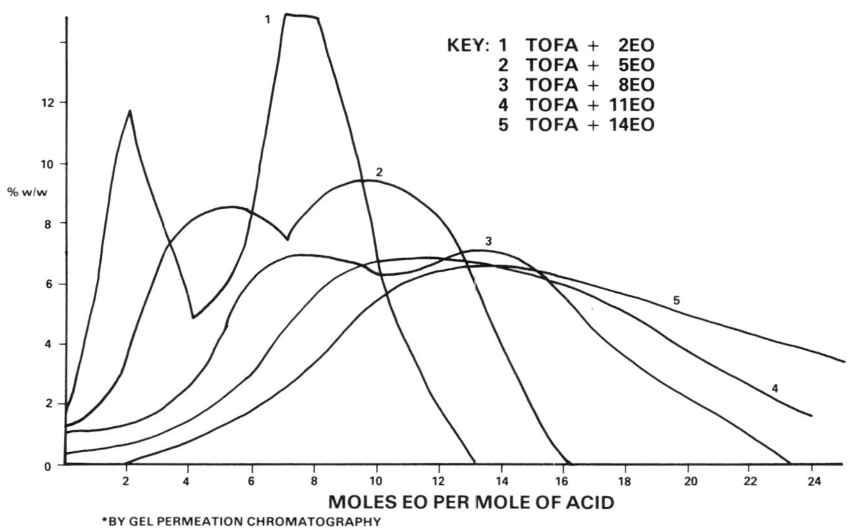

Fig. 5. Homologue distribution of tall oil fatty acid ethoxylates (obtained using gel permeation chromatography).

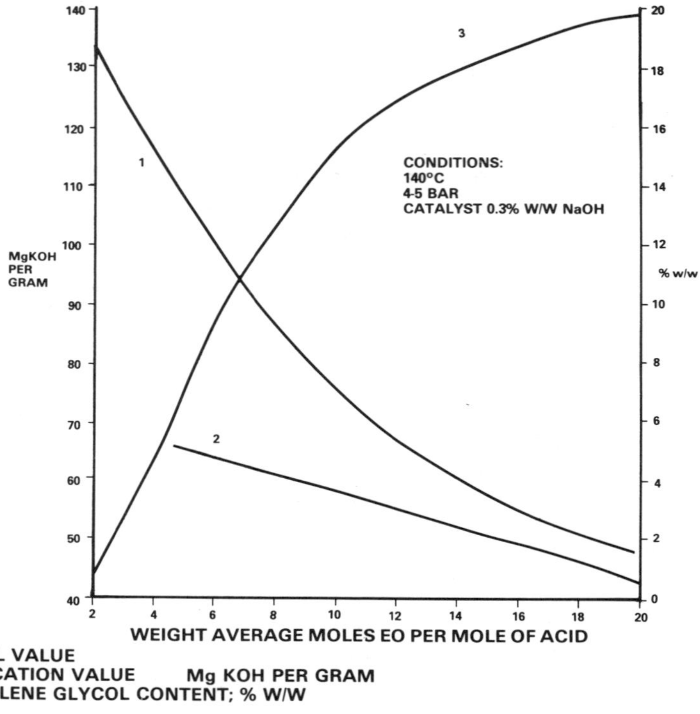

Fig. 6. Ethoxylation of tall oil (M.W. 289) fatty acid (chemical analysis).

Fatty Acid Alkanolamides

Reaction of a fatty methyl ester with mono- or diethanolamine with an alkali catalyst (e.g. NaOMe) yields an alkanolamide in up to 90% pass yield.

$$RCOOMe + H_2NCH_2CH_2OH \xrightarrow[160\,°C]{NaOMe} RCONHCH_2CH_2OH + MeOH + RCOOCH_2CH_2NH_2 + etc.$$

Amino esters and higher condensates are also produced. Simple alkanolamides can be further reacted with ethylene oxide under similar conditions to those used for fatty alcohols, increasing their water solubility and electrolyte tolerance.

Ethylene Oxide-Propylene Oxide Block Copolymers

Propylene oxide is condensed with propylene glycol or dipropylene glycols using an alkali catalyst to give complex mixtures of polypropylene glycols (PPG) typically of molecular weight 1500 to 2500. The terminal hydroxyl groups are secondary.

These PPGs are water-insoluble but soluble in polar hydrocarbons, forming the hydrophobe portion of the resultant surfactant formed by further condensation of ethylene oxide on to each end of the chain. The surfactant is thus difunctional (as are amine and some amide ethoxylates). These molecules are more polar than fatty nonionics and their HLB number is best calculated as 1/3 of their weight percentage of ethylene oxide. Apart from dilution of their activity by PEG (formed as previously described) the main impurities in PPG-PEG copolymers are propionaldehyde and acetaldehyde and their connate acids. These can cause acute problems in flaked products by accelerating peroxidation, depolymerization and acidification. The problem can be eliminated by pretreatment with $NaBH_4$ and/or addition of an antioxidant such as butylated hydroxytoluene (BHT).

SURFACE PROPERTIES

General

Nonionic surfactants are used to produce the following interfacial effects: foaming; wetting; detergency; emulsification; dispersion; defoaming; demulsification; solubilization; emulsion stabilization. These effects are influenced by variables such as: temperature; electrolytes (type and concentration); cosurfactants; water insolubles (e.g. oils, solids); organic water solubles (e.g. EtOH); their concentration; pH.

It is thus not surprising that classical colloid science prefers the study of specific surface effects in response to changes in these variables. Such research is rarely of use to the technologist seeking to achieve an economic end-effect and faced with an array of anonymous surfactants often described only by trade names. Systematic study can prove impossible and scientific problem-solving is reduced to "bottle reaching". This can be avoided by using chemically characterized surfactants (such as ICI's "Synperonics" or Shell's "Dobanols") in combination with an understanding of their fundamental interfacial properties.

Choosing the appropriate surfactant can be resolved by considering two questions: Which surfactants will have the maximum concentration at the interface one wishes to modify, and what will be the properties of the interfacial film thus obtained?

The water solubility of a nonionic surfactant is dependent on its polarity or proportion of water-solvatable groups such as -(ethylene oxy)-, amines or hydroxy. Its HLB number is a way of quantifying this polarity. When a nonionic is mixed into water these groups hydrate to form a voluminous sheath around their polar ends. The size of this sheath and tenacity with which this water is held determines at which interface the surfactant will concentrate and the strength of the interfacial film thus formed. Heating or the addition of water-soluble electrolytes causes dehydration of these polar groups and reduces the surfactant's solubility. The temperature at which this dehydration becomes visible in the form of solution turbidity (caused by surfactant molecules aggregating into larger micelles) is its "cloud point" (Chapter 1). An understanding of this characteristic property and its variation with chemical structure and solution variables is essential to the scientific application of nonionic surfactants.

As is illustrated in Fig. 7, the cloud point increases with polarity/ethylene oxide content/HLB and is depressed by the addition of NaCl and similar electrolytes. It is raised by addition of water- and oil-soluble solvents such as EtOH, isopropanol or "Ethoxol" (ICI trademark) and depressed by water-insoluble oils. A nonionic surfactant is most effective when used up to about 10°C below and 20°C above its cloud point under its conditions of use. In this temperature range its interfacial concentration is maximized and polar groups are still sufficiently hydrated to form a viscous interface-modifying film structured by lateral hydrogen bonding via intermediate water molecules.

These concepts can be used to understand and predict nonionics' properties as hereinafter described, bearing in mind the additional changes superimposed by pH decreases (enhances solubility of amine ethoxylates) or changes in hydroxyl: polyoxyethylene ratios (OH groups are more temperature and electrolyte resistant).

Gas - Liquid Interface

This encompasses the effects of surface tension reduction, foaming and defoaming.

Surface tension Reduction of surface tension from 72 dynes cm^{-1} to 28-29 dynes cm^{-1} (at best) can be achieved by concentrations of nonionic surfactants as low as 0.001% w/w. This is achieved as a result of ejection of the surfactant hydrophobe from the aqueous phase to form a hydrocarbon film on the water surface where the molecules are oriented vertically

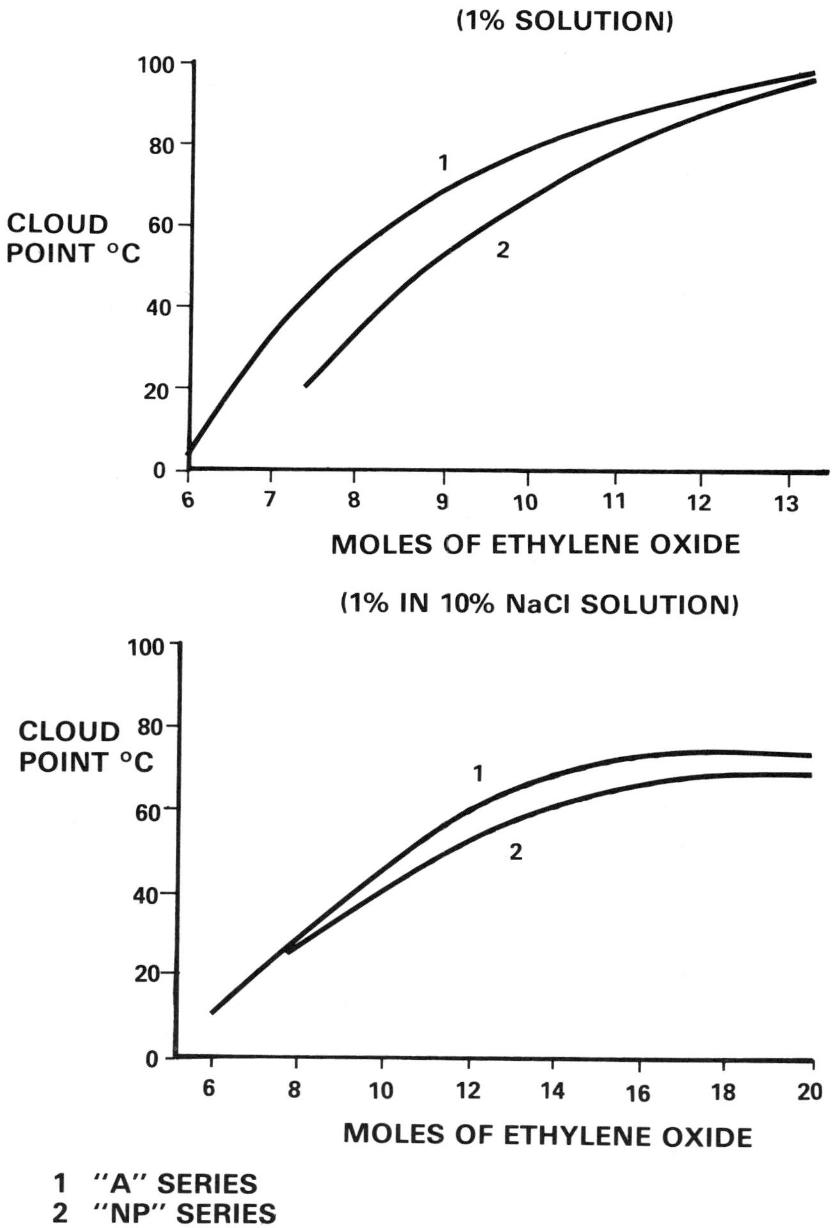

Fig. 7. Cloud points of "Synperonic A" and "NP" series.

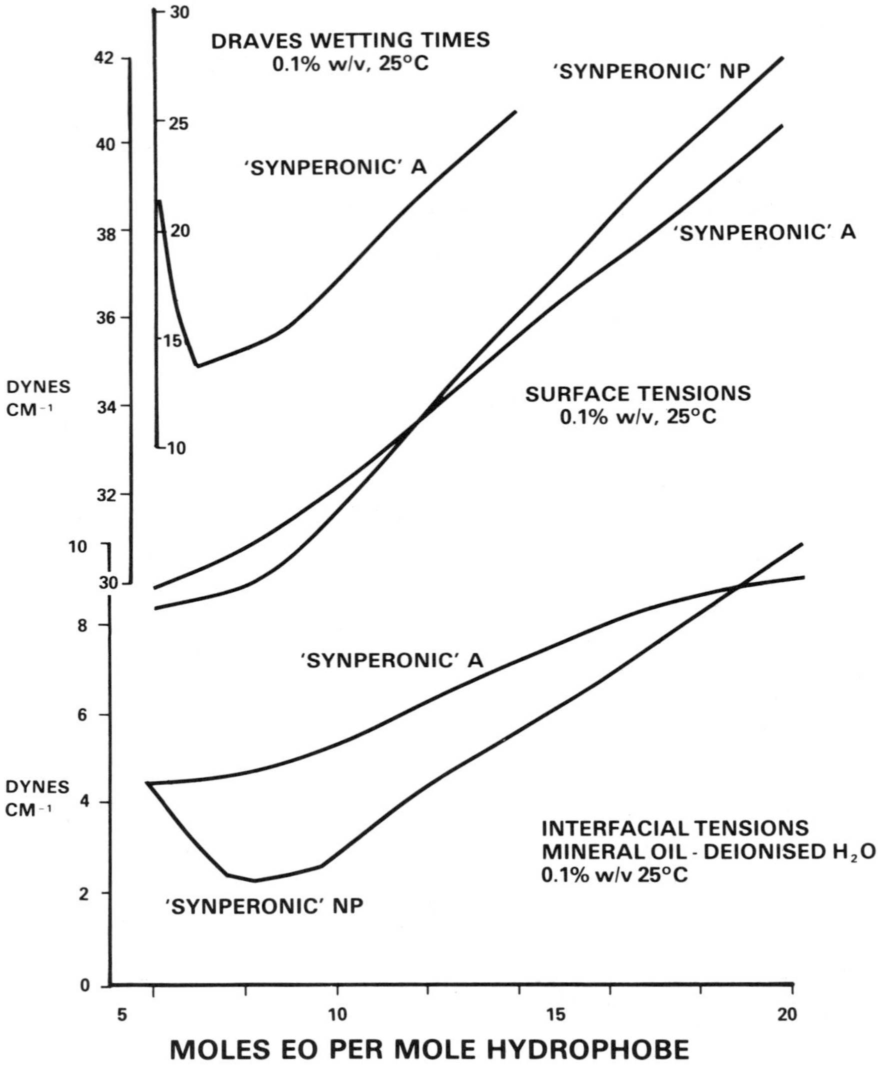

Fig. 8. Variation of surface tension, interfacial tension and wetting with moles EO for "Synperonic" A and NP series.

when the surface is covered by a monomolecular film. Their polar ends remain hydrated by the aqueous phase. The net effect is to reduce the asymmetric hydrogen bonding of the water molecules at the surface and reduce the energy needed for its deformation. Foam formation and wetting are thus eased.

The relationship between surface tension, structure and polarity for alcohol and alkylphenol ethoxylates is illustrated in Fig. 8 where it is apparent that at 25°C and in distilled water the lowest surface tensions are achieved with surfactants which are barely water-soluble, have cloud points of between 15 and 35°C and HLB numbers of between 11 and 12.5. For higher temperatures or with electrolyte solutions this optimum shifts towards higher HLB. Thus the lowest surface tensions are produced by the shortest ethoxylate chain which imparts water-solubility under the conditions of use. Lower surface tensions can be obtained with siloxane or fluorocarbon surfactants (down to 20-25 dynes cm^{-1}) or C_{10} alcohol ethoxylates (25-28 dynes cm^{-1}) but these have deficiencies which restrict their use.

As is also apparent from Fig. 8, cotton wetting is most rapid when surface tension is minimized but this requires a reduction of solid - liquid interfacial tension by concurrent coordination of surfactant hydrophobe to the solid surface.

Foam stabilization and defoaming A surfactant which lowers surface tension reduces the energy needed to create a foam but the parameter which dictates foam volume is its stability. Thus both defoamers and foam stabilizers can create low air - water surface tension but a foam stabilizer also forms a dense, ordered, viscous but mobile film. Defoamers form open, low viscosity surface films. Thus large volumes of stable foams are generated by application of mechanical energy to solutions of nonionics with 7-20 moles ethylene oxide. Maximum foam stability is at 12-16 moles EO where the ethoxylate chains can form dense closely-packed hydrated bi-layers crosslinked by interstitial water molecules. Foam stability reduces as the temperature (and electrolyte) concentration is increased.

Thus at or above their cloud points foam stability is greatly reduced, especially for alkylphenol ethoxylates on account of their narrower homologue distribution. This is illustrated in Fig. 9 which shows Ross-Miles foam heights after 5 minutes for alcohol and nonyl phenol ethoxylates between 25 and 65°C.

Nonionic surfactants are also widely used to enhance wetting or detergency whilst producing little or no foam even under conditions where much mixing energy is applied, as in bottle-washing plants or automatic dish-washing machines. Low foaming nonionics are invariably of low aqueous solubility under their conditions of use thus ensuring that they are concentrated at the gas - liquid interface. They usually give low surface tensions and have low CMC and they may by good wetting agents. Although they are interfacially active they

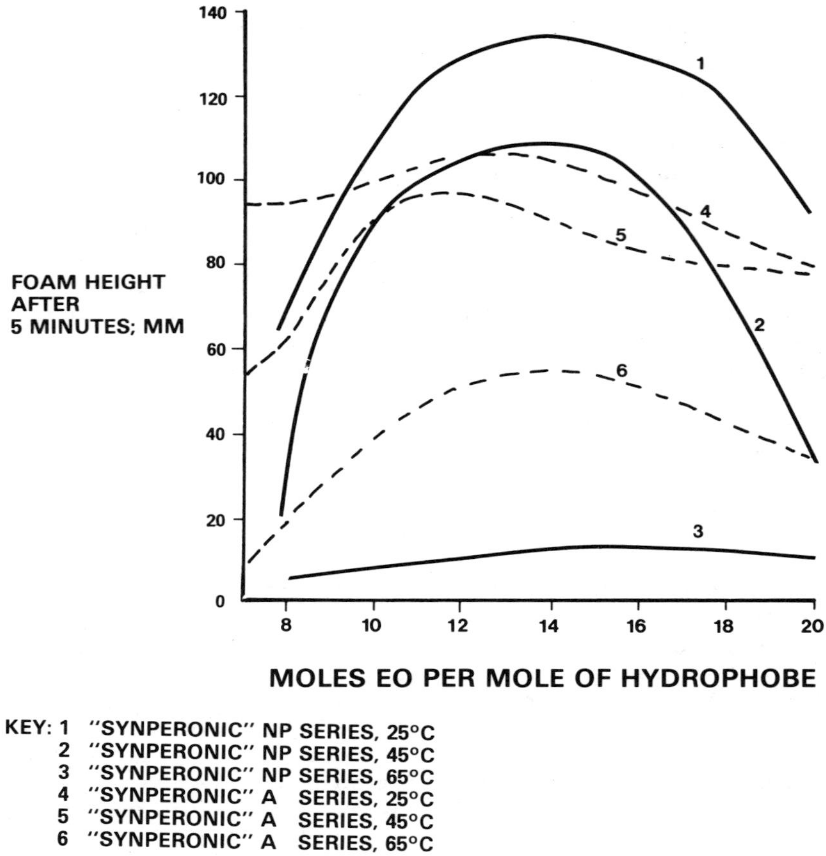

Fig. 9. Foaming properties of "Synperonic" A and NP Series (using Ross-Miles static method).

form low viscosity, mobile surface films which form areas of weakness in the bubble surface thereby aiding its rupture. These films are laterally weak because little hydrogen bonding is possible (via water molecules) between adjacent defoamer molecules as their ethoxylate chains are short or "end-capped" with bulky water-insoluble groups. Typical nonionic defoamers/low foam surfactants are polypropylene glycols (MWts 1500-2500), propylene oxide-ethylene oxide copolymers such as "Synperonic" PE30/10 or 30/20, fatty alcohol propoxylates, fatty alcohol ethoxypropoxylates, benzyl-"capped" alcohol or alkylphenolethoxylates, difatty PEG esters or PEG-polysiloxane copolymers such as the "Silcolapse" (ICI trademark) series. Defoamers can be effective at concentrations as low as 5 ppm. Simple nonionic surfactants are also effective defoamers if used above their cloud points.

Solid - Liquid Interface

This includes the phenomena of wetting, detergency, and dispersion.

Wetting Efficient wetting is obtained when the gas - liquid and solid - liquid interfacial tensions are minimized. Wetting efficiency can be quantified by contact angle determination (Chapter 10) or more usually by measuring the sinking time of a weighted cotton hank (the Draves-Clarkson method). Wetting times for the "Synperonic" A series are shown in Fig. 8 where it can be seen that the shortest wetting time is obtained with "Synperonic" A7, which also gives a very low surface tension. Likewise "Synperonic" NP8 and 9 are optimal in the corresponding nonylphenolethoxylate series.

As well as requiring surface tension reduction, surfaces are most efficiently wetted when the nonionic is barely soluble (i.e. on or above its cloud point) under the conditions of use. It is thus first rejected to the gas - liquid interface and thence also to the liquid - solid interface where the hydrophobe may weakly coordinate to the surface.

Detergency This is a complex effect composed of wetting, emulsification/emulsion stabilization and dispersion, each of which will be discussed separately. Suffice to say that although the detergent performance of a nonionic will obviously depend on other formulation variables such as inorganic builder type and concentration as well as cosurfactants, selection of the optimum product is amenable to certain semi-empirical rules. Thus the best nonionic detergents contain between 60% and 70% w/w ethylene oxide (or have HLB numbers of 12-14). Hydrophobe molecular weight should be between 200 and 350. The ethylene oxide content must be high enough to ensure the surfactant does not phase separate under wash conditions but not so great as to reduce its interfacial activity. To ensure oily soil emulsification/solubilization the hydrophobe should be above its melting point at the wash temperature. The additional requirements of good biodegradability and resistance to alkaline or acid hydrolysis further restrict the available options such that natural or synthetic fatty alcohol ethoxylates are the choice of most responsible detergent formulators.

Historically tallow (C_{16}-C_{18}) alcohol ethoxylates containing 11-20 moles EO were used at low levels (2-4% w/w) in powder formulations to enhance oily soil detergency. However, the reduction in wash temperatures (for energy conservation, increased use of synthetic fibres and development of non-iron, drip-dry, crease-resistant finishes) and phosphate builder levels coupled with improved formulation technology has resulted in moves to shorter hydrophobe chains (C_{12}-C_{15}) containing 7-9 moles EO at higher (5-7%) usage levels.

Dispersion and deflocculation Nonionic surfactants will disperse and deflocculate finely divided organic and inorganic solids in water. The hydrophobe is rejected from the aqueous phase on to any available surface where it may also be absorbed if the surface is hydrophobic (e.g. wax, PVC, polystyrene, coal). The effect is fairly general although longer ethoxylate chains (e.g. 15-20 moles EO) are usually more efficient. Polymeric surfactants such as the "Synperonic" DA or PE (Propylene-oxide-ethyleneoxide copolymers) series or ethoxylated alkyl-phenol-formaldehyde resins are especially effective as their larger hydrophobes permit the desirable combination of firmer anchoring (entropic and/or chemisorption) and low water solubility/polarity with long ethoxylate chains which thicken the flocculation resisting steric barrier. They also retain their effectiveness at high electrolyte concentrations. Unlike anionic dispersing agents (such as the lignosulphonates) which

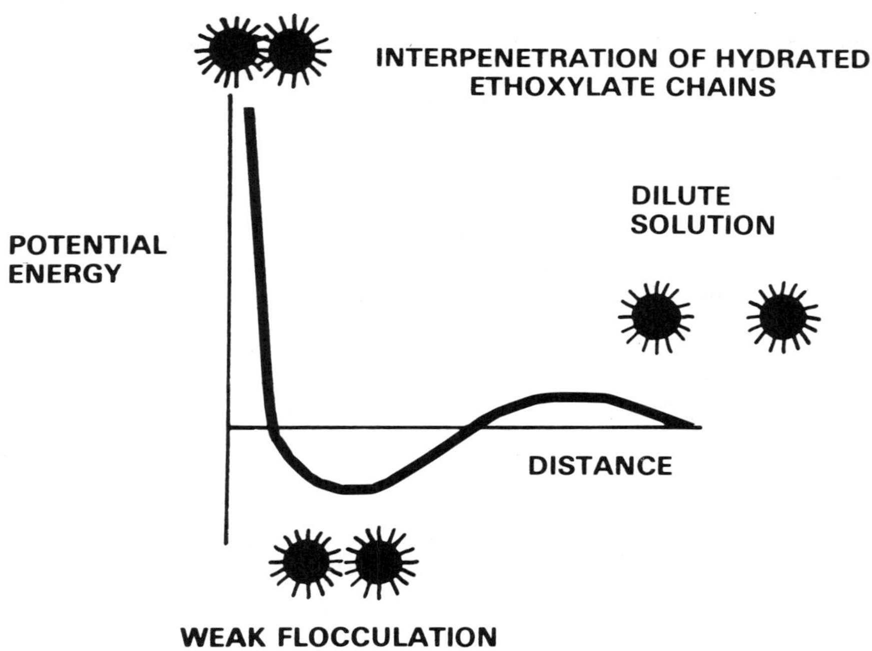

Fig. 10. Schematic representation of energy - distance curve for sterically stabilized dispersions.

often cause sedimentation by deflocculating suspensions, nonionic dispersants can enhance dispersion stability by forming a thermodynamically stable, weakly flocculated system.

Flotation Minerals can be concentrated for further processing by their preferential removal from an aqueous slurry of finely divided ore as a floated froth. This is achieved by firstly adding an organic chemical collector to the slurry which chemisorbs or reacts with the enriched ore particles making them hydrophobic. A surfactant designed to produce a foam (or froth) of controlled stability but not wet the ore surface is then added and air injected. Air bubbles preferentially adhere to the hydrophobic ore, floating it to the surface where the enriched froth is removed and dewatered. Typical collectors are xanthates (copper extraction) and hydrocarbon oils (coal flotation). Water-soluble collectors are typified by the ICI "Teefroth" range of methanol and n-butanol propoxylates and ethoxy-propoxylates which are used in each of these applications.

Liquid - Liquid Interface

This includes the processes of emulsification, emulsion stabilization, solubilization and demulsification.

Emulsification, emulsion stabilization, microemulsions and solubilization There is a mystique about emulsification which has been fostered, to some degree, by the HLB system of categorizing emulsifiers. Although a useful tool it inhibits a more general understanding of the above processes.

Nonionic emulsifiers are very important in almost all aspects of our daily lives and in industry. They are used in such widely diverse products as bread, salad cream, cosmetics, herbicides and plastics. Their use flexibility is a consequence of the control of their composition/polarity/EO content which is possible by their method of synthesis. Also hydrophobe types are available which are compatible with most water-insoluble materials and most of these surfactants are of low toxicity/good biodegradability.

An emulsion is a dispersion of fine droplets of a liquid within another liquid with which it is immiscible (Chapter 8). In the absence of a suitable surfactant(s) the emulsion can only be formed by vigorous mixing (mechanical or ultrasonic) and will quickly "break" by coalescence and phase separation of the internal phase. An emulsifier can significantly reduce the energy needed to form an emulsion by absorbing at oil - water interfaces (because of its simultaneous solubility in each phase) where it may reduce the oil - water interfacial tension from 20 dynes cm^{-1} to <1 or <0.1 dynes cm^{-1}. It thus follows that an emulsifier is most effective when its interfacial concentration is at a maximum. It should therefore be solvated by each phase but dissolved in neither. Hence the concept of hydrophobe - lyophobe "balance". However, agitation combined with a low interfacial tension will create both oil-in-water and water-in-oil emulsions. Both may be equally stable and a complex emulsion results or one or the other type predominates. This happens because of the relative rates of oil - oil drop coalescence or water - water coalescence. If the former is rapid a water-in-oil emulsion results and vice versa. It is the strength of the interfacial film formed by the emulsifier which dictates the ultimate emulsion type. Thus simple surfactants with ethylene oxide contents of up to 30% (HLB up to 6) form water-in-oil emulsions and those with 40-80% (HLB 8-16) form oil-in-water emulsions.

The only effective way to identify an emulsifier for a specific system is to screen candidates initially selected on the basis of polarity with a practical test controlling phase types/volumes, temperature, electrolyte or cosurfactant concentration, hydrophobe type, shape and molecular weight, emulsifier concentration and mixing energy. The objective is always to create an elastic, robust and coherent interfacial film which inhibits collision-induced droplet coalescence. Thus the above variables may alter surfactant interfacial concentration, interfacial molecular packing, interfacial inter-ethoxylate hydrogen bonding and hydrophobe - hydrophobe van der Waals forces.

When emulsifiers are screened empirically according to polarity (HLB) and hydrophobe several likely candidates will be quickly identified. These may then be fine-tuned by mixing, and varying their concentrations and emulsion preparation methods. Emulsion stability is assessed by rate of "creaming"/ phase separation at ambient and elevated temperatures and during freeze - thaw cycling. Centrifuging accelerates these processes.

A typical "coarse" milky appearing oil-in-water emulsion has an internal phase droplet size of 0.1-1.0µm. By carefully blending two or more emulsifiers around the optimum polarity a "microemulsion" region can usually be located (Chapter 5). In this range of compositions the oil - water interfacial tension sharply drops to <0.1 or even <0.01 dynes cm^{-1} with consequent reduction in dispersed phase droplet size below 0.1µm yielding a clear, usually opalescent homogeneous mixture of indefinite stability. This process is also known as solubilization and is due to the incorporation of the oil phase into expanded mixed micelles of uniformly graded polarity. The interfacial region is probably composed of "liquid crystals".

High surfactant : oil ratios (e.g. 1:1 to 1:2) are necessary. Microemulsions are easily generated by identifying the "best" single emulsifier from an homologous series, creating a "coarse" emulsion with an analogous emulsifier of lower polarity and then adding small increments of a more polar emulsifier until the mixture clears. Microemulsions are of increasing importance for enhanced oil recovery, and surfactant solubilization of insoluble drugs is an efficient way of introducing them into the body, minimizing dosage and reducing adverse side effects.

Most emulsifiers are also emulsion stabilizers. However, many polymeric materials (including polymeric surfactants) or solids do not markedly reduce oil - water interfacial tension but are very effective at forming a viscous ordered interface which prevents dispersed phase droplet coalescence once they have been formed. This may be achieved by high shear or ultrasonic mixing together with a low MWt emulsifier. Thus polyvinyl alcohol, sodium alginate, xanthan gum, waxes, asphaltenes or clays act as emulsion stabilizers. The first three for oil-in-water emulsions, the latter three for water-in-oil (especially crude oil).

Demulsifiers Demulsification is one of the least understood surfactant processes. Its greatest relevance is in the crude oil production and chemical industries. The former industry requires the removal of brines emulsified into crude oil by wellhead letdown equipment and stabilized by crystalline waxes/asphaltenes or partially oil/water wetted clays. The latter usually requires removal of emulsified oil in aqueous effluents. Apart from mechanical treatments (e.g. baffled oil - water separation tanks and electrical heater-treater units) or pH adjustment (to acidify alkali metal soaps) the petrochemical industry has a wide range of surfactant chemical demulsifiers available. These are commonly block copolymers of propylene and ethylene oxides combined with initiators such as glycerol, phenol-formaldehyde resins, melamine-formaldehyde resins, siloxanes, polyamines and polyols.

They are all of high molecular weight (3-100,000) with flexible, mobile "backbones" and usually contain from 5-50% w/w polyethylene oxide. Demulsifiers for oil-in-water contain 5-20% polyethylene oxide and for water-in-oil 20-50%. Their high molecular weight and limited water/oil solubility ensures that they concentrate at the oil - water interface creating mobile, low viscosity zones reducing the collision energy needed for droplet coalescence and eventual gravity-driven phase separation. They are commonly formulated into multicomponent solutions in aromatic solvents and used at levels of 5 to 100 ppm.

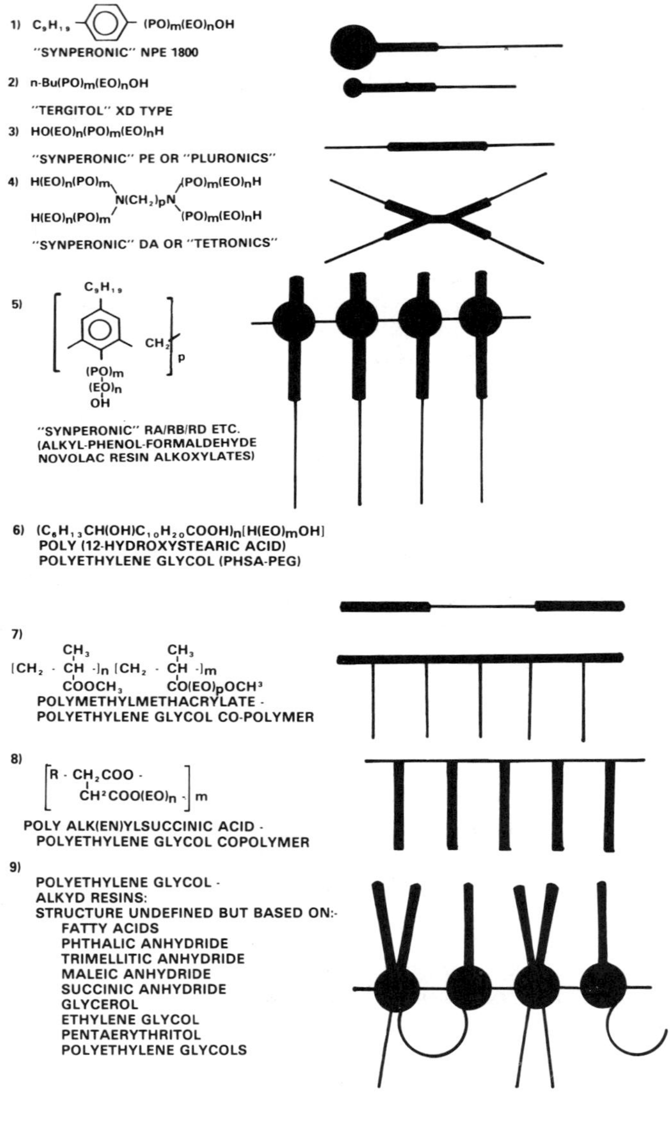

Fig. 11. Structure of some polymeric nonionic surfactants.

POLYMERIC NONIONIC SURFACTANTS

The formulae and chemical structures of most of the commercially available polymeric nonionic surfactants (excluding polysiloxanes) are illustrated in Fig. 11. Polymeric surfactants are useful mainly as demulsifiers (as discussed above), dispersing agents and emulsion stabilizers. Their wetting and detergency is much inferior to simpler nonionics, the ester type is easily hydrolysed by alkalis and the polyether types are not biodegradable.

However, their high molecular weight ensures maximum interfacial concentration by a significant increase in the energy of adsorption. This also gives them a better all-round performance as they are less affected by electrolytes, temperature or composition of dispersed phase and effective at lower concentrations. They may also be designed with longer and hence more efficient polyglycol chains to create a very robust steric barrier inhibiting disperse phase coalescence or flocculation without making them too water-soluble.

Thus products such as "Synperonic" NPE1800 or DA60/30 find applications in agrochemical emulsions containing high concentrations of electrolytes, stabilization of coal in water slurries, coal-in-oil dispersions and oil slick dispersant concentrates. Polyhydroxystearic acid - PEG copolymers are useful in emulsion polymerization and PEG-alkyd resins are used for producing water-emulsifiable gloss paints and industrial degreaser concentrates.

APPLICATIONS

The applications of nonionic surfactants are very diverse and impossible to discuss in a short review. The summary list in Table 2 is offered as a starting point for further investigations.

THERMAL AND OXIDATIVE STABILITIES

In the absence of oxygen, ethoxylate surfactants are high boiling involatile materials. Under high vacuum (e.g. <10m. bar) and temperatures in excess of 150°C it is possible to distil one- and two-mole adducts but at atmospheric pressure their initial boiling points are in excess of 300°C. Lower alcohol ethoxylates, however, contain appreciable levels of unreacted fatty alcohol which readily vaporizes during spray-drying nonionic containing detergent powders. This "pluming" problem is exacerbated by simultaneous oxidation and shortening of the ethoxylate chain which also yields volatile organic products such as acetaldehyde and acetic acid (plus water and CO_2). The relative susceptibility to oxidation and volatility

TABLE 2

APPLICATION	INDUSTRY	SUGGESTED "SYNPERONIC" SURFACTANTS
Agrochemicals	Manufacturers of pesticide and herbicide self-emulsifiable concentrates and wettable powders	Syn NP9, 13, 15, 20, NPE 1800 A7, A11
Alkaline cleaning of aluminium and zinc	Sheet aluminium and zinc production	Syn NP15, A11, A20
Aluminium stamping and drawing lubes	Aluminium forming industry	Syn A2, A3, NP4, NP6, PE30/40
Bath salts and bubble bath formulations	Cosmetic and perfumery companies	Syn A3 sulphate Synprol sulphate
Bottle washing	Alkaline formulations for bottle washing and dairy cleaning; janitorial formulators	Syn NP10, OP10, A11
Bottle washing, glass washing and sanitizing	Brewing industry; iodophore formulations	Syn NP8, NP15, 87K PE30/20
Brick and stone work cleaner	Contract cleaning companies	Syn NP9, A7
Bubble baths	Cosmetic formulation companies	Syn A2, A3 sulphates
Carboxymethylate manufacture	Feedstock for reaction with sodium chloracetate to give dye levelling agent and oil recovery detergent	Syn A2, A3, A6
Car shampoo, external and internal	Car accessory producers	Syn A3 sulphate

Cement foam insulation panels	Building industry	"Synprol" alcohol sulphate, Syn A3 sulphate
Concrete production	Foaming agent - Plasticiser -	Syn NP15 Syn NP8, NP9, NP12
Corrosion inhibitors for iron and steel	Oil and gas pipelines	Syn A2/A3 amine neutral-ized ether phosphates
Cotton desizing	Cotton industry	Syn NP6, NP8, K87
Cotton dye levelling	Cotton industry	Syn A7, A11
Cotton finishing (with polythene emulsions)	Cotton industry	Syn NP4, NP6, NP8
Cotton kier boiling	Cotton industry	Syn NP9, NP10
Cutting oils for metal working - soluble and emulsion types	Engineering industry	Syn A11
Dairy cleansing	Iodophore formulations for milking parlours and bottling plants	Syn NP8, NP15, 87K PE30/20
Defoaming aqueous solutions		Syn PE30/10, 30/20
Degreasing and deoiling of vehicles	Road, rail and aircraft industries. Chassis and engine cleaning. Runway cleaning. Solvent based formulations	Syn NP6, NP8, NP9, TC 30C, OP10
Demulsification of crude oil	For wellhead or refinery desalting; oil production industry	Syn RA, RB, RC, PE, DA series
Detergent laundry liquids	Manufacturers of heavy duty domestic detergent liquids for general purpose laundering	Syn A7, A11, 87K, A3 sulphate

TABLE 2 (continued)

APPLICATION	INDUSTRY	SUGGESTED "SYNPERONIC" SURFACTANTS
Detergent laundry powders	High, medium and controlled foam domestic detergent powder manufacturers	Syn A7, 87K, A11 "Synprol" alcohol sulphate
Speckled detergent washing powder production	Detergent powder formulators: nonionic used as dye carrier	Syn A20
Drilling mud dispersants and emulsifiers	Oil production and exploration	Syn NP9, NP10, NP15
Drug solubilization. Low irritancy hand cleaners	Pharmaceutical formulations	Syn PE39/70
Drum cleaning and reclamation	Solvent based formulations for pressure washing	Syn NP6, A7, K87
Dust suppression as wetting agent with $CaCl_2$	Mining/coal industry	Syn NP9, A7, OP10
Emulsifier for oil and waxes into water	Various industries, especially production of household cleaners	Syn A2, A3, NP4
Emulsifier for water into chlorinated solvent	Dry-cleaning	Syn A2, A3, NP4, NP6, NP8, NP10
Emulsion polymerization	PVC, PVA, polyacrylics, polypropylene, polyethylene	Syn OP10, 11, 16.5, 40, Syn NP8, 9, 10, 13, 15, 30, 40, Syn PE30/80
Enhanced oil recovery	By surfactant injection. Oil companies, drilling companies	All "Synperonic" surfactants, depends on well conditions

Enzyme coating	Coating enzymes for detergent powders	Syn A20, A50, A80
Fibre antistats	Synthetic fibre spinning industries	Syn NP4, A2, A3, PE30/20
Fibre lubricant and antistatic	Polyester and nylon spinning, weaving and knitting	Syn A2, A3, NP4, PE30/20
Fibre lubricants - mineral oil emulsifiers	Nylon, polyester, polyacrylic textile spinning	Syn NP6, NP8, A7, A11
Filter cake dewatering	Chemical process industries	Syn NP8 or 9, Syn A7, PE30/20, PE30/40
Fire fighting foams	Fire extinguisher producers	"Synprol" alcohol sulphate, Syn A3 sulphate
Floor and wall cleaning formulations - alkaline	Janitorial supply industry	Syn NP10, NP13, NP15, A11
Fluon dispersions	Fluon/Teflon coating of non-stick cooking utensils	Syn OP10
Fungicide for fruit trees	Pre-emergent spray: agrochemical formulators	Syn NP4, NP6, NP8, NP9
Fur dyeing and oiling	Fur industry	Syn NP8, NP9, NP15
Hand cleaning gels	Kerosene-water gels for oil and grease removal. Industrial cleaner formulators	Syn A7, NP4, NP6
Hard surface cleaner formulations	Manufacturers of domestic detergent formulations	Syn A7, 87K, NP9
Iron and steel acid cleaning and pickling	Steel industry prior to electroplating or painting	Syn NP8, NP9, OP10
Iron and steel sheet degreasing	Steel industry prior to painting or coating	Syn NP6, NP8, NP9, OP10

TABLE 2 (continued)

APPLICATION	INDUSTRY	SUGGESTED "SYNPERONIC" SURFACTANTS
Leather processing	Tanning and hide industries	Syn NP8, NP9, 87K
Newsprint deinking	Paper reclamation and recycling	Syn NP6-12, A7, A11
Nylon dyeing	Synthetic fibre industry	Syn A20
Oil slick dispersants	Local government. Oil companies. Shipping companies. Oil industry supply companies	Syn OSD 20, 41, TO5
Oil well acidizing	Oil production industry	Syn NP8, NP9, NP10
Oil well waterflooding viscosity improver	Oil production industry	Syn A3 sulphates
Oil well wettability improver	Oil production industry	Syn NPE 1800
Paint remover formulations	Solvent cleaner producers. Paint formulators and producers	Syn NP9
Phenol-formaldehyde foams	Insulating foam producers. Resin producers	Syn NP13, NP15
Phosphate ester production	Feedstocks for phosphate ester manufacturers. Oil industry chemical suppliers	Syn A2, A3, NP4
Phosphating of steel	Iron and steel industry	Syn NP8, NP9, OP10, A7, A11
Photographic wetting agents	Film industry, photographic industry	Syn NP9/10, OP10, A7

Polishes for floors, furniture and cars	Wax emulsifier. Household chemical manufacturers	Syn NP4, 6, 8, 9, OP10, A7
Polyester dyeing	Synthetic fibre industry	Syn A11
Polyethylene emulsions	Plastic coatings	Syn NP4, A3/A7 mixtures
Polystyrene latexes	Plastic forming industries	Syn A20, NP15, NP20, NPE 1800
Polyvinylacetate emulsion paints	Paint industry	Syn NP8, NP9, NP20, NP30
Printed circuit boards	Heat transfer fluid for electronic component producers	Syn NP9, NP15, NP20
PVC foam production	For carpet backing foams	Syn A11, A3 sulphate "Synprol" alcohol sulphate
Resin removal from wood pulp	Newsprint production and paper industry	Syn NP9, 10, 12, A7, A11
Rubber latex stabilization	Natural rubber industry	Syn NP20, A20, "Synprol" alcohol sulphate, Syn A3 sulphate
Sanitizing and cleaning of food processing equipment	Food and catering industries. Formulations with quaternary ammonium biocides	Syn NP8, NP9, OP10, A7, 87K
Shampoos	Cosmetic formulation companies	Syn A2, A3 sulphates
Silicone emulsions	Polydimethyl siloxane emulsions for defoaming aqueous processes	Syn NP4, NP6, A3, A4
Tank cleaning	Solvent based formulations for pressure cleaning of oil storage tanks, road and rail tank cars	Syn NP6, NP8, NP9, TC30C, Syn PE30/40, OP10

TABLE 2 (continued)

APPLICATION	INDUSTRY	SUGGESTED "SYNPERONIC" SURFACTANTS
Tin plating	Production of tinned steel sheeting	Syn NP8, NP9, OP10, A7
Toilet blocks	Production of household cleaners	Syn A20
Washing liquids for delicate fabrics	Domestic detergent formulators	Syn A3 sulphate, Syn A7, Syn A11
Washing-up liquid formulations	Domestic detergent formulators	Syn A2, A3 sulphates, Syn A7, A11, NP9, "Synprol" alcohol sulphate
Wetting agent for paper and cellulose	Paper industry	Syn A7, NP9
Wetting agent - low foaming	Textile, detergent, food industries	Syn A6, 87K, NP6, Syn PE30/20, 30/40
Window cleaning formulations	Aerosol or liquid domestic cleaners. Detergent formulators	Syn A7, 87K, A3 sulphates
Windscreen washing detergents	Car accessory producers	Syn A3 sulphate
Wool carbonizing (with dilute acid)	Wool industry	Syn NP8, NP9, 87K
Wool dye levelling	Wool industry	Syn NP8, NP9, 87K
Wool scouring	Raw wool processing industry. Sheepskin processing	Syn NP8, NP9, OP10, 87K, A7

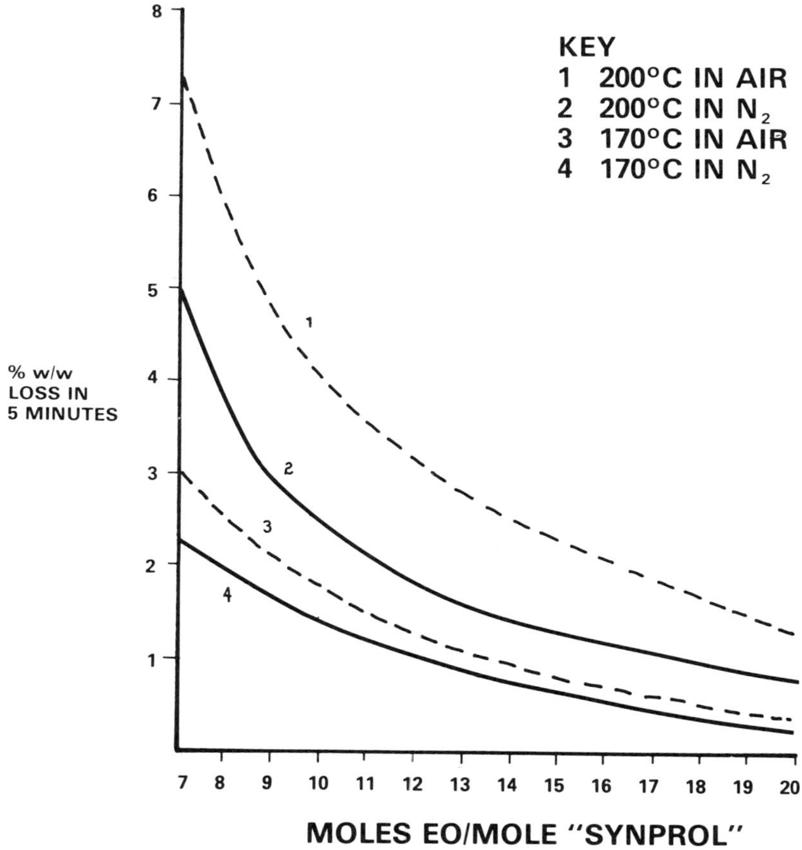

Fig. 12. Thermal and oxidative stability of "Synprol" alcohol ethoxylates.

of ethoxylates can be measured using a thermal analyser. Results of such measurements are illustrated in Fig. 12 which shows the maximum relative weight losses for "Synperonic" surfactants A7 to A20 over 5 minutes (upper limit of residence time in a detergent powder spray drier). Nonylphenol ethoxylates show much lower weight losses due to their narrower homologue distribution and absence of unreacted hydrophobe.

The weight losses displayed by the detergent alcohol ethoxylate series are proportional to the square root of their unethoxylated alcohol content and independent of the alcohol's structure and chain length. Thus excessive pluming is best reduced by using a higher ethoxylate, reducing its concentration in the formulation or the spray drier's air inlet temperature.

On prolonged heating exothermic oxidation of a neat ethoxylate initiates (in air) at a threshold temperature of about 130°C. The heat evolved is typically 600-800 calories/gram. The "induction period" for this reaction is 20-25 minutes at 140°C reducing to <2 minutes at 170°C. In a detergent powder such decomposition is retarded and may take up to two hours (for a 10% active nonionic formulation). Its consequences are seen as charred deposits within the spray tower and black speckles in the final products. This reduces on-line time and increases the frequency of tower cleaning and hence production costs. Addition of an antioxidant such as "Topanol" CA (ICI trademark) at a level of 0.1 to 1.0% w/w on the nonionic largely overcomes this problem.

BIODEGRADABILITY AND TOXICOLOGY

General

The increasing use of nonionic surfactants, especially in retail products, requires that they should be environmentally acceptable and nonhazardous to their users. Thus a range of tests have been developed which permit meaningful assessments of key parameters such as biodegradability, oral toxicity, skin and eye irritation and fish toxicities. Ideally a surfactant should be fully biodegradable under all conditions, nontoxic and nonirritant. Alcohol ethoxylates offer the best overall compromise as is discussed below.

Biodegradability

Nonionic biodegradability may be determined by a range of tests, some of which measure primary degradation (or loss of surface activity) and more recent protocols which seek to identify the fate of all the molecular fragments. The OECD has recently recommended two standard tests. The "Die Away" test where the surfactant is added to an aqueous solution of inorganic nutrients, inoculated with aerobic micro-organisms and its residual concentration determined by precipitation with $KBiI_4/BaCl_2/CH_3COOH$ (modified Dragendorff reagent).

Alternatively, a continuous flow activated sludge test (Husmann) may be used. Both of these tests measure only primary degradation but ultimate degradation (into CO_2, H_2O and mineralized fragments) may be estimated using the ratio of a surfactant's biological oxygen demand (BOD) to its chemical oxygen demand (COD) measured by dichromate oxidation.

Figures 13 and 14 illustrate how BOD/COD ratio decreases

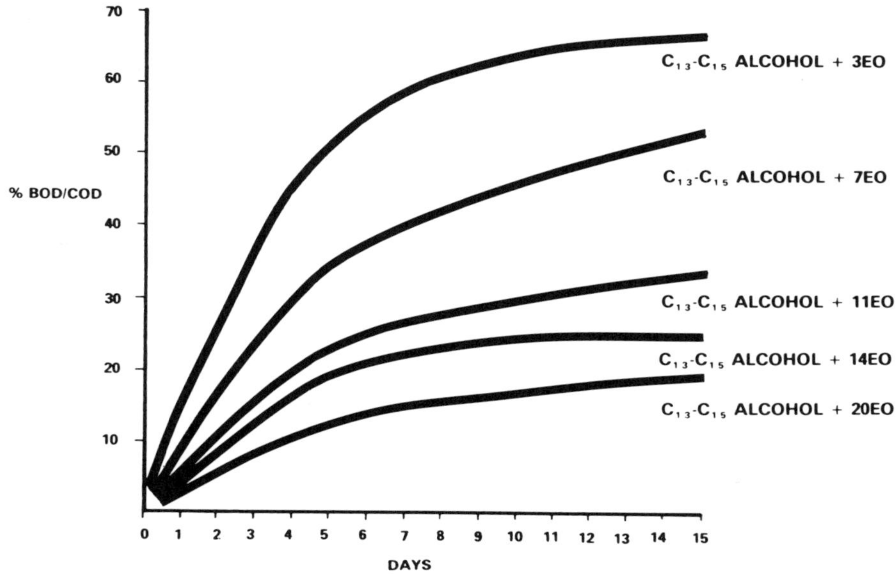

Fig. 13. Biodegradability of "oxo" alcohol ethoxylates (BOD/COD ratio).

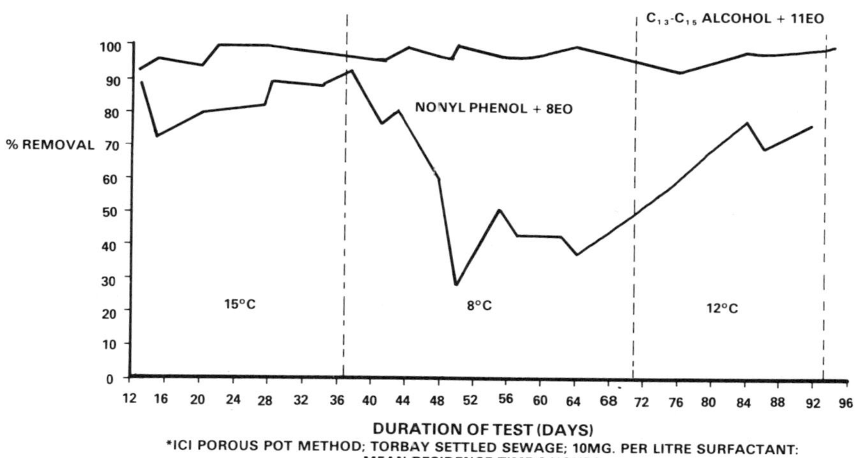

Fig. 14. Biodegradability as assessed by the continuous activated sludge method, (ICI Porous Pot method; Torbay settled sewage; 10 mg per litre surfactant; mean residue time 6 hours) at various temperatures.

as ethylene oxide content increases in the "Synperonic" A series (typical of alcohol ethoxylates) and the variation in percentage removal with temperature in a continuous flow activated sludge test. The difference between the degradability of the nonylphenol ethoxylate and the "oxo" alcohol ethoxylate is marked as the temperature is reduced. Geographical variations in degradability are also found, presumably due to differences in sewage micro-organisms. Currently, both alcohol and alkylphenol ethoxylates are accepted as greater than 80% biodegradable using the OECD protocols, although the former degrade more rapidly and completely.

During degradation polyethyleneglycol residues are generated which are not surface active but degrade relatively slowly, especially from ethoxylates containing more than 20 moles EO. Nonylphenol ethoxylates degrade initially to a 2-mole ethoxylate and its carboxylate derivative whereas alcohol ethoxylates degrade to the parent alcohol.

Linear fatty alcohols degrade faster than "oxo" alcohols containing 2-methyl isomers and degradability is further reduced as numbers and length of alkyl side chains are increased. Incorporation of propylene oxide into the ethoxylate chain reduces biodegradability, although terminal blocks of up to 4-5 moles do not prevent it proceeding beyond 80% under favourable conditions. However, high molecular weight EO/PO copolymers such as the "Synperonic" PEs are not biodegradable.

Toxicity

Oral toxicity Nonionic surfactants are of low acute and subacute oral toxicity. Fatty alcohol and alkylphenol ethoxylates typically have LD_{50} of 2-10 g/kg of body weight (rats). Fatty acid esters of PEG, sorbitol or sorbitol ethoxylates are less toxic with LD_{50} of 20-60 g/kg. Propylene oxide-ethylene oxide block copolymers have toxicities in the range 5-15 g/kg. None of these products thus presents a significant toxic hazard in normal use. The least toxic are approved as food additives. Subacute oral testing shows broadly similar trends.

Skin irritation and sensitization Alcohol and alkylphenol ethoxylates and fatty acid ethoxylates which are good emulsifiers and detergents can degrease skin both when neat and in concentrated aqueous solutions. This may result in drying and cracking leading to bacterially induced dermatitis. However, at normal dilutions they are non irritant nor are they skin sensitizers. Polypropylene oxide-polyethylene oxide copolymers have no irritant action on the skin.

Eye irritation Using the Draize test on rabbits' eyes has shown that neat or concentrated (10-25% w/v) aqueous solutions of alkylphenol ethoxylates give moderate to severe corneal damage. Dilute solutions (0.5% w/v) cause slight irritation. Dilute alcohol ethoxylate solutions are non irritant and slight irritants when neat.

Fish toxicity The 96-hour LC_{50} of alcohol and alkylphenol ethoxylates to rainbow trout is in the range 1-10 mg/l. This is similar to that of common anionic surfactants such as alkyl benzene sulphonates and alcohol ether sulphates. The nonionics which are most lethal to fish are those which effect the greatest reduction in the surface tension of water at low concentrations. Fortunately concentrations of nonionics in surface waters rarely approach even 1/10 of the LC_{50} value as most European domestic and industrial sewage is now processed through biological treatment plants which largely degrade them to less toxic intermediates and eventually CO_2, water and mineralized derivatives.

GENERAL REFERENCES

Davidsohn, A. and Mildwidsky, B.M. (1967). "Synthetic Detergents". Leonard Hill, London.

Griffin, W.C. (1949). *J. Soc. Cosmet. Chemists* 1, 311.

Schick, M.J. (1966). "Nonionic Surfactants". Marcel Dekker Inc., New York.

Schonfeldt, N. (1969). "Surface Active Ethylene Oxide Adducts". Pergamon Press, Oxford.

Schwartz, A.M., Perry, J.W. and Berch, J. (1958). "Surface Active Agents and Detergents", Vol.II. Interscience Publishers Inc., New York.

SOME APPLICATIONS OF SURFACTANTS IN BIOLOGICAL SYSTEMS

Th. F. Tadros

ICI Plant Protection Division,
Jealott's Hill Research Station,
Bracknell, Berkshire RG12 6EY, England.

INTRODUCTION

In previous chapters, examples have been given to illustrate the role of surface active agents in the preparation of disperse systems of various types, e.g. emulsions, suspensions, etc. Many chemicals used in pharmaceuticals, agriculture, cosmetics, etc., are formulated using surface active agents and/or macromolecules to produce systems such as tablets, granules, suspensions, emulsions, etc. which are convenient for application. Apart from this essential role in the preparation of the formulation and the maintenance of shelf life, surface active agents also influence the biological efficacy of the drug or pesticide.

Several examples may be quoted to illustrate the role of surfactants in biological systems. Thus, many poorly soluble drugs and pesticides are administered in a solubilized form using micellar solutions in order to increase the bioavailability and penetration of the chemical. Moreover, the use of surfactants as emulsifying agents, suspension stabilizers, wetting agents, etc., can coincidentally lead to significant changes in the biological activity of the active agent in the formulation. In addition, surface active agents may be needed for targetting the drug or pesticide to the site of action. They can also be used for the control of various surface phenomena such as adhesion promotion or prevention, spreading or deposit formation, etc.

A comprehensive description of the role of surface active agents in such complex phenomena is impossible within the limited space available in the present text. However, it was

SURFACTANTS
ISBN 0-12-682180-1

thought useful to give a few illustrations in order to indicate the role of surfactants in biological systems. The examples chosen are arbitrary and serve as an introduction to readers interested in such topics.

MICELLAR SOLUBILIZATION

Solubilization is usually described as the process of incorporation of an "insoluble" substance (usually referred to as the substrate or solubilizate) into surfactant micelles (the solubilizer). Broadly speaking solubilization may be referred to as the formation of a thermodynamically stable, isotropic solution of a substance, normally insoluble or slightly soluble in a given solvent, by the introduction of an additional amphiphilic component or components. Solubilization is usually reflected in a rapid increase in the "solubility" of the substrate above a critical concentration of the surfactant in solution, which is usually close to the c.m.c. of the surfactant. One may differentiate four different locations of the substrate in micelles: (i) in the hydrocarbon core; (ii) as a "sandwich" between the surfactant chains in the micelle, i.e. in the palisade layer, in which one may distinguish between deep and short penetration; (iii) in the outer hydrophilic region of a surfactant, e.g. in the ethylene oxide region of an ethoxylated surfactant; (iv) adsorbed on the micelle surface.

Several factors affect solubilization, of which the structure of the surfactant and solubilizate, temperature and addition of electrolytes and non-electrolytes are probably the most important (Attwood and Florence, 1983). Generalization about the manner in which the structural characteristics of the surfactant affect its solubilizing capacity are complicated by the existence of the different solubilization sites within the micelles. For solubilizates which are located either in the micelle core or in a position of deep penetration within the micelle, solubilization is expected to increase with increase in the alkyl chain length of the surfactant. On the other hand, if solubilization occurs in the hydrophilic portion of the surfactant, this will increase with increase in the size of that group. The solubilizate structure can also play a major role. For example, polarity and polarizability, chain branching, molecular size, shape and structure have been shown to have various effects.

Temperature has an effect on the extent of micellar solubilization which is dependent on the structure of the solubilizate and of the surfactant (Attwood and Florence, 1983). In most cases the amount of solubilization increases with increase in temperature. This is usually due to the increase in solubility of the solubilizate and increase of micellar size (particularly with nonionic ethoxylated surfactants).

Addition of electrolytes to ionic surfactants usually causes an increase in micelle size and a reduction in the c.m.c., and hence an increase in the solubilizing capacity. Non-electrolytes which are capable of being incorporated in the micelle, e.g. alcohols, lead to an increase in the micelle size and hence lead to an increase in solubilization.

The presence of micelles will have significant effects on the biological efficacy of an insoluble drug or pesticide. In the first instance, it will affect the rate of solution of the material. Below the critical micelle concentration, surfactant adsorption can aid wetting of the particles and consequently increase the rate of dissolution of particle agglomerates (Attwood and Florence, 1983). Above the c.m.c., the rate of dissolution is affected as a result of solubilization. According to the Noyes-Whitney relation, the rate of dissolution is directly related to the surface area of the particles, A, and the saturation solubility, c_s, i.e.

$$\frac{dc}{dt} = k\, A\, (c_s - c) \qquad (1)$$

where c is the concentration of the solute.

Higuchi (1964; 1967) assumed that an equilibrium exists between the solute and solution at the solid - solution interface and that the rate of movement of solute into the bulk is governed by the diffusion of the free and solubilized solute across a stagnant diffusion layer. The solute solubilized in the micelle will have a lower diffusion coefficient than the free solute. Thus, the effect of surfactant on dissolution rate will be related to the dependence of dissolution rate on the diffusion coefficients of the diffusing species and not on their solubilities as suggested in equation (1). However, experimental results have not confirmed this hypothesis and it was concluded that the effect of solute solubilization involves more steps than a simple effect on the diffusion coefficient. For example, it has been argued that the presence of surfactants may facilitate the transfer of solute molecules from the crystal surface into solution, since the activation energy for this process was found to be lower in surfactant than in water (Elworthy and Lipscomb, 1968). On the other hand, Chan *et al.* (1976) considered a multistage process in which surfactant micelles diffuse to the surface of the crystal, become adsorbed (as hemi-micelles) and form mixed micelles with the solubilizate. The latter is dissolved and it diffuses away into bulk solution releasing the solute from the crystal surface. This multistage process, which directly involves surfactant micelles, will probably enhance the dissolution rate.

Apart from this effect on dissolution rate, surfactant micelles also affect membrane permeability of the solute. Their exact role is difficult to predict without knowledge of the detailed interactions between the surfactant molecules and the membrane components. Experiments on real membrane systems have been few, due to the complexity of performing such measurements with isolated systems which remain undamaged. However, work using artificial membranes has shown that the effect of surfactants on membrane permeability is rather complex and depends on various factors. It has been claimed that certain surfactants have the ability to increase the permeability of some bacterial cell walls, and hence are synergistic with some antibacterial agents.

AGRICULTURAL SPRAYS

Surfactants play a role in the application of pesticides and herbicides by spraying, since they affect the various processes involved, namely droplet formation, their impaction and adhesion, their retention, and wetting and spreading. Control of all these processes is important for maximum control of pests and plant diseases by a particular spray (Hartley and Graham-Bryce, 1980). Below, a brief account of the role of surfactants in the above-mentioned processes is given.

Droplet Formation

In a hydraulic nozzle, the spray liquid ejects as jets which break into droplets as a result of surface tension effects. The surface area, and consequently the surface free energy, of a sphere is smaller than that of a less symmetrical body of the same volume. Hence small liquid volumes of other shapes tend to give rise to small spheres (Bickerman, 1958). A liquid cylinder becomes unstable and divides into two spherical droplets as soon as the length of the liquid is greater than its circumference. A very long cylinder tends to divide into drops of volume $(9/2)\pi r^3$, where r is the radius of the cylinder. As the surface area of two unequal spheres is less than that of two equal drops, the formation of a polydisperse spray is more probable.

Liquid jets ejecting from a spray nozzle usually become unstable by a process of wave formation at the surface of the liquid, which grow in amplitude until drops are formed. The driving force for the growth of this instability is governed by surface tension gradients (the so-called Gibbs-Marangoni effect) which depend on the type and concentration of surfactant. Clearly the process of liquid disruption occurs over short periods of time and hence dynamic effects are very important. It is therefore essential to know the dynamic surface tension of the spray solution under the practical conditions involved. It is clear that the droplet size distribution of a spray depends on the factors that govern diffusion, convection and adsorption of the surfactant at the liquid/air interface. However, analysing the various processes involved is beyond the scope of this review, although it must be realized that the adsorption and concentration of surfactant molecules at the surface of the liquid jet must play a major role in its disruption and atomization.

Spray Impaction and Adhesion

When a drop of a liquid falls on a solid surface, one of several situations may arise depending on the conditions: (a) the drop bounces back; (b) the drop undergoes fragmentation into two or more smaller droplets which bounce back and may return to the surface with a lower kinetic energy; (c) the drop adheres to the surface; this process involves several stages whereby the drop flattens, retracts, spreads and finally rests to form a hemispherical cap; (d) the drop does not adhere immediately but floats (for a fraction of a second to several seconds) before either adhering to the surface or leaving it again.

The most important parameters that determine what happens are: (a) the mass (volume) of the drop, its velocity in flight and the distance between the orifice and target; (b) the difference between the surface energy of the drop in flight, E_o, and its surface energy after impact, E_s; (c) displacement of air between the drop and the leaf surface. Clearly both (b) and (c) depend on the nature and concentration of surfactant used in the spray solution.

The question of whether a drop adheres to the surface or not depends on the balance between its kinetic energy in flight and the difference in surface energy between free and attached droplets. For adherence to take place E_o-E_s must exceed the kinetic energy. E_o is given by the equation,

$$E_o = 4\pi r^2 \gamma_{LV} \tag{2}$$

where r is the radius of the drop and γ_{LV} is the surface tension of the liquid. E_s is given by (Hartley and Brauskill, 1958),

$$E_s = \gamma_{LV}\pi r^2 (4/3)^{2/3} [2(1-\cos\theta) - \sin^2\theta \cos\theta] [1 - \cos\theta + 1/3 (\cos^3\theta - 1)]^{-2/3} \tag{3}$$

where θ is the contact angle of the liquid drop on the surface.

From equations (2) and (3), a critical contact angle may be calculated (at a given surface tension) for various droplet sizes below which adhesion of the droplets with various velocities occurs. For example, for water droplets of 100 μm diameter (surface tension ∿70 mNm^{-1}) falling with a velocity of 25 cm s^{-1}, the contact angle should be less than 160° for adhesion to take place. On the other hand, for water droplets of 200 μm, the contact angle should be less than 100° for adhesion to take place. Thus, at a given surface tension of the liquid, the contact angle of the liquid below which adhesion occurs is smaller, the larger the droplet size and the higher their velocities. In typical sprays, droplets of several hundred μm diameter are formed at high velocities and this necessitates the use of surfactants for droplet adhesion to occur on hydrophobic surfaces.

Spray Retention

After the droplets adhere to the leaf surface, it is essential for them to be retained in order to prevent sliding under gravity for low volume sprays, and "run-off" under high volume application. The problem of droplet sliding has been analysed by Furmidge (1962) who considered drops with mass m on an inclined surface with an angle α to the horizontal. To prevent sliding under gravity, the difference in work of wetting and dewetting must balance the gravitational force, i.e.,

$$\frac{mg\sin\alpha}{\omega} = \gamma_{LV}\ (\cos\theta_R - \cos\theta_A) \tag{4}$$

where ω is the droplet "width", θ_A and θ_R are the advancing and receding contact angles respectively, and g is the acceleration due to gravity. Thus, resistance to sliding depends on the surface tension of the liquid and the difference between θ_R and θ_A (the so-called contact angle hysteresis) both of which depend on the nature and concentration of the surfactant and the texture of the surface.

Under conditions of high volume spray, retention of spray droplets and prevention of "run-off" also depends on the contact angle and its hysteresis. For that purpose, Furmidge (1962) defined a retention factor, F (see Chapter 10) given by the following expression,

$$F = \theta_M\ [\gamma_{LV}\ (\cos\theta_R - \cos\theta_A)/\rho]^{\frac{1}{2}} \tag{5}$$

where θ_M is the mean contact angle and ρ is the density of the liquid. In general, addition of surface active agents tends to reduce γ_{LV}, θ_A and θ_R. The relative extent to which these three values are affected depends on the properties of the leaf surface and the chemical structure and concentration of surfactant at the surface. This is a very complex problem and predictions are often impossible.

Wetting and Spreading

For increased coverage, which may be an important factor under certain circumstances for pest control, spreading is required. Leaves have one of the most difficult surfaces to wet. This is due to their microroughness because they are usually covered with acicular crystals of straight chain paraffinic alcohols, etc. in the range of 24-35 carbon atoms. The crystals may be less than 1 μm thick and only a few μm apart. Thus, the area of the solid covered can be several times the gross area, whereas the liquid/air surface remains smooth. A roughness factor, r, should be defined (see Chapter 10) and depending on the magnitude of the contact angle, roughness may enhance or reduce wetting. If $\theta>90°$, roughness will reduce wetting (θ is larger on a rough surface), whereas if $\theta<90°$, roughness will enhance wetting, i.e., θ is smaller on a rough surface.

The wetting and spreading of a liquid on a leaf surface may be defined in terms of a spread factor, S_F, which is the ratio of the diameter of area wetted, D, to the diameter of the drop, d. Provided θ is not too small,

$$S_F = D/d = [4 \sin^3\theta/(1 - \cos\theta)^2 \ (2 + \cos\theta)]^{1/3} \tag{6}$$

Equation (6) shows that S_F increases greatly when $\theta<35°$. Alternatively, spreading may be defined in terms of Harkin's (1952) spreading coefficient, S, i.e.,

$$S = \gamma_{LV} \ (\cos\theta - 1) \tag{7}$$

which means that S is negative for values of $\theta>0$, reaching a value of 0 when $\theta = 0$.

It is clear from equations (6) and (7) that spreading of a liquid on a solid surface depends on the magnitude of the contact angle, which, in turn, is determined by the nature of the surface and the nature and concentration of the surface active agent used. This immediately shows the major role played by surfactants in agricultural sprays which in many cases need to provide maximum "coverage" by maximum spreading. In this context, it is perhaps useful to use Zisman's (1963) definition of the critical surface tension for liquids below which spreading is spontaneous (see Chapter 10). By proper choice of surfactant and concentration, it is possible to match the critical surface tension of the surface thus ensuring spreading.

ENHANCEMENT AND PREVENTION OF PARTICLE ADHESION

There are many biological processes in which adhesion of particles to each other or to surfaces is desirable, whereas in other cases adhesion has to be prevented. For example, the sticking together of like cells (usually referred to as cohesion) or unlike cells (adhesion) is essential in tissue formation and morphogenesis. An example where prevention of adhesion of organisms to surfaces could be advantageous is in the prevention of dental plaque. The forces responsible for adhesion are of an electromagnetic nature and it is often convenient to distinguish three types of interactions: those due to van der Waals forces between neutral atoms; those due to electrostatic forces between dipoles or free electrons and ions; chemical, hydrogen and hydrophobic bonding (Tadros, 1980).

In vitro, the forces of interaction between cells in a dispersion or between cells and surfaces appear to be similar to those which act between colloidal particles. If this is the case, it is easy to describe such interactions in terms of the well-known theories of colloid stability, with the limitation that with cells the geometry and surface groups are much less well characterized than model colloid particles. As is commonly discussed with colloids, the interaction can be described in terms of the double layer repulsion and van der Waals attraction in the so-called DLVO theory (see Chapter 9). Other shorter range types of interactions which occur at separations less than 1.5 nm, are not considered in this theory. However, these shorter range interactions, e.g. ion - dipole, dipole - dipole interactions, chemical bonds, hydrogen bonds, etc. should be considered when dealing with close contact between the plasma membranes of cells or between the cells and the surfaces. Moreover, if macromolecules are present in the gap between the cell membrane or the cell and surface, one has to consider steric repulsive forces and polymer bridging (see Chapter 9).

All the above types of interactions control adhesion, and such interactions are considerably modified by the addition of surfactants which may interact with cell surfaces by electrostatic forces, or simply by hydrophobic bonding. For example, many bacterial cells contain acid groups which are ionized above a specific pH value giving a negatively charged surface. Addition of a cationic surfactant to such systems causes charge neutralization and, above a certain surfactant concentration, charge reversal may take place. This surfactant adsorption certainly modifies the interaction between the cells and with any surface. Under certain conditions, addition of the surfactant may enhance the adhesion of the cells to walls, either by reducing the repulsive forces or by short-range interactions resulting from hydrophobic bonding. On the other hand, addition of surfactants may be applied to prevent adhesion by simultaneous adsorption on the cell and surface. This may find application in prevention of dental plaque.

CONTROLLED AND TARGETTED DELIVERY SYSTEMS

In many applications of pharmaceutical and agricultural systems, it is essential to control the release of the chemical on application. This is achieved by creating a "barrier" between the chemical and its environment, a process usually referred to as "microencapsulation". For example, when a water-soluble core substance is microencapsulated with a water-insoluble wall material, the release of the core material to an external aqueous phase is controlled by the rates of diffusion and dissolution. Various types of capsule systems may be applied of which polymeric walls are the most commonly used materials. A liquid membrane may also be used as, for example, in the case of the so-called multiple emulsions (see Chapter 8). In targetted delivery, the drug or pesticide is incorporated in a material that can interact specifically with the site of action. An example of this is the use of liposomes as drug delivery systems. Liposomes are smectic mesophases of phospholipids organized into bilayers which assume a multilamellar or unilamellar structure. The latter is usually obtained by sonication of the multilamellar structure. Below, a brief account of the various delivery systems that are commonly used in drugs and pesticides is given, with particular attention to the role of surface active agents in their preparation and subsequent stability.

Microcapsules

Several methods may be used for microencapsulation of drugs and pesticides and these have been described in a book by Kondo (1979). Three major classes of processes may be identified: (i) chemical processes, namely interfacial polymerization, *in situ* polymerization and rapid insolubilization of polymer; (ii) physico-chemical processes such as phase separation from aqueous or non-aqueous solution; and (iii) mechanical processes, e.g. air suspension coating, spray drying, electrostatic aerosol, etc. In this review only interfacial polymerization and phase separation (coacervation) types of microencapsulation will be considered to illustrate the role of surfactants.

In interfacial polymerization, a hydrophobic monomer A is placed in the oil phase, and a hydrophilic monomer B is placed in the aqueous phase. The two monomers interact at the interface between the oil and the aqueous solution forming a capsule wall. Two main types of systems may be identified. For example, if the material to be encapsulated is oil-soluble, oil-dispersable or an oil itself, an oil-in-water (o/w) emulsion is prepared. In this case the hydrophobic monomer is dissolved in the oil phase which forms the disperse phase. The role of surfactants in this process is obvious since an o/w emulsifier will be needed. Such an emulsifier will control the droplet size distribution and the size of the capsules formed. On the other hand, if the material to be encapsulated is water-soluble, a water-in-oil (w/o) emulsion is prepared with the hydrophilic monomer dissolved in the internal phase. Again, the type and concentration of emulsifier is critical in controlling the size distribution of the microcapsules formed.

In interfacial polymerization, monomers A and B are polyfunctional monomers capable of causing a polycondensation or a polyaddition reaction between them (Kondo, 1979). As mentioned above, they are selected so that one monomer is oil-soluble and the other is water-soluble. Examples of oil-soluble monomers are polybasic acid chloride, bis-haloformate and polyisocyanate, whereas soluble monomers could be polyamine, or glycol polyphenol. Thus, a capsule wall of polyamide, polyurethane or polyurea may be formed. As mentioned above, the ultimate capsule size is determined by the size of the dispersed droplet of the first monomer and this size is determined by the nature and concentration of emulsifier used. The latter also affects the physical stability of the resulting capsule dispersion and hence the role of the surfactant is crucial.

There are four types of encapsulation utilizing the system of phase separation from an aqueous solution (Kondo, 1979). These are: (i) complex coacervation or phase separation resulting from two oppositely charged colloids neutralizing one another; (ii) simple coacervation where a non-electrolyte, such as alcohol, causes formation of a separate polymer-rich phase; (iii) salt coacervation where a polymer separates as a result of a salting-out process; and (iv) precipitation and insolubilization of a polymer by changing the pH of the aqueous solution system. An example of complex coacervation is the interaction between gelatin and gum arabic. In this case a dispersion of oil in a dilute aqueous solution of gelatin - gum arabic is prepared. Gelatin usually has an iso-electric point (i.e.p.) at about pH = 4.8, whereas gum arabic, which contains only carboxylic groups, is usually negative over a wide range of pH values. Thus, by lowering the pH to a value below the i.e.p. of gelatin, say to pH 4, the gelatin acquires a positive charge and becomes coacervated with gum arabic forming the capsules. Various other polyanionic polymers may be used such as sodium alginate, agar, polyvinylbenzenesulphonic acid, etc. In general, effective materials include polymers, surface active agents and organic compounds which have acid groups in the molecule (Kondo, 1979).

Encapsulation by phase separation can also be applied to non-aqueous media. This is particularly suitable for encapsulating water-soluble materials. In this case, phase separation of the oil-soluble polymer may be induced by addition of another polymer, non-solvent or by changing the temperature.

Apart from their direct role in the preparation of microcapsules and their stabilization, surfactants can also be used to control the release of chemical from microcapsule dispersions. For example, Wada *et al.* (1983) have shown that the efficacy of an edifenphos capsule suspension can be improved by addition of a surfactant either to the aqueous medium or to the core. This was attributed to the possible solubilization of the pesticide by the surfactant, thus increasing its release rate.

Multiple Emulsions (Liquid Capsules)

Multiple emulsions are emulsion systems in which the disperse phase contains dispersed droplets of the external phase (Attwood and Florence, 1983). Water-in-oil-in-water (w/o/w) type multiple emulsions are oil-in-water emulsions in which the dispersed oil drops themselves contain smaller dispersed aqueous droplets. On the other hand, oil-in-water-in-oil (o/w/o) multiple emulsions are those in which the dispersed water droplets themselves contain smaller oil droplets.

Multiple emulsions are usually prepared using a two-stage emulsification process. For example, a w/o/w multiple emulsion is prepared by firstly making a stable w/o emulsion (the primary emulsion) followed by further emulsification (secondary process) into water to form the multiple emulsion. It is essential to control the stability of the primary emulsion and the process of the secondary emulsification, otherwise coalescence of the multiple droplets takes place. The success of preparation of a given multiple emulsion system depends on the proper choice of emulsifier for the primary and secondary processes and their ratios (Attwood and Florence, 1983).

The potential of w/o/w systems for prolonged release of drugs rests on the presence of an active ingredient in the disperse phase droplets. For example, a water-soluble drug in a non-aqueous droplet has to diffuse out through an oil layer of disperse droplets. However, if the multiple emulsion is not stable, the drug may be released by gradual coalescence of the droplets. Thus, in principle, it is possible to control the release of the drug either by controlling diffusion through an oil-barrier or by controlling the stability. In practice, however, such control is not easy and a great deal of work is needed before such systems can be applied.

Liposomes

As mentioned above, unilamellar liposomes are produced by sonication of dispersions of phospholipids and cholesterol into water. The net charge of the liposome can be varied by incorporation of a long chain amine (e.g. stearylamine) to give positively charged vesicles, or a phosphate surfactant (e.g. dicetyl phosphate) to give a negatively charged system (Attwood and Florence, 1983). Both lipid-soluble and water-soluble drugs can be entrapped in liposomes. In the first case the drug is solubilized in the hydrocarbon interior of the lipid bilayers, whereas in the latter case the compound is simply intercalated in the aqueous layer. The use of liposomes as drug carriers has been reviewed by Fendler and Romero (1977) to which the reader should refer for further details. Liposomes can be administered intravenously, orally or intramuscularly, thus offering various means of specificity, uptake and in some cases, controlled release. Unfortunately, the vesicles are very sensitive to interactions with various components of serum and with surfactants and this limits their application in many cases. However, by proper choice of surfactant systems it is possible, in certain cases, to enhance their stability and specificity. Thus, the liposomes offer a potential system for application in the future, although it must be realized that a great deal of study is needed in order to understand their stability and specificity.

REFERENCES

Attwood, D. and Florence, A.T. (1983). "Surfactant Systems, Their Chemistry, Pharmacy and Biology". Chapman and Hall, London.

Bickerman, J.J. (1958). "Surface Chemistry". 2nd edn, Academic Press, New York.

Chan, A.F., Evans, D.F. and Cussler, E.L. (1976). *A. I. Chem. E.J.* 22, 1006.

Elworthy, P.H. and Lipscomb, F.J. (1968). *J. Pharm. Pharmocol.* 20, 923.

Fendler, J.H. and Romero, A. (1977). *Life Sci.* 20, 1109.

Furmidge, C.G.L. (1962). *J. Colloid Sci.* 17, 309.

Harkins, M.D. (1952). "The Physical Chemistry of Surface Films". Reinhold Publishing Corp., New York.

Hartley, G.S. and Brauskill, R.T. (1958). "Surface Phenomena

in Chemistry and Biology". Pergamon Press, New York.
Hartley, G.S. and Graham-Bryce, I.J. (1980). "Physical Principles of Pesticide Behaviour". Academic Press, New York.
Higuchi, W.I. (1964). *J. Pharm. Sci.* 53, 532.
Higuchi, W.I. (1967). *J. Pharm. Sci.* 56, 315.
Kondo, A. (1979). "Microcapsule Processing and Technology". Marcel Dekker Inc., New York.
Tadros, Th. F. (1980). *In* "Microbial Adhesion to Surfaces" (Eds R.C.W. Berkeley, J.M. Lynch, J. Melling, P.R. Rutter and B. Vincent) Chapter 5, p.93. Ellis Hoorwood Publishers, Chichester.
Wada, Y., Nakahara, T., Orrii, T., Okano, Y., Aya, M., Kasui, K., Kamochi, A., Yamada, Y., Katsumata, O., Sakawa, S. and Kurahashi, Y. (1983). "Proc. 5th Int. Congress Pesticide Chem." Vol. 4. Pergamon Press, Oxford.
Zisman, M. (1963). *Ind. Eng. Chem.* 55, 18.

SUBJECT INDEX

Adhesion
enthalpies of, 155
of particles, 330
of spray, 327
tension, 228, 230
work of, 156, 228, 229, 230
Adsorption
at air/liquid interface, 4, 153
at liquid/liquid interface, 153, 282
at solid/liquid interface, 154, 166, 170, 282
determination of, 157, 168
enthalpies of, 155
free energies of, 32-37, 160, 161
model of, 22
of sodium dodecyl sulphate on alumina, 170
of surfactants at liquid interfaces, 27
Adsorption isotherm
Gibbs equation, 4, 23, 27-29, 113, 159
surface equations of state and, 29-31
Agricultural sprays, 326
Amphiphilic compounds
possible states in solution of, 20
space filling models of, 84
Association process, 83, 86, 90
Bicontinuous structure
schematic representation of, 178
Binding
of counterions to micelles, 93
Biological systems
application of surfactants in, 323

Caking of suspensions
control of, 212
Capillary phenomena, 202
Capillary pressure
in liquid films, 191
Cohesion
enthalpies of, 155
work of, 228, 230
Comminution of particles, 203
Conductivity
of microemulsions, 128
of surfactant solutions, 5-7
Contact angle, 201, 221
calculation of, 233
dynamic, 260
effect of surface roughness on, 255
effect of surface tension on, 262
effect of surfactant adsorption on, 242-254
effect of viscosity on, 262

Contact angles (contd.)
hysteresis, 222, 254
kinetic factors, 258
practical aspects of, 267
Counterion binding
to micelles, 93
Critical micelle concentration
changes of properties at, 89
prediction of values of, 45
values of, 8, 91
variation with temperature of, 8
Critical surface tension of wetting, 238-241
measurement of, 241
Crystal Growth, 213
Cubic phase, 16, 70

Deflocculation, 304
Demulsifiers, 307
Detergency, 303
Diffusion coefficient
of counterions, 94, 95
of micelles, 94, 95
of surfactant ions, 94, 95
Dispersion forces, 232
Dispersion methods
for preparation of suspensions, 200
of finely divided solids, 304
Double layer
interaction, 205
model for, 167
Droplet formation, 326
Dynamic measurement,
of concentrated systems, 142

Electrical double layer
at solid/liquid interface, 167, 205
Electrical interactions
at solid/liquid interface, 169
Electrostatic effects
in ionic surfactants, 93, 98
Electroviscous effects
primary, 136
secondary, 138
Emulsification, 305
spontaneous, 280
Emulsions, 175 - 196
breakdown of, 181, 182
coalescence of, 190
flocculation of, 186-189
formation of, 176
Ostwald ripening of, 183-186
sedimentation of, 189
stability of, 181, 182, 305
Enhanced oil recovery, 277-285

Films
instabilities in, 192
Flow behaviour
of suspensions, 215
of coagulated suspensions, 218
Foams, 175 - 196
breakdown of, 181, 182, 301
coalescence of, 190
flocculation of, 186-189
formation of, 176
Ostwald ripening of, 183-186
sedimentation of, 189
stability of, 181, 182 301
Free energy of adsorption, 32-34

Gibbs equation, 4, 23, 27-29, 113, 159
Gibbs-Marangoni effect, 193
Gibbs phase rule, 56

Hemimicelles, 171
Hexagonal phase, 15, 16, 69
Hydrocarbon-water interaction, 153
Hydrophilic-lipophilic balance (HLB), 178, 289
Hydrophobic effect, 36

Impaction of spray, 327
Interaction energy-distance curves
  for electrostatically stabilised dispersions, 209
  for sterically stabilised dispersions, 211, 304
Interfacial rheology, 280
Interfacial tension,
  as a function of surfactant concentration, 113, 115
  as a function of temperature, 165
  as a function of time, 181
  between water and pure liquids, 25
  determination of, 127, 157
  ultralow, 164
Intermolecular forces, 231
  macroscopic approach, 236

Kinetics
  of micellization, 15, 93
  of wetting, 260, 263
Krafft temperature
  location of, 8, 72

Lamellar phase, 16, 69
Laplace pressure, 176
Lever rule, 60
Light scattering
  dynamic, 128
  of microemulsions, 127
  of surfactant solutions, 5, 6
Line tensions, 228
Liposomes, 334
Liquid crystals, 68
  boundaries of, 73
  lyotropic, 16, 69
  thermotropic, 68
Liquid interface
  models of, 21, 22
Liquid surfaces, 20

Mesophases, 16, 53, 68, 73
Micellar solubilization, 324
Micelles
  aggregation of, 39
  aggregation number of, 92
  counterion binding to, 93
  dynamics of, 104
  formation of, 90
  internal structure of, 104
  models of, 10, 90, 98
  properties of, 9-11
  shape of, 46, 86
  size of, 46, 86, 88
  size distribution of, 92
  space filling model of, 85
  structural aspects of, 83-109
Micellization
  enthalpy of, 13, 14, 49, 92

Micellisation (contd.)
enthalpy of, 13, 14, 49, 92
entropy of, 49
free energy of, 11, 12
kinetics of, 15, 93
mass action model of, 40, 91, 92
multiple equilibrium model of, 42, 92
phase separation model of, 39, 90
thermodynamics of, 11, 12, 49, 91, 92
Microcapsules, 331
Microemulsions, 111-132, 164, 305
applications of, 129
model consideration of, 114
phase diagrams of, 117-121
picture of W/O and O/W, 114
techniques for characterisation of, 127
thermodynamics of, 122-125
Middle phase, 16, 17
Miscibility gaps, 74
Monolayers, 25
Multiple emulsions, 333

Neat phase, 16, 17
NMR
of surfactant systems, 13, 87, 103, 105
of microemulsion systems, 128
Nonionic surfactants
adsorption of at liquid interfaces, 27
application of, 309-317
biodegradability of, 318
classes of, 287-289
cloud points of, 299
in microemulsions, 125
micelles of, 102
Nonionic surfactants (contd.)
oxidative stability of, 309
phase diagrams of, 17, 101, 126, 179
structure of, 308
surface properties of, 297-298
thermal stability of, 309
toxicology of, 318
Nucleus
free energy of formation of, 199

Oil mobilization
mechanism of, 279
Oscillatory measurements, 142
Ostwald ripening, 183-186, 213-215

Particle aggregation
control of, 204
Phase diagrams
determination of, 65-67
four component, 121
of C_{12} E_6-water, 17
of C_{12} E_5-water
of C_{10} E_5-water, 58, 101
of C_{12} E_8-water, 101
of microemulsions, 117-121
of sodium palmitate-sodium chloride-water, 59
of sodium palmitate-water, 63
three-component, 62, 63, 74
two-component, 60
Phase equilibria, 53, 116
Phase information,
dimensions of, 55
experimental methods, 65
isoplethal methods, 55, 65
isothermal methods for, 65
utility of, 75

Phase rule
  Gibbs, 56
Phase separation
  thermodynamics of, 53
Polymer-surfactant
  interaction, 283
Polymeric surfactants
  application of, 309-318

Rheology
  and the microstructure, 147
  mechanical analogues of, 140
  models of, 148, 149
  of concentrated systems, 140
  of dilute systems, 134
  of surfactants and related systems, 133-151
  of suspensions, 215

Sedimentation
  of emulsions, 189
  of foams, 189
  of suspensions, 212
Self diffusion
  studies of in microemulsions, 128
  studies of in surfactants, 94, 103
Solid/liquid interface
  electrical double layer at, 167
  electrical nature of, 166
Solubilization, 305
  micellar, 324
Sprays
  adhesion of, 327
  impaction of, 327
  retention of, 328
Spreading, 329
Spreading coefficient, 228, 229
Steric interactions, 207, 304
Surface active agents (surfactants)
  adsorption of at liquid interfaces, 27
  anionic, 2
  applications of, 323
  biodegradibility of, 318
  bulk properties of, 6
  cationic, 2
  classification of, 2-4, 287-289
  cloud points of, 298, 299
  clouding of, 102
  compositions, 292-297
  critical micelle concentration of, 8 45, 91
  floods, 278
  foaming properties of, 302
  free energy of adsorption of, 32-37
  in enhanced oil recovery, 277
  interfacial properties of, 5
  ionic, 93
  macromolecular, 287-321
  microstructure of, 133
  nonionic, 3, 27, 291
  phase behaviour of, 53-81, 67-78
  polymer interactions with, 283
  properties of solutions of, 4, 5
  raw materials of, 289
  rheology of, 133
  self diffusion studies of, 94

Surface active agents (contd.)
solubility of, 7 - 9
solutions of, 15, 16
states of matter in, 67-71
structural aspects of, 83
structure of concentrated solutions of, 16
surface properties of, 4-6, 297
thermodynamics of solutions of, 19
toxicology of, 318
zwitterionic, 3
Surface equations of state, 30, 161,
Surface films, 32
Surface free energy, 23
Surface pressure, 161
as a function of area, 163
Surface tension, 23, 298
calculation of, 233
critical, 238
determination of, 157
of aqueous amino-acid solutions, 26
Suspensions, 197-220
claying (or caking) of, 212
control of physical stability of, 204
preparation of, 198

Tie-lines, 60

van der Waals interaction of spherical particles, 206
Viscosity
as a function of volume fraction, 139 147, 148, 216
determination of, 135
of concentrated systems, 140
of dilute surfactant systems, 134
of dispersions, 134, 215
of microemulsions, 128
stress relationship, 139

Wetting, 221, 303, 329
adhesional, 201
complete, 227, 250
critical surface tension of, 238
immersional, 201
kinetics of, 260
molecular basis for, 231
of solid by a liquid, 201, 251
practical aspects of, 265
spreading, 201
surface, 281
thermodynamics of, 224

Young's equation, 224
and intermolecular forces 231